Computability & Unsolvability

Computability & Unsolvability

MARTIN DAVIS

Department of Mathematics and Computer Sciences
Courant Institute of Mathematical Sciences
New York University

DOVER PUBLICATIONS, INC.
NEW YORK

Published in Canada by General Publishing Company, Ltd., 30 Lesmill Road, Don Mills, Toronto, Ontario.
Published in the United Kingdom by Constable and Company, Ltd.

This Dover edition, first published in 1982, is an enlarged version of the work originally published by McGraw-Hill Book Company, New York, in 1958. The Dover edition includes a new preface and a new appendix to the text. This appendix, "Hilbert's Tenth Problem Is Unsolvable," originally appeared in *The American Mathematical Monthly* (Vol. 80, No. 3, March 1973, pp. 233–269) and is reprinted here with the permission of The Mathematical Association of America, 1529 Eighteenth Street, N.W., Washington, D.C. 20036.

Manufactured in the United States of America
Dover Publications, Inc., 180 Varick Street, New York, N.Y. 10014

Library of Congress Cataloging in Publication Data

Davis, Martin, 1928-
Computability & unsolvability.

Reprint. Originally published : New York : McGraw-Hill, 1958. (McGraw-Hill series in information processing and computers)
Includes bibliographical references and index.
1. Recursive functions. 2. Unsolvability (Mathematical logic) 3. Computable functions. I. Title. II. Series: McGraw-Hill series in information processing and computers.
QA9.6.5.D38 1982 511.3 82-7287
ISBN 0-486-61471-9 (pbk.) AACR2

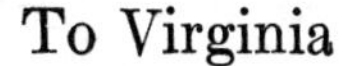
To Virginia

PREFACE TO THE DOVER EDITION

Work in computer science over the past two decades has amply justified the hope expressed in the original preface to this book that the general theory of computability, and in particular Turing's model of computation, would play an important role in the "theory of digital computers." In fact, much of the terminology introduced here has become a standard part of the terminology of theoretical computer science, and many computer scientists have told me that this book was their theoretical introduction. Naturally I am delighted that my first book continues to find an audience, and that it will live on in this Dover edition.

One of the great pleasures of my life came in February 1970, when I learned of the work of Yuri Matiyasevič which completed the proof of the crucial conjecture in Chapter 7 and thereby showed that Hilbert's tenth problem is recursively unsolvable. My paper on the subject from the March 1973 *American Mathematical Monthly* is reprinted with kind permission as Appendix 2, and I have taken the opportunity to correct several minor errors in it. Readers can either regard this appendix as a substitute for Chapter 7 or can simply apply Theorem 5.1 of Appendix 2 to Theorem 3.2 of Chapter 7 to obtain the conjectured converse of Corollary 2.3 of Chapter 7.

MARTIN DAVIS

1982

PREFACE TO THE FIRST EDITION

This book is an introduction to the theory of computability and noncomputability, usually referred to as the theory of recursive functions. This subject is concerned with the existence of purely mechanical procedures for solving various problems. Although the theory is a branch of pure mathematics, it is, because of its relevance to certain philosophical questions and to the theory of digital computers, of potential interest to nonmathematicians. The existence of absolutely unsolvable problems and the Gödel incompleteness theorem are among the results in the theory of computability which have philosophical significance. The existence of universal Turing machines, another result of the theory, confirms the belief of those working with digital computers that it is possible to construct a single "all-purpose" digital computer on which can be programmed (subject of course to limitations of time and memory capacity) any problem that could be programmed for any conceivable deterministic digital computer. This assertion is sometimes heard in the strengthened form: anything that can be made completely precise can be programmed for an all-purpose digital computer. However, in this form, the assertion is false. In fact, one of the basic results of the theory of computability (namely, the existence of nonrecursive, recursively enumerable sets) may be interpreted as asserting the possibility of programming a given computer in such a way that it is impossible to program a computer (either a copy of the given computer or another machine) so as to determine whether or not a given item will be part of the output of the given computer. Another result (the unsolvability of the halting problem) may be interpreted as implying the impossibility of constructing a program for determining whether or not an arbitrary given program is free of "loops."

Because it was my aim to make the theory of computability accessible to persons of diverse backgrounds and interests, I have been careful (particularly in the first seven chapters) to assume no special mathematical training on the reader's part. For this reason also, I was very pleased when the McGraw-Hill Book Company suggested that the book be included in its Information Processing and Computers Series.

Although there is little in this volume that is actually new, the expert will perhaps find some novelty in the arrangement and treatment of certain topics. In particular, the notion of Turing machine has been made central in the development. This seemed desirable, on the one hand, because of the intuitive suggestiveness of Turing machines and the analogies between them and actual digital computers and, on the other hand, because combining Turing's approach with the powerful syntactic methods of Gödel and Kleene makes it possible to present the various aspects of the theory of computability, from Post's normal systems to the Kleene hierarchy, in a unified manner.

Some of the material in this book was used in a graduate course, given by me, at the University of Illinois and in lecture series at the Control Systems Laboratory of the University of Illinois and at the Bell Telephone Laboratories. Part of the work for this book was done while the author was at the Institute for Advanced Study on a grant from the Office of Naval Research. The book could be used as a text, or supplementary text, in courses in mathematical logic or in the theory of computability.

I should like to thank Mr. Donald Kreider, Professor Hilary Putnam, Professor Hartley Rogers, Jr., and Dr. Norman Shapiro for suggesting many corrections and improvements.

MARTIN DAVIS

CONTENTS

GLOSSARY OF SPECIAL SYMBOLS

References in this glossary are to the page(s) where the symbol in question is introduced.

$\mathfrak{x}^{(n)}$	xx
$\mathfrak{y}^{(n)}$	xx
$\mathfrak{z}^{(n)}$	xx
$m \in S$	xx
$m \notin S$	xx
$R \cup S$	xx
$R \cap S$	xx
$\bar{R}$	xx
ϕ	xxi
$R \subset S$	xxi
$C_S(x_1, \ldots, x_n)$	xxii
$p \wedge q$	xxii
$p \vee q$	xxii
$\sim p$	xxii
$\{\mathfrak{x}^{(n)} \mid P(\mathfrak{x}^{(n)})\}$	xxii
$P(\mathfrak{x}^{(n)}) \leftrightarrow Q(\mathfrak{x}^{(n)})$	xxiii
$C_P(x_1, \ldots, x_n)$	xxiii
$\bigvee_{y=0}^{z} P(y, \mathfrak{x}^{(n)})$	xxiv
$\bigwedge_{y=0}^{z} P(y, \mathfrak{x}^{(n)})$	xxiv
$\bigvee_{y} P(y, \mathfrak{x}^{(n)})$	xxiv
$\bigwedge_{y} P(y, \mathfrak{x}^{(n)})$	xxiv
$q_i\, S_j\, S_k\, q_l$	5
$q_i\, S_j\, R\, q_l$	5
$q_i\, S_j\, L\, q_l$	5
$q_i\, S_j\, q_k\, q_l$	5, 21
B	5
1	5
$\alpha \to \beta \quad (Z)$	6
$\mathrm{Res}_Z\,(\alpha_1)$	7
$\bar{n}$	9
$\overline{(n_1, n_2, \ldots, n_k)}$	9
$\Psi_Z{}^{(n)}(x_1, x_2, \ldots, x_n)$	9
$\Psi_Z(x)$	9
$S(x)$	12
$x \dot{-} y$	15
$U_i{}^n(x_1, x_2, \ldots, x_n)$	16
$\alpha \underset{A}{\longrightarrow} \beta \quad (Z)$	21
$\mathrm{Res}_Z{}^A\,(\alpha_1)$	22
$\Psi_{Z;A}^{(n)}(x_1, x_2, \ldots, x_n)$	22
$\Psi_Z{}^A(x)$	22
$\theta(Z)$	25
$Z^{(n)}$	26
$\min_y\, [f(y, \mathfrak{x}^{(n)}) = 0]$	38
$N(x)$	42
$\alpha(x)$	42
$[\sqrt{x}]$	42
$[x/y]$	42
$R(x, y)$	43
$J(x, y)$	43
$K(z)$	44
$L(z)$	44
$T_i(w)$	46
$\underset{y=0}{\overset{z}{\mathfrak{m}}}\, P(y, \mathfrak{x}^{(n)})$	52
Prime (x)	54
Pr (n)	54

INTRODUCTION

1. Heuristic Remarks on Decision Problems. The primary task of present-day mathematicians is that of determining whether various propositions concerning mathematical objects (e.g., integers, real numbers, continuous functions, etc.) are true or false. Our principal concern, however, will be with another kind of mathematical task, one which was considered of great importance during the earliest phases of the development of mathematics and which still yields problems of considerable mathematical interest. That is, we shall be concerned with the problem of the existence of *algorithms* or *effective computational procedures* for solving various problems. What we have in mind are sets of instructions that provide mechanical procedures by which the answer to any one of a class of questions can be obtained. Such instructions are to be conceived of as requiring no "creative" thought in their execution. In principle, it is always possible to construct a machine for carrying out such a set of instructions or to prepare a program by means of which a given large-scale digital computer will be enabled to carry them out.

As an example, consider the problem of obtaining the sum of two positive integers given in the ordinary decimal notation. An algorithm that can be used will be found in any textbook of arithmetic. Intuitively, we are at once prepared to admit that this procedure is purely mechanical, that it can be carried out by a human computer who does nothing but obey direct and elementary commands. Moreover, there do exist adding machines which accomplish exactly this.

A more sophisticated example is given by the theory of linear diophantine equations. Let a, b, c be given integers positive, negative, or zero. Suppose we wish to know whether or not there exist integers x, y such that

$$ax + by = c. \tag{1}$$

If $a = 2$, $b = -1$, $c = 1$, it is easily seen that $x = 1$, $y = 1$ provides a solution of the desired sort. If $a = 2$, $b = -6$, $c = 1$, it is easy to see that there are *no integers* x, y that provide a solution of (1) [for, in this case, the left-hand side of (1) would always be even, whereas the right-hand side is simply 1, which, of course, is odd]. Is there an *algorithm* which will enable us to decide, for given values of a, b, c, whether or not there are integers x, y that satisfy (1)? A basic result in the elementary theory of numbers asserts that (1) has an integral solution if and only if the largest positive integer that is a divisor of both a and b is also a divisor of c. It is not difficult to convince oneself that this criterion actually does furnish an algorithm of the desired kind.

An immediate generalization of the problem just considered is obtained by replacing linear polynomials in two variables by polynomials of arbitrary degree in arbitrarily many variables. That is, the problem is that of determining, of a given polynomial equation:

$$\sum_{i_1, \ldots, i_k = 0}^{n} a_{i_1, \ldots, i_k} x_1^{i_1} \cdots x_k^{i_k} = 0,$$

where the $a_{i_1, \ldots, i_k}$'s are integers, whether or not there exist integers $x_1, \ldots, x_k$ which satisfy it. However, *no algorithm is known* for solving this problem. Moreover, we shall see (cf. Chap. 7) that, for a *suitable further generalization* of this problem, not only is no algorithm known, but *none is possible.*

Problems of this kind, which inquire as to the existence of an algorithm for *deciding* the truth or falsity of a whole class of statements, are called *decision problems* to distinguish them from ordinary mathematical questions concerning the truth or falsity of single propositions. A *positive solution* to a decision problem consists of giving an algorithm for solving it; a *negative solution* consists of showing that no algorithm for solving the problem exists, or, as we shall say, that the problem is *unsolvable.* Positive solutions to decision problems occur quite frequently in classical mathematics. To recognize such a solution as valid, it suffices to verify that the alleged algorithm really is an algorithm; this is ordinarily taken for granted and remains on the level of intuition. However, in order to obtain the *unsolvability* of a mathematical decision problem, this does not suffice. It becomes necessary to give an exact mathematical definition of the term "algorithm." This will be done in Chap. 1. The rest of the present section is devoted to showing, still on the intuitive level, that the problem, mentioned above, of verifying that an alleged algorithm is indeed an algorithm is not so simple as might be supposed and is, in fact, itself *unsolvable.*

We consider functions of a single variable $f(x)$, defined on the positive integers (that is, for $x = 1, 2, 3$, etc.) and whose values are positive integers. Examples of such functions are x^2, 2^x, the xth digit in the decimal expansion of π, etc. We shall say that such a function $f(x)$ is *effectively calculable* if there exists a definite algorithm that enables us to compute the functional value corresponding to any given value of x.

Let us assume that such an algorithm can be expressed as a set of instructions in the English language. Furthermore, let us imagine all such sets of instructions ordered according to the number of letters they contain: first, those (if any) that consist of a single letter; then those that employ two letters; etc. Where there is more than one set of instructions consisting of the same number of letters, they are to be ordered among themselves, alphabetically, like the entries in a dictionary. Thus, there will be a first set of instructions, a second set of instructions, a third, etc. With each positive integer i, there is associated the ith set of instructions in this list, E_i, which tells us how to compute the values of some function. The function associated in this way with E_i we will call $f_i(x)$.

Now, let

$$g(x) = f_x(x) + 1. \tag{2}$$

Then, $g(x)$ is a perfectly good function. Its value for a given integer x is obtained by finding the xth set of instructions E_x, then applying it to the number x as argument, and finally increasing this result by 1. We have:

I. *For no value of i is it the case that $g(x) = f_i(x)$.*

PROOF. Suppose that $g(x) = f_{i_0}(x)$ for some integer i_0. Then, by (2),

$$f_{i_0}(x) = f_x(x) + 1$$

for all values of x. In particular, this equation would have to hold for $x = i_0$, yielding

$$f_{i_0}(i_0) = f_{i_0}(i_0) + 1.$$

But this is a contradiction.

Now, from the manner of choice of the E_i, the functions $f_i(x)$ were to include *all* effectively calculable functions. This yields:

II. *$g(x)$ is not effectively calculable.*

In spite of the interdiction which II seems to impose, let us try to develop an algorithm for computing $g(x)$. A first attempt might run as follows:

"Given a value x_0, in order to compute the number $g(x_0)$, begin by generating the list E_i, until E_{x_0} is obtained. Having E_{x_0}, apply its very instructions to the number x_0. Finally, add 1 to the number thus obtained."

Since we know, from II, that this set of instructions cannot really be an algorithm, let us see wherein it fails to qualify. Clearly, it must be in our gratuitous assumption that the list E_i can be generated in a purely mechanical way. But what is wrong with the following attempt at a mechanical procedure for generating E_i?

"Begin by generating a list of all possible pieces of English writing, whether meaningful or not. This can be done by taking all possible permutations of the letters and punctuation marks which make up the English alphabet (including the space between words!) that have a given length (first one letter, then two letters, etc.). After obtaining each such piece of writing, check to see whether it is a set of instructions for computing a numerical function $f(x)$. If it is, place a $\checkmark$ before it in the list.

"Then the list of E_i can be generated by simply crossing off, from the list obtained above, all pieces of writing not preceded by a $\checkmark$."

Here, the difficulty is not quite so easy to find. It lies in our assumption that one can determine mechanically whether or not an *alleged* algorithm for computing a function $f(x)$ is indeed such an algorithm. If we believe II (and, as we shall see later, it can be quite rigorously justified), our only recourse is to conclude:

III. *There is no algorithm that enables one to decide whether an alleged algorithm for computing the values of a function whose domain of definition is the set of positive integers, and all of whose values are positive integers, is indeed such an algorithm.*

If the methods by which these results have been inferred seem somewhat questionable, it is only because the notion of algorithm has not yet been accurately defined. We shall remedy this situation in Chap. 1.

2. Suggestions to the Reader. For the most part, this book is self-contained. Although the reader should possess the ability to follow a detailed proof, no specific knowledge of advanced mathematics is necessary. It has become quite usual to employ some of the notation of symbolic logic in various mathematical subjects. This notation is not used in any essential way, but rather serves to abbreviate conveniently certain frequently occurring logical notions. This has been particularly true of the theory of recursive functions, in part because it was developed primarily by professional logicians and in part because of the nature of the subject. Section 3 of this Introduction is devoted to the special notation (largely borrowed from symbolic logic) that we find convenient. We suggest that the reader use this material for reference, referring to it only when necessary in following the text. It should be emphasized that no formal knowledge of symbolic logic is necessary on the reader's part.

Chapters 1 to 7 constitute a general introduction to recursive-function theory. Chapters 1 to 5, i.e., Part 1, develop the theory to the point where unsolvable problems can be produced. Chapters 6 and 7 give applications to other branches of mathematics, Chap. 6 to algebra and Chap. 7 to number theory. In these first seven chapters, we have kept particularly in mind the reader who has had little experience with professional mathematics; that is, we have made a special effort to avoid leaving details as "exercises for the reader" and to include enough intuitive explanations so that the reader who wishes to skip many of the proofs, or at least not to follow them in detail, will not be entirely lost.

Some of the material in the earlier chapters consists of detailed proofs that certain rather complicated procedures can be accomplished by suitable iteration of more basic procedures. As might be supposed, such proofs are very much like programs for use with a digital computer. The details tend to become both routine and tedious, and the reader may well wish merely to skim much of it. Most of this material is to be found in Chap. 1, Sec. 3; in Chap. 2; and in Chap. 4, Sec. 1.

In Chap. 3, we have need of several notions (e.g., prime number, congruence) and results (i.e., the unique-factorization theorem and the Chinese remainder theorem) from the elementary theory of numbers. For the benefit of the reader who is not familiar with these results and who does not wish to take them on faith, we have included a brief appendix on number theory, where they are proved.

The most spectacular applications of recursive-function theory, thus far, have been to mathematical logic. These applications are discussed in Chap. 8. In their most general form the results have little meaning except to specialists; their interest stems, to a large extent, from their application to specific systems of logic. Unfortunately, however, a detailed development of the properties of several of the better-known systems of logic would in itself require a book larger than the present volume. The compromise which we have adopted should present no difficulty to one who is familiar with Church [4], Church [5], or Hilbert and Ackermann [1].

The material in Part 3 is of rather specialized interest, and we have felt free to make our exposition there more terse than that in the earlier chapters.

The numbering of theorems, definitions, etc., begins anew with each section. Thus, Theorem 3.2 is the second theorem in Sec. 3. When referring to an item from a previous chapter, we precede the item's number with that of the chapter. Thus, in Chap. 4, a reference to Theorem 1.7 is a reference to Theorem 1.7 of Chap. 4, Sec. 1, whereas (still in Chap. 4) a reference to Theorem 2-3.2 is a reference to Theorem 3.2 of Chap. 2, Sec. 3.

3. Notational Conventions. The following discussion of the special notation we employ is intended largely for reference. It might be well, however, to read Secs. 3.1, 3.2, and 3.3 before beginning Chap. 1.

3.1. *Numbers and n-tuples.* Ordinarily, we shall be dealing with the *natural numbers* 0, 1, 2, 3, ... The words "number" and "integer" are to be regarded as synonymous with "natural number" *unless the contrary is explicitly stated.*

We shall also be concerned with *ordered n-tuples* ($n \geqq 1$) of natural numbers. Ordinarily the word "ordered" will be omitted, and we shall speak simply of *n-tuples.* An n-tuple will be denoted by such an expression as $(a_1, a_2, \ldots, a_n)$. It is, of course, by no means required that the a_i's be distinct. Thus,

$$\begin{aligned} (3, 2, 7) &\text{ is a 3-tuple,} \\ (6) &\text{ is a 1-tuple,} \\ \text{and } (5, 4, 1, 4) &\text{ is a 4-tuple.} \end{aligned}$$

A 2-tuple is called a *pair;* a 3-tuple is called a *triple.* We shall not ordinarily distinguish between the 1-tuple (a) and the number a itself.

For our purposes, the characteristic feature of n-tuples is given by the relation

$$(a_1, a_2, \ldots, a_n) = (b_1, b_2, \ldots, b_n)$$

if and only if

$$a_1 = b_1, a_2 = b_2, \ldots, \text{ and } a_n = b_n.$$

In dealing with n-tuples the following abbreviative convention will be useful:

$$\begin{aligned} \mathfrak{x}^{(n)} &\text{ is to mean } x_1, x_2, \ldots, x_n, \\ \mathfrak{y}^{(n)} &\text{ is to mean } y_1, y_2, \ldots, y_n, \\ \text{and } \mathfrak{z}^{(n)} &\text{ is to mean } z_1, z_2, \ldots, z_n. \end{aligned}$$

3.2. *Sets.* We shall ordinarily designate *sets* or *collections* of numbers by capital letters of the Roman alphabet. If m is a number and S is a set, we write

$$m \in S$$

to indicate that m is a member of S, and we write

$$m \notin S$$

to indicate that m is not a member of S.

We shall also consider sets which consist of n-tuples (usually for fixed n; thus, we shall not ordinarily consider sets which contain both pairs and triples). We indicate that the n-tuple $(a_1, \ldots, a_n)$ is or is not a member of the set S of n-tuples by writing $(a_1, \ldots, a_n) \in S$ or $(a_1, \ldots, a_n) \notin S$, respectively.

If R and S are sets, then $R \cup S$ is the *union* of R and S, that is, the set of all numbers (or of n-tuples of numbers, as the case may be) that belong to either R or S *or both.* Also, $R \cap S$ is the *intersection* of R and S, that is, the set of all numbers (or of n-tuples of numbers, as the case may be) that belong to both R and S. Finally, $\bar{R}$ is the *complement* of R, that is, the set of all numbers (or of n-tuples of numbers, as the case may be) that do not belong to R. In order that every set of numbers (or of n-tuples of numbers) may have a complement, it is necessary that we allow the *empty* or *null set*, which has no members, as a legitimate set. The null set is written ϕ. We write

$$R \subset S$$

if each element of R is also an element of S. In particular,

$$R \subset R$$
$$\phi \subset R$$

for any set R whatever.

We shall, on occasion, employ the notation of this section even when dealing with sets whose elements are not natural numbers or n-tuples of natural numbers.

3.3. *Functions.* Let $n \geqq 1$ be fixed in the following discussion. Let D be some set of n-tuples. Then, by a *function on* D or a *function whose domain is* D is meant a definite correspondence by which there is associated with each element of D a single natural number, called the value of the function for this element of D. Such a function is called *n-ary* or a *function of n variables.* A 1-ary function is called *singulary;* a 2-ary function is called *binary;* a 3-ary function is called *ternary.*

A function will ordinarily be designated by a single letter, such as f, g, h, or a lower-case Greek letter. If f is an n-ary function and if $(a_1, \ldots, a_n)$ is in the domain of f, then we write $f(a_1, \ldots, a_n)$ to designate the number associated with the n-tuple $(a_1, \ldots, a_n)$ by the function f. We shall tolerate the "abuse of language" whereby an n-ary function f is written $f(x_1, \ldots, x_n)$.

An n-ary function whose domain is the set of *all* n-tuples is called *total.*

To say of the n-ary functions f, g that

$$f = g$$

means that f and g have the same domain D and that, for every n-tuple $(a_1, \ldots, a_n)$ for which $(a_1, \ldots, a_n) \in D$, we have

$$f(a_1, \ldots, a_n) = g(a_1, \ldots, a_n).$$

The set of all numbers that are values of a function f is called the *range* of f.

Thus, the process of squaring a number gives rise to a singulary function which we write x^2. The domain of x^2 is the set of all integers; its range is the set of all perfect squares. x^2 is a total function.

The process of division by 2 gives rise to a singulary function which we write $x/2$. The domain of $x/2$ is the set of even numbers; its range is the set of all numbers.

The process of subtraction gives rise to a binary function which we write $x - y$. The domain of $x - y$ is the set of all pairs (x, y) for which $x \geqq y$; its range is the set of all integers.

Let S be a set of n-tuples. Then, by the *characteristic function of the set* S, written $C_S(x_1, \ldots, x_n)$, is to be understood the total n-ary function whose value, for a given n-tuple $(a_1, \ldots, a_n)$, is 0 if $(a_1, \ldots, a_n) \in S$ and is 1 if $(a_1, \ldots, a_n) \notin S$. Setting $M = R \cup S$ and $N = R \cap S$, we have

$$C_M = C_R \cdot C_S,$$
$$C_N = C_R + C_S - (C_R \cdot C_S),$$
$$C_{\bar{R}} = 1 - C_R.$$

3.4. *Statements.* The fundamental property of a statement is that it asserts a proposition that must be either true or false.

If p, q are given statements, then:

$p \wedge q$ (read "p and q") is the statement that asserts that *both* p and q hold. Thus, $p \wedge q$ is true if and only if both p and q are true.

$p \vee q$ (read "p or q") is the statement that asserts that *either* p or q *or both* hold. Thus, $p \vee q$ is true if and only if at least one of p, q is true.

$\sim p$ (read "not p") is the statement that asserts that p does not hold. Thus, $\sim p$ is true if and only if p is false.

3.5. *Predicates.* Consider the expression

$$x + y = 5.$$

As it stands, this is clearly not a statement. (For how can we say that it is true or that it is false?) If, however, we replace the letters x, y by numbers, a definite statement, true or false, is obtained. Thus,

$$2 + 3 = 5 \text{ is true,}$$
$$\text{but } 2 + 6 = 5 \text{ is false.}$$

An expression that contains lower-case letters of the Roman alphabet (with or without subscripts) and that becomes a statement when these letters are replaced by any numbers whatever (always assuming that the same letter, at two different occurrences in the expression, is replaced by the same number) is called a *predicate*. We shall usually employ upper-case letters of the Roman alphabet, such as P, Q, R, S, to designate predicates. The lower-case letters that occur in a predicate are called its arguments. If a predicate P has the arguments $x_1, x_2, \ldots, x_n$, then the statement which results on replacing x_1 by the number a_1, x_2 by the number a_2, . . . , and x_n by the number a_n is written $P(a_1, a_2, \ldots, a_n)$. Such a predicate is called n-ary. As with functions, "1-ary" and "2-ary" may be replaced by "singulary" and "binary," respectively. We shall tolerate the "abuse of language" whereby an n-ary predicate is written $P(x_1, \ldots, x_n)$.

Let $P(x_1, \ldots, x_n)$ be an n-ary predicate. Then, by the *extension* of P, written

$$\{x_1, x_2, \ldots, x_n \mid P(x_1, x_2, \ldots, x_n)\},$$

we shall mean the set of all n-tuples $(a_1, \ldots, a_n)$ for which $P(a_1, \ldots, a_n)$ is true.

Hence,

$$(a_1, \ldots, a_n) \in \{x_1, \ldots, x_n \mid P(x_1, \ldots, x_n)\}$$

if and only if $P(a_1, \ldots, a_n)$ is true.

Thus, if we let

$$S = \{x, y \mid x + y = 5\},$$

we have

$$(2, 3) \in S, \quad (4, 1) \in S, \quad (6, 2) \notin S, \quad (5, 0) \in S.$$

Two predicates are said to be *equivalent* if they have the same extension. We write

$$P(x_1, \ldots, x_n) \leftrightarrow Q(x_1, \ldots, x_n)$$

to indicate that P and Q are equivalent. Thus,

$$P(x_1, \ldots, x_n) \leftrightarrow Q(x_1, \ldots, x_n)$$

if and only if

$$\{x_1, \ldots, x_n \mid P(x_1, \ldots, x_n)\} = \{x_1, \ldots, x_n \mid Q(x_1, \ldots, x_n)\}.$$

For example,

$$x + y = 5 \leftrightarrow x + y + 1 = 6.$$

By the *characteristic function* of an n-ary predicate $P(x_1, \ldots, x_n)$ is understood the characteristic function of its extension, $\{x_1, \ldots, x_n \mid P(x_1, \ldots, x_n)\}$. The characteristic function of $P(x_1, \ldots, x_n)$ is written $C_P(x_1, \ldots, x_n)$. We have

$$\begin{aligned} C_P(a_1, \ldots, a_n) &= 0 \text{ if } P(a_1, \ldots, a_n) \text{ is true,} \\ C_P(a_1, \ldots, a_n) &= 1 \text{ otherwise.} \end{aligned}$$

It is clear that the connectives "$\vee$", "$\wedge$", and "$\sim$" can be applied to predicates to obtain new predicates. Thus, for example,

$$\begin{aligned} &\sim (x + y = 5) \leftrightarrow x + y \neq 5, \\ &(x < y \vee x = y) \leftrightarrow x \leqq y. \end{aligned}$$

3.6. *Quantifiers.* Let $P(y, x_1, \ldots, x_n)$ [or, as we may write, following the convention of Sec. 3.1, $P(y, \mathfrak{x}^{(n)})$] be an $(n + 1)$-ary predicate. Then the expression

$$P(0, \mathfrak{x}^{(n)}) \vee P(1, \mathfrak{x}^{(n)}) \vee \cdots \vee P(z, \mathfrak{x}^{(n)})$$

is another $(n + 1)$-ary predicate, which we may write $Q(z, \mathfrak{x}^{(n)})$. The statement obtained by inserting definite numbers z, $\mathfrak{x}^{(n)}$ for the letters in the expression is true if and only if there is a number $y \leqq z$ for which

$P(y, \mathfrak{x}^{(n)})$ is true. We designate this predicate by

$$\bigvee_{y=0}^{z} P(y, \mathfrak{x}^{(n)}).$$

That is,

$$\bigvee_{y=0}^{z} P(y, \mathfrak{x}^{(n)}) \leftrightarrow P(0, \mathfrak{x}^{(n)}) \vee P(1, \mathfrak{x}^{(n)}) \vee \cdots \vee P(z, \mathfrak{x}^{(n)}).$$

$\bigvee_{y=0}^{z} P(y, \mathfrak{x}^{(n)})$ is read, "There exists a y between 0 and z such that $P(y, \mathfrak{x}^{(n)})$." Similarly, we write

$$\bigwedge_{y=0}^{z} P(y, \mathfrak{x}^{(n)}) \leftrightarrow P(0, \mathfrak{x}^{(n)}) \wedge P(1, \mathfrak{x}^{(n)}) \wedge \cdots \wedge P(z, \mathfrak{x}^{(n)}).$$

$\bigwedge_{y=0}^{z} P(y, \mathfrak{x}^{(n)})$ is read, "For all y between 0 and z, $P(y, \mathfrak{x}^{(n)})$." We have

$$\bigwedge_{y=0}^{z} P(y, \mathfrak{x}^{(n)}) \leftrightarrow \sim \bigvee_{y=0}^{z} \sim P(y, \mathfrak{x}^{(n)}).$$

The symbols "$\bigvee_{y=0}^{z}$" and "$\bigwedge_{y=0}^{z}$" are referred to as a *bounded existential quantifier* and a *bounded universal quantifier*, respectively.

$\bigvee_{y} P(y, \mathfrak{x}^{(n)})$ may be regarded as an abbreviation of the "infinite expression"

$$P(0, \mathfrak{x}^{(n)}) \vee P(1, \mathfrak{x}^{(n)}) \vee \cdots .$$

More accurately, $\bigvee_{y} P(y, \mathfrak{x}^{(n)})$ is an n-ary predicate which gives rise to a true statement, for a given choice of values for the arguments $\mathfrak{x}^{(n)}$, if and only if there is a number y_0 such that $P(y_0, \mathfrak{x}^{(n)})$ is a true statement for the given choice of values.

Similarly, $\bigwedge_{y} P(y, \mathfrak{x}^{(n)})$ may be regarded as an abbreviation of the "infinite expression"

$$P(0, \mathfrak{x}^{(n)}) \wedge P(1, \mathfrak{x}^{(n)}) \wedge \cdots .$$

More accurately, $\bigwedge_{y} P(y, \mathfrak{x}^{(n)})$ is an n-ary predicate which gives rise to a true statement, for a given choice of values for the arguments $\mathfrak{x}^{(n)}$, if and only if, for every number y_0, $P(y_0, \mathfrak{x}^{(n)})$ is a true statement for the given choice of values.

$\bigvee_y P(y, \mathfrak{x}^{(n)})$ is read, "There exists a y such that $P(y, \mathfrak{x}^{(n)})$."

$\bigwedge_y P(y, \mathfrak{x}^{(n)})$ is read, "For all y, $P(y, \mathfrak{x}^{(n)})$."

"$\bigvee_y$" and "$\bigwedge_y$" are referred to as an *existential quantifier* and a *universal quantifier*, respectively. We have

$$\bigwedge_y P(y, \mathfrak{x}^{(n)}) \leftrightarrow \sim \bigvee_y \sim P(y, \mathfrak{x}^{(n)}).$$

The following equivalences serve to illustrate the above concepts:

$$\bigvee_y (x + y = 5) \leftrightarrow x \leqq 5,$$

$$\bigvee_y (x + y = 5) \leftrightarrow \bigvee_{y=0}^{5} (x + y = 5),$$

$$\bigwedge_y (xy = 0) \leftrightarrow x = 0,$$

$$\bigvee_{y=0}^{z} (x + y = 5) \leftrightarrow [(x + z \geqq 5) \wedge (x \leqq 5)].$$

PART 1

THE GENERAL THEORY OF COMPUTABILITY

CHAPTER 1

COMPUTABLE FUNCTIONS

1. Turing Machines. We shall proceed to define a class of functions which we propose to identify with the *effectively calculable functions*, i.e., with those functions for which an algorithm that can be used to compute their values exists (cf. Introduction, Sec. 1). Our point of departure is the remark that, if an algorithm for performing a task exists, then, at least in principle, a computing machine for accomplishing this task can be constructed. Such a computing machine is *deterministic* in the sense that, while it is in operation, its entire future is completely specified by its status at some one instant.[1]

Thus, we shall give a mathematical characterization of a class of objects which we shall call *Turing machines*.[2] These will be defined by analogy with physical computers of a certain kind. With this will come a mathematical characterization of a class of numerical functions, the functions "computed" by these Turing machines. These functions will be called *computable functions*, and it will be proposed to identify the intuitive concept of effectively calculable function with the new precise concept of computable function. Discussion of whether this identification is too narrow (i.e., omits functions that should be considered effectively calculable) or too wide (i.e., includes functions that should not be considered effectively calculable), or perhaps both, had best await the detailed definition.

It is important that, in the following, the reader carefully distinguish between formal definitions and more or less vague explanations intended to appeal to intuition, and, particularly, that he satisfy himself that only formal definitions are employed in actual proofs.

Whereas physical computing machines are handicapped by having only a finite region available for storing input data and the various intermediate expressions formed in the course of a computation (e.g., the partial products in a multiplication), we shall imagine a computing machine that prints symbols on a linear tape, assumed to be *infinite* in both directions, ruled into a two-way infinite sequence of boxes (cf.

[1] This excludes from our consideration both analogue computers and computers that contain "random" elements.

[2] After A. M. Turing. Cf. Turing [1].

Fig. 1). We assume that the machine is capable of only a finite number of distinct internal states or configurations and that its next immediate operation at a given moment is determined by its internal configuration at that moment, taken in conjunction with the (finite) expression that then appears on the tape. If the machine is regarded as "sensitive" to only one of the squares on the tape at a time (the square "scanned" by the machine), then its future behavior must be determined by its internal configuration taken in conjunction with the symbol that appears on the scanned square. For, although a physical computing machine will consist of parts such as gears or vacuum tubes (so that, at any point in its operation, the internal configuration of such a machine is the actual arrangement of these parts), in effect the only function of the internal

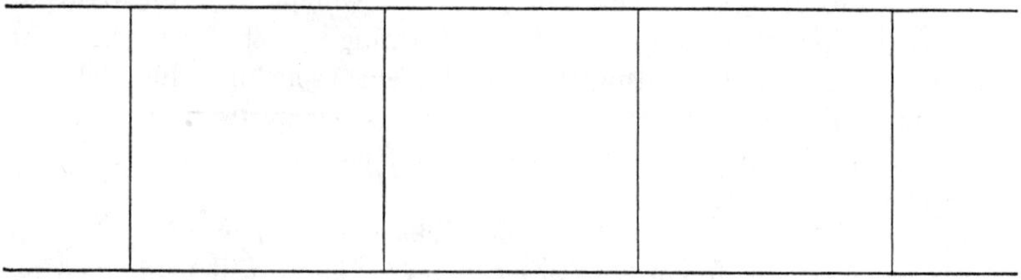

FIG. 1

configuration is to specify the next act of the computer, given knowledge of the symbol that appears on the scanned square. We shall assume that this next act must be one of the types that we now indicate. It can be a complete halt of operations; or it can be either a specified change in the symbol that appears on the scanned square or a change of the scanned square itself to the square one to the left or one to the right, followed by a specified change of internal configuration. Thus we may regard an internal configuration to be determined mathematically by a set of rules that specify, for each symbol the machine is capable of printing, what the next act of the machine (in the sense of the previous sentence) will be, given that the symbol in question is the one that appears on the scanned square.

For the purpose of symbolic representation of these concepts, we introduce the following notational conventions.

The symbols $q_1, q_2, q_3, q_4, \ldots$ will be regarded as denoting internal configurations; the symbols $S_0, S_1, S_2, \ldots$ will be regarded as symbols which various machines may be capable of printing; the symbols R and L will represent a move of one square to the right and one square to the left, respectively.

The above material is to be considered a preliminary explanation intended to help make the following more formal material readily comprehensible.

DEFINITION 1.1. *An* **expression** *is a finite sequence (possibly empty) of symbols chosen from the list:* $q_1, q_2, q_3, \ldots ; S_0, S_1, S_2, \ldots ; R, L$.

DEFINITION 1.2. *A* **quadruple** *is an expression having one of the following forms:*

(1) $q_i\ S_j\ S_k\ q_l$.
(2) $q_i\ S_j\ R\ q_l$.
(3) $q_i\ S_j\ L\ q_l$.
(4) $q_i\ S_j\ q_k\ q_l$.

A Turing machine will be defined so as to consist entirely of quadruples. A quadruple of the form 1, 2, or 3 above specifies the *next act* of a Turing machine when in internal configuration q_i and scanning a square on which appears the symbol S_j. Thus, quadruple 1 indicates that the next act is to replace S_j by S_k on the scanned square and to enter internal configuration q_l. Quadruple 2 indicates motion of one square to the right followed by entry into internal configuration q_l. Quadruples of type 3 similarly indicate motion leftward. Quadruples of the form 4 will enter our development only in Sec. 4 of this chapter, and discussion of their role is therefore deferred until that point.

DEFINITION 1.3. *A* **Turing machine** *is a finite (nonempty) set of quadruples*[1] *that contains no two quadruples whose first two symbols are the same.*[2]

The q_i's and S_i's that occur in the quadruples of a Turing machine are called its **internal configurations** *and its* **alphabet,** *respectively.*

If none of the quadruples of a Turing machine Z is of the type 4, *Z is called* **simple.**

In Secs. 1 to 3, we are concerned entirely with simple Turing machines, although our statements will hold for any Turing machines.

The symbols S_0 and S_1 will have a special role in our development. S_0 will also be written B, and S_1 will be written 1. S_0 will serve as a blank; so replacing a symbol by S_0 will amount to erasing it.

In order to bring change into our, at present, static picture of the Turing machine, we must indicate how to capture in a single expression an entire present state of a Turing machine. Our intuitive description indicates that such an expression should include:

(1) The expression on the tape.
(2) The internal configuration.
(3) The square scanned.

[1] Turing's original development employed *quintuples* rather than quadruples. The present formulation (including the use of a two-way infinite tape) follows Post [6].

[2] This last restriction is one of consistency. That is, it guarantees that no Turing machine will ever be confronted with two different instructions at the same time.

DEFINITION 1.4. *An* **instantaneous description** *is an expression that contains exactly one* q_i, *neither* R *nor* L, *and is such that* q_i *is not the rightmost symbol.*

If Z *is a Turing machine and* α *is an instantaneous description, then we say that* α **is an instantaneous description of** Z *if the* q_i *that occurs in* α *is an internal configuration of* Z *and if the* S_i*'s that occur in* α *are part of the alphabet of* Z.

DEFINITION 1.5. *An expression that consists entirely of the letters* S_i *is called a* **tape expression.**

In what follows, P and Q usually are tape expressions.

Now, we visualize each instantaneous description of a Turing machine as determining at most one immediately subsequent instantaneous description of this same Turing machine in a manner determined by an appropriate quadruple of this Turing machine. Thus, the instantaneous descriptions of a Turing machine may be regarded as an abstract substitute for successive moments of time. This gives rise to the following suggestive language:

DEFINITION 1.6. *Let* Z *be a Turing machine, and let* α *be an instantaneous description of* Z, *where* q_i *is the internal configuration that occurs in* α *and where* S_j *is the symbol immediately to the right of* q_i. *Then we call* q_i *the* **internal configuration of** Z **at** α, *and we call* S_j *the* **symbol scanned by** Z **at** α. *The tape expression obtained on removing* q_i *from* α *is called the* **expression on the tape of** Z **at** α.

At this point, we must remove a basic inconsistency between our rigorous definitions and our intuitive picture. We have visualized our machine as possessing an *infinite* tape; yet, formally, the expression on the tape of a Turing machine at an instantaneous description α is always *finite*, and we have required that there always be a symbol scanned at any instantaneous description. This difficulty is removed by means of the special symbol S_0 or B. We change our intuitive picture to that of a machine with a tape that is always *finite* but that can be extended. The machine is to have the property that, whenever it is about to rush off an end of the tape, a new square, on which appears a B, is "spliced" onto that end of the tape.

We now indicate the precise manner in which an instantaneous description of a Turing machine is replaced by a succeeding instantaneous description.

DEFINITION 1.7. *Let* Z *be a Turing machine, and let* α, β *be instantaneous descriptions. Then we write* $\alpha \rightarrow \beta \quad (Z)$, *or (when no ambiguity can result) simply* $\alpha \rightarrow \beta$, *to mean that one*[1] *of the following alternatives holds:*

[1] By the definition of Turing machine, at most one of the alternatives can hold.

(1) *There exist expressions P and Q (possibly empty) such that*

$$\alpha \text{ is } Pq_iS_jQ,$$
$$\beta \text{ is } Pq_lS_kQ,$$

where

$$Z \text{ contains } q_i\ S_j\ S_k\ q_l.$$

(2) *There exist expressions P and Q (possibly empty) such that*

$$\alpha \text{ is } Pq_iS_jS_kQ,$$
$$\beta \text{ is } PS_jq_lS_kQ,$$

where

$$Z \text{ contains } q_i\ S_j\ R\ q_l.$$

(3) *There exists an expression P (possibly empty) such that*

$$\alpha \text{ is } Pq_iS_j,$$
$$\beta \text{ is } PS_jq_lS_0,$$

where

$$Z \text{ contains } q_i\ S_j\ R\ q_l.$$

(4) *There exist expressions P and Q (possibly empty) such that*

$$\alpha \text{ is } PS_kq_iS_jQ,$$
$$\beta \text{ is } Pq_lS_kS_jQ,$$

where

$$Z \text{ contains } q_i\ S_j\ L\ q_l.$$

(5) *There exists an expression Q (possibly empty) such that*

$$\alpha \text{ is } q_iS_jQ,$$
$$\beta \text{ is } q_lS_0S_jQ,$$

where

$$Z \text{ contains } q_i\ S_j\ L\ q_l.$$

The following theorems are immediate consequences of this definition.

THEOREM 1.1. *If* $\alpha \rightarrow \beta \quad (Z)$ *and* $\alpha \rightarrow \gamma \quad (Z)$, *then* $\beta = \gamma$.

THEOREM 1.2. *If* $\alpha \rightarrow \beta \quad (Z)$, *and* $Z \subset Z'$,† *then* $\alpha \rightarrow \beta \quad (Z')$.

DEFINITION 1.8. *An instantaneous description* α *is called* **terminal**[1] **with respect to** *Z if for no* β *do we have* $\alpha \rightarrow \beta \quad (Z)$.

DEFINITION 1.9. *By a* **computation of a Turing machine** *Z is meant a finite sequence* $\alpha_1, \alpha_2, \ldots, \alpha_p$ *of instantaneous descriptions such that* $\alpha_i \rightarrow \alpha_{i+1} \quad (Z)$ *for* $1 \leqq i < p$ *and such that* α_p *is terminal with respect to Z. In such a case, we write* $\alpha_p = \text{Res}_Z(\alpha_1)$ *and we call* α_p *the* **resultant** *of* α_1 *with respect to Z.*

† For the meaning of the symbol "$\subset$" of set inclusion, see the Introduction, Sec. 3.2.

[1] Thus, the machine interprets the absence of an instruction as a "stop" order.

Ordinarily, the internal configuration at α_1 will be taken as q_1.

By way of illustrating some of these concepts, let Z consist of the following quadruples:

$$\begin{array}{l} q_1\ S_0\ R\ q_1 \\ q_1\ S_2\ R\ q_1 \\ q_1\ S_5\ R\ q_1 \\ q_1\ S_3\ S_5\ q_2 \\ q_2\ S_5\ L\ q_3 \\ q_3\ S_0\ S_5\ q_3 \\ q_3\ S_2\ S_5\ q_3 \\ q_3\ S_3\ S_5\ q_3. \end{array}$$

Z has the *internal configurations* q_1, q_2, q_3 and the *alphabet* S_0, S_2, S_3, S_5. The following are *computations* of this Z:

(1)
$$\begin{aligned} S_2q_1S_0S_5S_3 &\rightarrow S_2S_0q_1S_5S_3 \\ &\rightarrow S_2S_0S_5q_1S_3 \\ &\rightarrow S_2S_0S_5q_2S_5 \\ &\rightarrow S_2S_0q_3S_5S_5; \end{aligned}$$

hence, $\text{Res}_Z\ (S_2q_1S_0S_5S_3) = S_2S_0q_3S_5S_5$.

(2)
$$\begin{aligned} q_1S_3 &\rightarrow q_2S_5 \\ &\rightarrow q_3S_0S_5 \\ &\rightarrow q_3S_5S_5; \end{aligned}$$

hence, $\text{Res}_Z\ (q_1S_3) = q_3S_5S_5$. Thus the "effect of Z" is to hunt for the symbol S_3, looking first in the square initially scanned (i.e., scanned at α_1) and then moving steadily to the right. If an S_3 is located, it is replaced by S_5, and another S_5 is then placed in the square immediately to the left.

Now, what if α_1 is so chosen that there is no occurrence of S_3 to the right of q_1? Then the computation will go on "forever," or, more accurately, there will be no computation beginning with α_1 in the sense of Definition 1.9. Thus, if $\alpha_1 = S_2q_1S_0S_5S_2$, then we may write

$$\begin{aligned} S_2q_1S_0S_5S_2 &\rightarrow S_2S_0q_1S_5S_2 \\ &\rightarrow S_2S_0S_5q_1S_2 \\ &\rightarrow S_2S_0S_5S_2q_1S_0 \\ &\rightarrow S_2S_0S_5S_2S_0q_1S_0 \\ &\rightarrow \cdots; \end{aligned}$$

so one never arrives at a terminal instantaneous description. Hence, $\text{Res}_Z\ (S_2q_1S_0S_5S_2)$ is undefined.

2. Computable Functions and Partially Computable Functions. In the previous section we introduced Turing machines and showed how they could be used to perform symbolic computations. In order to have Turing machines perform *numerical* computations, it is necessary that we introduce a suitable symbolic representation for numbers. The simplest way of doing this, and the best for our purposes, is to choose one symbol as basic—we choose S_1—and to symbolize a number by an expression consisting entirely of occurrences of this symbol (recall that we have agreed to write 1 for S_1). If n is a positive integer, we write S_i^n for the expression $\underbrace{S_iS_i \cdots S_i}_{n}$ that consists of n occurrences of S_i. For completeness, we take S_i^0 to be the null expression. Then, we may write

DEFINITION 2.1. *With each number n we associate the tape expression* $\bar{n}$ *where* $\bar{n} = 1^{n+1}$.

Thus $\bar{3} = 1111$ and, in general, $\bar{n} = \underbrace{11 \cdots 1}_{n+1}$.

DEFINITION 2.2. *With each k-tuple* $(n_1, n_2, \ldots, n_k)$ *of integers we associate the tape expression* $\overline{(n_1, n_2, \ldots, n_k)}$, *where*

$$\overline{(n_1, n_2, \ldots, n_k)} = \bar{n}_1 B \bar{n}_2 B \cdots B \bar{n}_k.$$

Thus, $\overline{(2, 3, 0)} = 111B1111B1$.

This notation is convenient in connection with initial data or inputs. For outputs, we use

DEFINITION 2.3. *Let M be any expression. Then* $\langle M \rangle$ *is the number of occurrences of* 1 *in M.*

Thus, $\langle 11BS_4q_3 \rangle = 2$; $\langle q_3q_2S_5 \rangle = 0$. Note that $\langle \overline{m - 1} \rangle = m$ and that $\langle PQ \rangle = \langle P \rangle + \langle Q \rangle$.

DEFINITION 2.4. *Let Z be a Turing machine. Then, for each n, we associate with Z an n-ary function*[1]

$$\Psi_Z^{(n)}(x_1, x_2, \ldots, x_n)$$

as follows.

For each n-tuple $(m_1, m_2, \ldots, m_n)$, *we set* $\alpha_1 = q_1\overline{(m_1, m_2, \ldots, m_n)}$ *and we distinguish between two cases:*

(1) *There exists a computation of* Z, $\alpha_1, \ldots, \alpha_p$. *In this case we set*

$$\Psi_Z^{(n)}(m_1, m_2, \ldots, m_n) = \langle \alpha_p \rangle = \langle \mathrm{Res}_Z(\alpha_1) \rangle.$$

(2) *There exists no computation* $\alpha_1, \ldots, \alpha_p$ [*that is,* $\mathrm{Res}_Z(\alpha_1)$ *is undefined*]. *In this case we leave*

$$\Psi_Z^{(n)}(m_1, m_2, \ldots, m_n)$$

undefined. We write $\Psi_Z(x)$ *for* $\Psi_Z^{(1)}(x)$.

[1] For a discussion of "n-ary," "total functions," etc., cf. Introduction, Sec. 3.3.

DEFINITION 2.5. *An n-ary function* $f(x_1, \ldots, x_n)$ *is* **partially computable** *if there exists a Turing machine Z such that*

$$f(x_1, \ldots, x_n) = \Psi_Z^{(n)}(x_1, \ldots, x_n).$$

In this case we say that Z computes f. If, in addition, $f(x_1, \ldots, x_n)$ *is a total function, then it is called* **computable.**

It is the concept of *computable function* that we propose to identify with the intuitive concept of effectively calculable function. A partially computable function may be thought of as one for which we possess an algorithm which enables us to compute its value for elements of its domain, but which will have us computing forever in attempting to obtain a functional value for an element not in its domain, without ever assuring us that no value is forthcoming. In other words, when an answer is forthcoming, the algorithm provides it; when no answer is forthcoming, the algorithm has one spend an infinite amount of time in a vain search for an answer. We shall now comment briefly on the adequacy of our identification of effective calculability with computability in the sense of Definition 2.5. The situation is quite analogous to that met whenever one attempts to replace a vague concept, having a powerful intuitive appeal, with an exact mathematical substitute. (An obvious example is the area under a curve.) In such a case, it is, of course, pointless to demand a mathematical proof of the equivalence of the two concepts; the very vagueness of the intuitive concept precludes this. However, it is possible to present arguments, having strong intuitive appeal, which tend to make this identification extremely reasonable. We shall outline several arguments of this sort.

Historically, proposals were made by a number of different persons at about the same time (1936), mostly independently of one another, to identify the concept of effectively calculable function with various precise concepts. In this connection we may mention Church's notion of λ-*definability*,[1] the Herbrand-Gödel-Kleene notion of *general recursiveness*,[2] Turing's notion of *computability*[3] (defined in a manner differing somewhat from that of the present work), Post's notion of 1-*definability*,[4] and Post's notion of *binormality*.[5] These notions, which (except for the third and fourth) were quite different in formulation, have all been proved equivalent[6] in the sense that the classes of functions obtained are the

[1] Cf. Church [1, 3].

[2] Defined in Gödel [2]. The proposal to identify with effective calculability first appeared in Church [1]. Cf. also Kleene [1, 4, 6].

[3] Cf. Turing [1].

[4] Cf. Post [1].

[5] Cf. Post [2, 3]; Rosenbloom [1].

[6] The equivalence of λ-definability with general recursiveness is proved in Kleene [2]. The equivalence of λ-definability with Turing's notion of computability is

same in each case. Now, the fact that these different concepts have turned out to be equivalent tends to make the identification of them with effective calculability the more reasonable.

Next, we may note that every computable function must surely be regarded as effectively calculable. For let $f(m)$ (for simplicity, we consider a singulary function) be computable, and let Z be a Turing machine which computes $f(m)$. Then, if we are given a number m_0, we may begin with the instantaneous description $\alpha_1 = q_1\overline{m_0} = q_1\underbrace{11 \cdots 1}_{m_0+1}$ and successively obtain instantaneous descriptions $\alpha_2, \alpha_3, \ldots, \alpha_p$, where $\alpha_1 \to \alpha_2 \to \alpha_3 \to \cdots \to \alpha_p$ and where α_p is terminal. Since $f(m)$ is computable, such a terminal α_p must be obtainable in a finite number of steps. But then $f(m_0) = \langle\alpha_p\rangle$, and $\langle\alpha_p\rangle$ is simply the number of 1's in α_p. That this procedure would ordinarily be regarded as "effective" is clear when one realizes that to decide, for a given instantaneous description α of Z, whether or not there exists an instantaneous description β such that $\alpha \to \beta \quad (Z)$, and, if the answer is affirmative, to determine which β satisfies this condition, it suffices to write α in the form Pq_iS_jQ and to locate that quadruple of Z, if one exists, that begins $q_i\, S_j$.

This indicates, at any rate, that our definition is not too "wide." Is it, perhaps, too "narrow"? An answer as satisfying as the one given for the previous question is, presumably, not to be expected. For how can we ever exclude the possibility of our being presented, some day (perhaps by some extraterrestrial visitors), with a (perhaps extremely complex) device or "oracle" that "computes" a noncomputable function? However, there are fairly convincing reasons for believing that this will never happen. It is possible to show directly, for various possible theoretical computing devices which seem to possess greater power than Turing machines, that any functions computed by them are, in fact, computable. Thus, we might consider machines which could move any number of squares to the right or left, or which operate on a two-, three-, or one-hundred-dimensional tape, or which are capable of inserting squares into their own tape. Now, it is not very difficult to show that any computation that could be carried out by such a machine can also be performed by a Turing machine. This will be more readily

---

proved in Turing [2]. It follows at once, from the results of our Chap. 4 and the main result of Kleene [1], that our present notion of computability is equivalent to general recursiveness. Equivalence proofs for Post's notion of 1-definability and binormality do not appear in the published literature. 1-definability is quite similar, conceptually, to computability, and an equivalence proof is quite easy. A direct proof of the equivalence of binormality with general recursiveness is given in the author's dissertation (Davis [1]). This equivalence is, in fact, an immediate consequence of the results of our present Chap. 6.

apparent in the light of the results of Chap. 4, at which point we shall resume this discussion.

3. Some Examples. In this section we shall consider several computable and partially computable functions. These examples are introduced at this point primarily for their illustrative value. However, the fact that the functions chosen are computable will be employed later.

Example 3.1. Addition. Let $f(x, y) = x + y$. We shall construct a Turing machine Z which computes $f(x, y)$, that is, such that

$$\Psi_Z^{(2)}(x, y) = x + y.$$

We take Z to consist[1] of the quadruples

$$\begin{array}{llll} q_1 & 1 & B & q_1 \\ q_1 & B & R & q_2 \\ q_2 & 1 & R & q_2 \\ q_2 & B & R & q_3 \\ q_3 & 1 & B & q_3. \end{array}$$

Let $\alpha_1 = q_1(\overline{m_1, m_2}) = q_1\overline{m_1}B\overline{m_2}$. Then

$$\begin{aligned} \alpha_1 &= q_1 1 1^{m_1} B 1 1^{m_2} \\ &\rightarrow q_1 B 1^{m_1} B 1 1^{m_2} \\ &\rightarrow B q_2 1^{m_1} B 1 1^{m_2} \\ &\rightarrow \cdots \\ &\rightarrow B 1^{m_1} q_2 B 1 1^{m_2} \\ &\rightarrow B 1^{m_1} B q_3 1 1^{m_2} \\ &\rightarrow B 1^{m_1} B q_3 B 1^{m_2}, \end{aligned}$$

which is terminal. Thus,

$$\begin{aligned} \Psi_Z^{(2)}(m_1, m_2) &= \langle \mathrm{Res}_Z\,(\alpha_1)\rangle \\ &= \langle B1^{m_1}Bq_3B1^{m_2}\rangle \\ &= m_1 + m_2. \end{aligned}$$

Example 3.2. The Successor Function. Let $S(x) = x + 1$. We shall show that $S(x)$ is computable.

In fact, let Z be any Turing machine with respect to which $q_1\overline{m}$ is terminal for all m. Thus, Z may consist of the single quadruple $q_1\ B\ B\ q_1$. Then

$$\Psi_Z(m) = \langle q_1\overline{m}\rangle = m + 1.$$

Example 3.3. Subtraction. Let $f(x, y) = x - y$. This function is defined only for $x \geqq y$. We shall show that $f(x, y)$ is a partially com-

[1] The purpose of Z is simply that of erasing two 1's. (Recall that for "input" the number n is represented by a sequence of 1's of length $n + 1$, whereas for "output" it is represented by a tape expression containing exactly n 1's.)

putable function; that is, we shall construct a Turing machine Z such that

$$\Psi_Z^{(2)}(x, y) = x - y.$$

Z will operate by successively canceling 1's from both ends of the tape expression $1^{m_1+1}B1^{m_2+1}$ until one of the two groups of 1's is exhausted. As might be expected, this Turing machine is considerably more complicated than those we have considered so far. To aid the reader, various of the quadruples (or groupings of quadruples) which make up Z will be followed by brief comments suggesting their function. Such comments (e.g., the first sentence of this paragraph) unavoidably tend to become animistic in nature. Of course, such remarks are not part of our formal development, and they may be ignored by the reader who is so minded.[1] We take Z to consist of the following quadruples:

$q_1\ 1\ B\ q_1$ (erase 1 on the left)
$q_1\ B\ R\ q_2$

$q_2\ 1\ R\ q_2$ (locate the separating B)
$q_2\ B\ R\ q_3$

$q_3\ 1\ R\ q_3$ (locate the right-hand end)
$q_3\ B\ L\ q_4$

$q_4\ 1\ B\ q_4$ (erase 1 on the right)
$q_4\ B\ L\ q_5$

$q_5\ 1\ L\ q_6$ (if m_2 has been exhausted, stop; otherwise continue)

$q_6\ 1\ L\ q_6$ (locate the separating B)
$q_6\ B\ L\ q_7$

$q_7\ 1\ L\ q_8$ (if m_1 has been exhausted, go to q_1; otherwise continue)
$q_7\ B\ R\ q_9$

$q_8\ 1\ L\ q_8$ (locate the left-hand end, and return to q_1)
$q_8\ B\ R\ q_1$

$q_9\ B\ R\ q_9$ (compute in an infinite cycle)
$q_9\ 1\ L\ q_9$.

To show formally that $\Psi_Z^{(2)}(m_1, m_2) = m_1 - m_2$, we proceed as follows.

[1] Other readers may well prefer to read only the informal remarks. Cf. the third paragraph of the Introduction, Sec. 2.

Let $\alpha_1 = q_1\overline{(m_1, m_2)} = q_1\overline{m_1}B\overline{m_2}$. First, suppose that $m_1 \geqq m_2$, and let $k = m_1 - m_2$. Then

$$\begin{aligned}
\alpha_1 &= q_1 1^{m_1+1} B 1^{m_2+1} \\
&= q_1 1 1^{m_2} 1^k B 1^{m_2} 1 \\
&\to q_1 B 1^{m_2} 1^k B 1^{m_2} 1 \\
&\to B q_2 1^{m_2} 1^k B 1^{m_2} 1 \\
&\to \cdots \\
&\to B 1^{m_2} 1^k q_2 B 1^{m_2} 1 \\
&\to B 1^{m_2} 1^k B q_3 1^{m_2} 1 \\
&\to \cdots \\
&\to B 1^{m_2} 1^k B 1^{m_2} 1 q_3 B \\
&\to B 1^{m_2} 1^k B 1^{m_2} q_4 1 B \\
&\to B 1^{m_2} 1^k B 1^{m_2} q_4 B B \\
&\to \cdots \\
&\to B 1^{m_2} 1^k B q_6 1^{m_2} B B \\
&\to B 1^{m_2} 1^k q_6 B 1^{m_2} B B \\
&\to \cdots \\
&\to q_8 B 1^{m_2} 1^k B 1^{m_2} B B \\
&\to B q_1 1^{m_2} 1^k B 1^{m_2} B B = \alpha_s.
\end{aligned}$$

Now, except for initial and final B's, α_s is like α_1 with a pair of 1's canceled. The process is now repeated. Eventually,

$$\begin{aligned}
\alpha_1 &\to \cdots \\
&\to \alpha_s \\
&\to \cdots \\
&\to \cdots \\
&\to B^{m_2} q_1 1 1^k B 1 B^{m_2+1} \\
&\to \cdots \\
&\to B^{m_2+1} 1^k q_2 B 1 B^{m_2+1} \\
&\to \cdots \\
&\to B^{m_2+1} 1^k B q_4 1 B^{m_2+1} \\
&\to B^{m_2+1} 1^k B q_4 B B^{m_2+1} \\
&\to B^{m_2+1} 1^k q_5 B B B^{m_2+1} = \alpha_p,
\end{aligned}$$

which is terminal. But $\langle \alpha_p \rangle = k = m_1 - m_2$. Hence, if $m_1 \geqq m_2$,

$$\Psi_Z^{(2)}(m_1, m_2) = m_1 - m_2.$$

Next, suppose that $m_1 < m_2$, and let $k = m_2 - m_1$. Then

$$\begin{aligned}
\alpha_1 &= q_1 1^{m_1+1} B 1^{m_2+1} \\
&= q_1 1 1^{m_1} B 1^k 1^{m_1} 1 \\
&\to \cdots \\
&\to B q_1 1^{m_1} B 1^k 1^{m_1} B B
\end{aligned}$$

$$\begin{aligned}
&\to \cdots \\
&\to \cdots \\
&\to B^{m_1}q_1 1B1^k 1B^{m_1+1} \\
&\to B^{m_1}q_1 BB1^k 1B^{m_1+1} \\
&\to B^{m_1}Bq_2 B1^k 1B^{m_1+1} \\
&\to B^{m_1}BBq_3 1^k 1B^{m_1+1} \\
&\to \cdots \\
&\to B^{m_1}BB1^k q_4 1B^{m_1+1} \\
&\to B^{m_1}BB1^k q_4 BB^{m_1+1} \\
&\to \cdots \\
&\to B^{m_1}Bq_6 B1^k BB^{m_1+1} \\
&\to B^{m_1}q_7 BB1^k BB^{m_1+1} \\
&\to B^{m_1}Bq_9 B1^k BB^{m_1+1} = \alpha_r \\
&\to B^{m_1}BBq_9 1^k BB^{m_1+1} \\
&\to B^{m_1}Bq_9 B1^k BB^{m_1+1} \\
&\to \cdots
\end{aligned}$$

Here, the last two instantaneous descriptions listed are transformed back and forth into each other; so no terminal instantaneous description is ever reached (that is, the Turing machine "moves" back and forth between two adjacent squares). So, if $m_1 < m_2$, $\Psi_{Z^{(2)}}(m_1, m_2)$ is undefined.

Hence,

$$\Psi_{Z^{(2)}}(x, y) = x - y.$$

EXAMPLE 3.4. PROPER SUBTRACTION. Consider the function $x \dot{-} y$, defined by

$$\begin{aligned}
x \dot{-} y &= x - y \text{ if } x \geqq y, \\
x \dot{-} y &= 0 \text{ if } x < y.
\end{aligned}$$

Thus, $x \dot{-} y$ is defined for all x, y. We shall show that it is computable.

Let Z' be obtained from the Z of Example 3.3 by deleting the quadruples $q_9\ B\ R\ q_9$ and $q_9\ 1\ L\ q_9$ and then adding the quadruples:

$$\begin{aligned}
&q_9\ B\ R\ q_{10} \qquad \text{(erase all 1's)} \\
&q_{10}\ 1\ B\ q_9.
\end{aligned}$$

Then it is clear that, if $m_1 \geqq m_2$, the analysis given for the Z of Example 3.3 holds just as well here, and that we have

$$\Psi_{Z'^{(2)}}(m_1, m_2) = \Psi_{Z^{(2)}}(m_1, m_2) = m_1 - m_2 = m_1 \dot{-} m_2.$$

Furthermore, if $m_1 < m_2$, then the analysis that led to the instantaneous description $\alpha_r = B^{m_1}Bq_9B1^kBB^{m_1+1}$ holds with respect to Z'. However, with respect to Z', α_r leads to a terminal instantaneous description as follows:

$$\begin{aligned}\alpha_r &= B^{m_1}Bq_9B1^kBB^{m_1+1}\\ &\rightarrow B^{m_1}BBq_{10}1^kBB^{m_1+1}\\ &\rightarrow B^{m_1}BBq_9B1^{k-1}BB^{m_1+1}\\ &\rightarrow B^{m_1}BBBq_{10}1^{k-1}BB^{m_1+1}\\ &\rightarrow \cdot\cdot\cdot\\ &\rightarrow B^{m_1+k+2}q_{10}BB^{m_1+1} = \alpha_s,\end{aligned}$$

which is terminal. Since $\langle\alpha_s\rangle = 0$,

$$\Psi_{Z'}^{(2)}(m_1, m_2) = 0$$

if $m_1 < m_2$.

Thus,

$$\Psi_{Z'}^{(2)}(x, y) = x \dot{-} y.$$

Hence $x \dot{-} y$ is a computable function.

Now, just as we have "completed" subtraction to obtain proper subtraction, we can complete any partially computable function $f(m)$ by setting

$$g(m) = f(m), \text{ where } f(m) \text{ is defined,}$$
$$g(m) = 0, \text{ elsewhere.}$$

However, it is possible to show (cf. Theorem 5-1.8) that there exists a partially computable function $f(m)$ whose completion $g(m)$ is *not* computable.

EXAMPLE 3.5. THE IDENTITY FUNCTION. Let $I(x) = x$. We show that $I(x)$ is computable.

Let Z consist of the single quadruple

$$q_1\ 1\ B\ q_1.$$

Then

$$q_1\bar{n} = q_1 1 1^n \rightarrow q_1 B 1^n,$$

which is terminal.

Hence,

$$\Psi_Z(n) = \langle q_1 B 1^n\rangle = n.$$

EXAMPLE 3.6. OTHER IDENTITY FUNCTIONS. We consider the n-ary functions $U_i^n(x_1, x_2, \ldots, x_n)$, $1 \leqq i \leqq n$, such that

$$U_i^n(x_1, x_2, \ldots, x_n) = x_i.$$

Thus, $U_1^1(x) = I(x)$. We proceed to show that these functions are all computable. That is, for each n and for each i, $1 \leqq i \leqq n$, there is a Turing machine Z such that

$$\Psi_Z^{(n)}(x_1, x_2, \ldots, x_n) = x_i.$$

We let Z consist of the following quadruples, where j runs over all integers $\neq i$ such that $1 \leqq j \leqq n$.

$$\left.\begin{array}{r}q_j\ 1\ B\ q_{2n+j}\\ q_j\ B\ R\ q_{j+1}\\ q_{2n+j}\ B\ R\ q_j\end{array}\right\}\qquad\text{(erase a block of 1's)}$$

$$\begin{array}{rl}q_i\ 1\ B\ q_i & \qquad\text{(erase the initial 1, only)}\\ q_i\ B\ R\ q_{2n+i} & \end{array}$$

$$\begin{array}{r}q_{2n+i}\ 1\ R\ q_{2n+i}\\ q_{2n+i}\ B\ R\ q_{i+1}.\end{array}$$

Then

$$\begin{aligned}
q_1 1^{m_1+1}B1^{m_2+1}B & \cdots B1^{m_i+1}B \cdots B1^{m_n+1}\\
&\rightarrow \cdots\\
&\rightarrow B^{m_1+1}Bq_2 1^{m_2+1}B \cdots B1^{m_i+1}B \cdots B1^{m_n+1}\\
&\rightarrow \cdots\\
&\rightarrow \cdots\\
&\rightarrow B^{m_1+1}BB^{m_2+1}B \cdots Bq_i 1^{m_i+1}B \cdots B1^{m_n+1}\\
&\rightarrow B^{m_1+1}BB^{m_2+1}B \cdots Bq_i B1^{m_i}B \cdots B1^{m_n+1}\\
&\rightarrow B^{m_1+1}BB^{m_2+1}B \cdots BBq_{2n+i}1^{m_i}B \cdots B1^{m_n+1}\\
&\rightarrow \cdots\\
&\rightarrow B^{m_1+1}BB^{m_2+1}B \cdots BB1^{m_i}Bq_{i+1} \cdots B1^{m_n+1}\\
&\rightarrow \cdots\\
&\rightarrow \cdots\\
&\rightarrow B^s 1^{m_i}B^t q_n 1^{m_n+1} \qquad \text{(for } s, t \text{ suitably chosen)}\\
&\rightarrow \cdots\\
&\rightarrow B^s 1^{m_i}B^t B^{m_n+1}Bq_{n+1}B,
\end{aligned}$$

which is terminal (note that, if we had used q_{n+j} rather than q_{2n+j}, this would not be terminal). Hence,

$$\Psi_{Z^{(n)}}(m_1, \ldots, m_n) = m_i.$$

EXAMPLE 3.7. MULTIPLICATION. As our final example we choose the function $(x + 1)(y + 1)$. Because of our notational convention, and also because $0 \cdot x = 0$, it is easier to show directly that $(x + 1)(y + 1)$ is computable than that xy is. However, the fact that xy is computable will appear as an immediate corollary of the fact that $(x + 1)(y + 1)$ is, in the light of the results of Chap. 2.

We shall construct a Turing machine Z such that

$$\Psi_{Z^{(2)}}(x, y) = (x + 1)(y + 1).$$

Our construction is based on the fact that

$$(x + 1)(y + 1) = \underbrace{(y + 1) + (y + 1) + \cdots + (y + 1)}_{x + 1 \text{ times}}.$$

That is, we shall let the block of 1's of length $y + 1$, which originally appears on the tape, be duplicated x times. The alphabet of Z will contain two symbols other than 1, B. They may be, say, S_2 and S_3. We write them ϵ and η, respectively. They play the role of counters. That is, they "mind the place" of the machine while it is off on an auxiliary mission.

We take Z to consist of the following quadruples:

$q_1\ 1\ B\ q_1$	(erase a single 1, leaving m_1 1's to be counted)	
$q_1\ B\ R\ q_2$		
$q_2\ 1\ \epsilon\ q_3$	(if the m_1 1's have all been counted, stop; otherwise count another one of them)	
$q_3\ \epsilon\ R\ q_3$	(go right until a double blank is reached)	
$q_3\ 1\ R\ q_3$		
$q_3\ B\ R\ q_4$		
$q_4\ 1\ R\ q_3$		
$q_4\ B\ L\ q_5$		
$q_5\ 1\ L\ q_6$		(duplicate one group of 1's of length $m_2 + 1$)
$q_5\ B\ L\ q_6$		
$q_6\ 1\ \eta\ q_6$	(count another 1 from a group of $m_2 + 1$ 1's)	
$q_6\ \eta\ R\ q_7$		
$q_6\ B\ B\ q_{10}$	($m_2 + 1$ 1's have been counted; prepare to repeat the whole process)	
$q_7\ 1\ R\ q_7$		
$q_7\ B\ R\ q_8$		
$q_8\ 1\ R\ q_8$		
$q_8\ B\ 1\ q_9$	(write another 1 to correspond to the 1 that has just been counted)	
$q_9\ 1\ L\ q_9$		
$q_9\ B\ L\ q_9$	(go left until η has been reached; prepare to count again)	
$q_9\ \eta\ 1\ q_5$		
$q_{10}\ 1\ L\ q_{10}$	(go left until ϵ is reached).	
$q_{10}\ B\ L\ q_{10}$		
$q_{10}\ \epsilon\ B\ q_1$		

The operation of this Turing machine is more complex than that of any considered hitherto. This is because there are two distinct processes involved—that of duplication and that of counting and regulating the total number of duplications—each of which is iterative in character. The situation is completely analogous to what digital-computer programmers call a "double inductive loop." This is in contradistinction to our "subtracter" (Example 3.3) which employs a "single inductive loop." (The term "inductive loop" refers to successive repetitions of a certain act until some counting operation signals a halt.)

It is convenient to begin our proof that

$$\Psi_{Z^{(2)}}(m_1, m_2) = (m_1 + 1)(m_2 + 1)$$

in the middle (that is, at the inner inductive loop), rather than at the beginning. Thus, we let

$$\alpha = PB1^{m_2+1}q_5BB,$$

where P is an arbitrary tape expression, and we note that

$$\begin{aligned}
\alpha &\to PB1^{m_2}q_61BB \\
&\to PB1^{m_2}q_6\eta BB \\
&\to PB1^{m_2}\eta q_7BB \\
&\to PB1^{m_2}\eta Bq_8B \\
&\to PB1^{m_2}\eta Bq_91 \\
&\to PB1^{m_2}\eta q_9B1 \\
&\to PB1^{m_2}q_9\eta B1 \\
&\to PB1^{m_2}q_51B1 \\
&\to PB1^{m_2-1}q_611B1 \\
&\to PB1^{m_2-1}q_6\eta 1B1 \\
&\to \cdots \\
&\to PB1^{m_2-1}\eta 1B1q_8B \\
&\to PB1^{m_2-1}\eta 1B1q_91 \\
&\to \cdots \\
&\to \cdots \\
&\to PB\eta 1^{m_2}B1^{m_2}q_91 \\
&\to \cdots \\
&\to PBq_9\eta 1^{m_2}B1^{m_2+1} \\
&\to PBq_51^{m_2+1}B1^{m_2+1} \\
&\to Pq_6B1^{m_2+1}B1^{m_2+1} \\
&\to Pq_{10}B1^{m_2+1}B1^{m_2+1}.
\end{aligned}$$

To recapitulate,

$$PB1^{m_2+1}q_5BB \to \cdots \to Pq_{10}B1^{m_2+1}B1^{m_2+1}; \tag{1}$$

so the effect has been to duplicate on the tape the expression 1^{m_2+1}.

Now, let

$$\alpha_1 = q_111^{m_1}B1^{m_2+1}.$$

Then,

$$
\begin{aligned}
\alpha_1 &\rightarrow q_1 B 1^{m_1} B 1^{m_2+1} \\
&\rightarrow B q_2 1^{m_1} B 1^{m_2+1} \\
&\rightarrow B q_3 \epsilon 1^{m_1 - 1} B 1^{m_2+1} \\
&\rightarrow \cdots \\
&\rightarrow B \epsilon 1^{m_1 - 1} q_3 B 1^{m_2+1} \\
&\rightarrow B \epsilon 1^{m_1 - 1} B q_4 1^{m_2+1} \\
&\rightarrow \cdots \\
&\rightarrow B \epsilon 1^{m_1 - 1} B 1^{m_2+1} q_3 B \\
&\rightarrow B \epsilon 1^{m_1 - 1} B 1^{m_2+1} B q_4 B \\
&\rightarrow B \epsilon 1^{m_1 - 1} B 1^{m_2+1} q_5 B B \\
&\rightarrow \cdots \\
&\rightarrow B \epsilon 1^{m_1 - 1} q_{10} B 1^{m_2+1} B 1^{m_2+1} \qquad \text{[by (1)]} \\
&\rightarrow \cdots \\
&\rightarrow B q_{10} \epsilon 1^{m_1 - 1} B 1^{m_2+1} B 1^{m_2+1} \\
&\rightarrow B q_1 B 1^{m_1 - 1} B 1^{m_2+1} B 1^{m_2+1} \\
&\rightarrow B B q_2 1^{m_1 - 1} B 1^{m_2+1} B 1^{m_2+1} = \alpha_k.
\end{aligned}
$$

Thus, the effect of our procedure is to delete a 1 from the expression 1^{m_1} and to duplicate 1^{m_2+1}. This process occurs again and again, until

$$
\begin{aligned}
\alpha_k &\rightarrow \cdots : \\
&\rightarrow B^{m_1} q_{10} \epsilon \underbrace{B 1^{m_2+1} B 1^{m_2+1} B \cdots B 1^{m_2+1}}_{m_1 + 1 \text{ times}} \\
&\rightarrow \cdots \\
&\rightarrow B^{m_1+1} q_2 \underbrace{B 1^{m_2+1} B 1^{m_2+1} B \cdots B 1^{m_2+1}}_{m_1 + 1 \text{ times}} = \alpha_p.
\end{aligned}
$$

But α_p is terminal, and

$$\langle \alpha_p \rangle = (m_1 + 1)(m_2 + 1).$$

Hence,

$$\Psi_Z^{(2)}(x, y) = (x + 1)(y + 1).$$

4. Relatively Computable Functions. Thus far, we have dealt only with closed computations. However, it is easy to imagine a machine that halts a computation at various times and requests additional information. We shall show how our formalism can be adapted to include such computations. We do this, not only because the extended concept possesses a certain interest in itself, but also because of the insight it will eventually afford into the relationships between computable and noncomputable functions. Our treatment of computable functions is interrupted at this point to allow for this extended kind of computability (which we shall call *relative computability*) because in developing the theory of relative computability we are enabled to obtain the theory of computability as a special case. Moreover, the complexity is increased but little thereby.

What sort of information can a Turing machine be expected to request? We shall restrict the interrogations permitted the machine to those of the form

$$\text{Is } n \in A?$$

where n is an integer and where A is a set of integers, fixed for a given context. Actually, as we shall see later, this limitation is not nearly so restrictive as might be supposed.

It is at this point that we make use of quadruples of type 4, that is, those of the form

$$q_i \; S_j \; q_k \; q_l.$$

Let A be a set of integers, to be thought of as remaining fixed in what follows.

DEFINITION 4.1. *Let α, β be instantaneous descriptions. Then we write $\alpha \underset{A}{\rightarrow} \beta$ (Z) if there exist expressions P and Q (possibly empty) such that*

$$\alpha \text{ is } Pq_iS_jQ,$$
$$Z \text{ contains } q_i \; S_j \; q_k \; q_l,$$

and either

(1) $\langle \alpha \rangle \in A$ *and* β *is* Pq_kS_jQ, *or*
(2) $\langle \alpha \rangle \notin A$ *and* β *is* Pq_lS_jQ.

(When no ambiguity can result, we shall sometimes omit explicit reference to the set A or to the Turing machine Z.)

This provides a Turing machine with a means of communication with "the external world." When a machine is at an instantaneous description Pq_iS_jQ, where $q_i \; S_j \; q_k \; q_l$ is a quadruple of the machine, the machine may be interpreted as inquiring:

If n is the number of occurrences of 1 on my tape at the present moment, is $n \in A$?

If the answer is yes, this is indicated by changing the machine's internal configuration to q_k; if the answer is no, by changing it to q_l.

DEFINITION 4.2. *Let α be an instantaneous description of the form Pq_iS_jQ. Then, α is* **final** *with respect to Z if Z contains no quadruple whose initial pair of symbols is $q_i \; S_j$.*

We have the following immediate corollaries:

THEOREM 4.1. *If Z is simple, then α is terminal with respect to Z if and only if it is final with respect to Z.*

THEOREM 4.2. *α is final with respect to Z if and only if*

(1) *α is terminal with respect to Z, and*

(2) *No matter what set A is chosen, there is no β such that $\alpha \underset{A}{\rightarrow} \beta \quad (Z)$.*

DEFINITION 4.3. *By an A-**computation** of a Turing machine Z is meant a finite sequence $\alpha_1, \alpha_2, \ldots, \alpha_p$ of instantaneous descriptions such that, for each i, $1 \leqq i < p$, either*

$$\alpha_i \rightarrow \alpha_{i+1} \quad (Z) \qquad \text{or} \qquad \alpha_i \underset{A}{\rightarrow} \alpha_{i+1} \quad (Z),$$

*and such that α_p is final. In this case, we write $\alpha_p = \mathrm{Res}_Z{}^A\,(\alpha_1)$, and we call α_p the A-**resultant** of α_1 with respect to Z.*

Note that, if Z is a simple Turing machine, then the requirement of Definition 4.3 is quite independent of A and that an A-computation in the sense of this definition is simply a computation in the sense of Definition 1.9. [This is true because $\alpha \underset{A}{\rightarrow} \beta \quad (Z)$ can *never* hold if Z is a *simple* Turing machine.]

DEFINITION 4.4. *Let Z be a Turing machine. Then, for each n, we associate with Z an n-ary function (which, in general, depends on the set A)*

$$\Psi_{Z;A}^{(n)}(x_1, x_2, \ldots, x_n)$$

as follows:

For each $(m_1, m_2, \ldots, m_n)$, we set $\alpha_1 = q_1(\overline{m_1, m_2, \ldots, m_n})$, and we distinguish between two cases:

(1) *There exists an A-computation of Z, $\alpha_1, \alpha_2, \ldots, \alpha_p$. In this case, we set*

$$\Psi_{Z;A}^{(n)}(m_1, \ldots, m_n) = \langle \alpha_p \rangle = \langle \mathrm{Res}_Z{}^A\,(\alpha_1) \rangle.$$

(2) *There exists no A-computation, $\alpha_1, \alpha_2, \ldots, \alpha_p$; that is, $\mathrm{Res}_Z{}^A\,(\alpha_1)$ is undefined. In this case, we leave*

$$\Psi_{Z;A}^{(n)}(m_1, \ldots, m_n)$$

undefined.

We write $\Psi_Z{}^A(x)$ for $\Psi_{Z;A}^{(1)}(x)$.

By the remark immediately following Definition 4.3, it is clear that, if Z is simple, then

$$\Psi_{Z;A}^{(n)}(x_1, \ldots, x_n)$$

is independent of A and that, in fact,

$$\Psi_{Z;A}^{(n)}(x_1, \ldots, x_n) = \Psi_Z{}^{(n)}(x_1, \ldots, x_n).$$

DEFINITION 4.5. *An n-ary function $f(x_1, x_2, \ldots, x_n)$ is **partially A-computable** if there exists a Turing machine Z such that*

$$f(x_1, \ldots, x_n) = \Psi_{Z;A}^{(n)}(x_1, \ldots, x_n).$$

*In this case, we say that Z A-**computes** f.*

If, in addition, $f(x_1, \ldots, x_n)$ is a total function, then it is called A- **computable.**[1]

COROLLARY 4.3. *If $f(x_1, \ldots, x_n)$ is (partially) computable, then it is (partially) A-computable for any set A.*[2]

PROOF. Immediate from the remark following Definition 4.4.

THEOREM 4.4. *To every Turing machine Z, there corresponds a simple Turing machine Z′ such that*

$$\Psi_{Z'^{(n)}}(x_1, \ldots, x_n) = \Psi^{(n)}_{Z;\phi}(x_1, \ldots, x_n).$$[3]

PROOF. Let Z' be obtained from Z by replacing each quadruple of Z having the form $q_i\ S_j\ q_k\ q_l$ by $q_i\ S_j\ S_j\ q_l$. The result follows at once.

COROLLARY 4.5. *If $g(x_1, \ldots, x_n)$ is (partially) ϕ-computable, then $g(x_1, \ldots, x_n)$ is (partially) computable.*

THEOREM 4.6. *$g(x_1, \ldots, x_n)$ is (partially) computable if and only if it is (partially) ϕ-computable.*

PROOF. This is an immediate consequence of Corollaries 4.3 and 4.5.

Thus, from each theorem we prove concerning A-computability, we can infer a comparable theorem concerning computability simply by taking $A = \phi$. That is, the theory of computability is a special case of the theory of relative computability.

DEFINITION 4.6. *We say that the set S is* **computable** (*A*-**computable**) *if the characteristic function*[4] *of S, $C_S(x)$, is computable (A-computable).*

THEOREM 4.7. *A is A-computable.*

PROOF. Let Z consist of the quadruples

$$\begin{array}{l} q_1\ 1\ B\ q_1 \\ q_1\ B\ q_2\ q_3 \\ q_2\ B\ R\ q_4 \\ q_4\ 1\ B\ q_2 \\ q_3\ B\ R\ q_5 \\ q_5\ 1\ B\ q_3 \\ q_5\ B\ 1\ q_3. \end{array}$$

[1] The basic idea of this definition goes back to Turing [3]. The theory of relative computability has been studied in Kleene [4, 6], Post [7], Kleene and Post [1], and Davis [1]. The first complete formulation of a definition equivalent to our 4.5 appears in Kleene [4].

[2] An assertion, such as this corollary, containing a word in parentheses, is intended to abbreviate two assertions, one containing the parenthetical word and one omitting it. Thus, the present corollary abbreviates the following pair of assertions:

If $f(x_1, \ldots, x_n)$ is computable, then it is A-computable for any set A.

If $f(x_1, \ldots, x_n)$ is partially computable, then it is partially A-computable for any set A.

[3] ϕ is the empty set. Cf. Introduction, Sec. 3.2.

[4] The characteristic function of a set is 0 on the set, 1 elsewhere. Cf. Introduction, Sec. 3.3.

Then, we assert that

$$\Psi_Z{}^A(m) = C_A(m).$$

Let $\alpha_1 = q_1\bar{m} = q_1 1^{m+1}$. Then $\alpha_1 = q_1 1 1^m \rightarrow q_1 B 1^m$. Now,

$$\langle q_1 B 1^m \rangle = m.$$

First, suppose that $m \in A$. Then

$$\begin{aligned} q_1 B 1^m &\underset{A}{\rightarrow} q_2 B 1^m \\ &\rightarrow \cdots \\ &\rightarrow B^{m+1} q_4 B, \end{aligned}$$

which is final. Also, $\langle B^{m+1} q_4 B \rangle = 0$. Thus, if $m \in A$,

$$\Psi_Z{}^A(m) = 0 = C_A(m).$$

Next, suppose that $m \notin A$. Then

$$\begin{aligned} q_1 B 1^m &\underset{A}{\rightarrow} q_3 B 1^m \\ &\rightarrow \cdots \\ &\rightarrow B^{m+1} q_3 B \\ &\rightarrow B^{m+2} q_5 B \\ &\rightarrow B^{m+2} q_3 1, \end{aligned}$$

which is final. Also, $\langle B^{m+2} q_3 1 \rangle = 1$. Thus, if $m \notin A$,

$$\Psi_Z{}^A(m) = 1 = C_A(m).$$

This proves the theorem.

CHAPTER 2

OPERATIONS ON COMPUTABLE FUNCTIONS

1. Preliminary Lemmas. As we have seen, a Turing machine that will compute so simple a function as $(x + 1)(y + 1)$ is rather complex. In what follows, we shall often wish to establish that some quite complicated functions are computable. This and the following chapter are devoted to the development of techniques which will make it possible to do this efficiently. In the present chapter, we introduce two operations by means of which new computable functions are obtained from given computable functions. These operations can then be repeatedly performed on the functions proved computable in Chap. 1, Sec. 3, and on the new functions thus obtained, to yield a huge class of computable functions. In Chap. 3, we shall study this class in detail, and in Chap. 4, we shall see that it actually includes *all* computable functions.

The procedures developed in the present chapter are analogous to subroutines for use with electronic digital computers.

We shall adopt the convention of systematically omitting final occurrences of a blank, B, in an instantaneous description. Thus, if Z contains the quadruple $q_3\ B\ L\ q_3$, we shall write

$$11S_3S_21q_3B \rightarrow 11S_3S_2q_31 \quad (Z).$$

On the other hand, *we shall not omit initial occurrences* of B. For example, if Z were to contain the quadruple $q_2\ B\ R\ q_2$, then we should write

$$q_2B11 \rightarrow Bq_211 \quad (Z).$$

The following definitions will prove convenient in our work with Turing machines (except for a brief occurrence in Chap. 9, their use is confined entirely to the present chapter).

DEFINITION 1.1. *If Z is a Turing machine we let $\theta(Z)$ be the largest number i such that q_i is an internal configuration of Z.*

DEFINITION 1.2. *A Turing machine Z is called n-**regular** $(n > 0)$ if*

(1) *There is an $s > 0$ such that, whenever* $\text{Res}_Z{}^A\ [q_1(\overline{m_1, \ldots, m_n})]$ *is defined, it has the form $q_{\theta(Z)}(\overline{r_1, \ldots, r_s})$ for suitable $r_1, \ldots, r_s$,* and

(2) *No quadruple of Z begins with $q_{\theta(Z)}$.*

n-regular Turing machines are useful because they present the results of a computation ("outputs") in a form suitable for use (as "inputs") at the beginning of a new computation by another Turing machine.

[Note that, by our convention permitting omission of B, the expression $q_{\theta(Z)}\overline{(r_1, \ldots, r_s)}$ may include additional occurrences of B on the right. However, $q_{\theta(Z)}$ must be the leftmost symbol.]

DEFINITION 1.3. *Let Z be a Turing machine. Then $Z^{(n)}$ is the Turing machine obtained from Z by replacing each internal configuration q_i, at all of its occurrences in quadruples of Z, by q_{n+i}.*

We begin with several lemmas. The first of these enables us to rewrite the numerical result of a computation in such a form that it is available for use as the beginning of a new computation.

LEMMA 1. *For every Turing machine Z, we can find a Turing machine Z' such that, for each n, Z' is n-regular, and, in fact,*

$$\mathrm{Res}_{Z'}{}^{A}\,[q_1\overline{(m_1, \ldots, m_n)}] = q_{\theta(Z')}\overline{\Psi_{Z;A}^{(n)}(m_1, \ldots, m_n)}.\dagger$$

PROOF. Our first attempt at a proof might well be as follows. We let Z' begin where Z halts, and we have it count up the number of 1's on its tape, assembling them into a block and erasing everything else. The difficulty with this plan is that we should not know how to tell Z' when its computation was at an end. No matter how long it had been searching unsuccessfully for more 1's, it could not be certain that it had located all of them. To circumvent this difficulty, we rearrange Z's computation so that it is carried out entirely between two special markers. These markers must be moved as more computing space is required. Then, at the end, all of the 1's on the tape will appear between the markers.

We now proceed formally. Let λ, ρ be the first two symbols in the list $S_2, S_3, S_4, \ldots$ that are not in the alphabet of Z. Let Z_1 consist of the quadruples

$q_1\ 1\ L\ q_1$
$q_1\ B\ \lambda\ q_1$ (print λ on the left)
$q_1\ \lambda\ R\ q_2$

$q_2\ 1\ R\ q_2$
$q_2\ B\ R\ q_3$ (move right until a double blank is reached)
$q_3\ 1\ R\ q_2$
$q_3\ B\ L\ q_4$

$q_4\ B\ \rho\ q_4$ (print ρ on the right)
$q_4\ \rho\ L\ q_5$

$q_5\ 1\ L\ q_5$
$q_5\ B\ L\ q_5$ (move left until λ is reached)
$q_5\ \lambda\ R\ q_6.$

† Equality between partial functions is understood to imply equality of their domains. Cf. Introduction, Sec. 3.3.

It is easy to see that, with respect to Z_1,

$$q_1(\overline{m_1, \ldots, m_n}) \to \cdots \to \lambda q_6(\overline{m_1, \ldots, m_n})\rho, \tag{1}$$

which is final. Thus, the effect of Z_1 is to seal the initial instantaneous description with the letters λ, ρ.

Now, $Z^{(5)}$ (cf. Definition 1.3) will behave precisely like Z except that it will begin in internal configuration q_6 instead of q_1 and the index of all of its other internal configurations will be similarly advanced. Thus, we set $K = \theta(Z^{(5)})$, and we let Z_2 consist of all of the quadruples of $Z^{(5)}$ and, in addition, the following quadruples, where q_i may be any internal configuration of $Z^{(5)}$:

$q_i\ \lambda\ B\ q_{K+i}$	(erase the marker λ)
$q_{K+i}\ B\ L\ q_{2K+i}$	
$q_{2K+i}\ B\ \lambda\ q_{2K+i}$	(print λ one square to the left)
$q_{2K+i}\ \lambda\ R\ q_i$	(return to the main computation)
$q_i\ \rho\ B\ q_{3K+i}$	(erase the marker ρ)
$q_{3K+i}\ B\ R\ q_{4K+i}$	
$q_{4K+i}\ B\ \rho\ q_{4K+i}$	(print ρ one square to the right)
$q_{4K+i}\ \rho\ L\ q_i$	(return to the main computation).

These last quadruples serve to define the actions of Z_2 with respect to the new symbols λ, ρ. Now, either $\text{Res}_Z{}^A\,[q_1(\overline{m_1, \ldots, m_n})]$ is defined, in which case we have, with respect to Z_2,

$$\lambda q_6(\overline{m_1, \ldots, m_n})\rho \to \cdots \to \lambda\alpha\rho, \tag{2}$$

which is final, where

$$\langle\alpha\rangle = \langle\text{Res}_Z{}^A\,[q_1(\overline{m_1, \ldots, m_n})]\rangle, \tag{3}$$

or $\text{Res}_Z{}^A\,[q_1(\overline{m_1, \ldots, m_n})]$ is undefined, in which case

$$\text{Res}_{Z_2}{}^A\,[\lambda q_6(\overline{m_1, \ldots, m_n})\rho]$$

is likewise undefined.

Let $L = 5K + 1$, and let Z_3 consist of all quadruples of the form

$$q_i\ S_j\ S_j\ q_L,$$

where q_i is any internal configuration of Z_2, where S_j belongs to the alphabet of Z_2, and where *no quadruple beginning with* $q_i\ S_j$ belongs to Z_2. Clearly, if $\lambda P q_i Q \rho$ is a final instantaneous description with respect to Z_2, we have

$$\lambda P q_i Q \rho \to \lambda P q_L Q \rho \quad (Z_3) \tag{4}$$

which last is final.

Next, let Z_4 consist of the following quadruples, where S may be any symbol in the alphabet of Z other than 1 and B:

$q_L\ 1\ L\ q_L$
$q_L\ B\ L\ q_L$ (move leftward looking for λ)
$q_L\ S\ L\ q_L$
$q_L\ \lambda\ R\ q_{L+1}$

$q_{L+1}\ S\ B\ q_{L+1}$
$q_{L+1}\ B\ R\ q_{L+1}$ (move rightward looking for a 1)
$q_{L+1}\ 1\ B\ q_{L+2}$
$q_{L+1}\ \rho\ B\ q_{L+4}$ (if ρ is reached without a 1, prepare to terminate)

$q_{L+2}\ B\ L\ q_{L+2}$
$q_{L+2}\ 1\ R\ q_{L+3}$ (locate the block of 1's)
$q_{L+2}\ \lambda\ R\ q_{L+3}$

$q_{L+3}\ B\ 1\ q_{L+3}$ (add 1 to the block of 1's)
$q_{L+3}\ 1\ R\ q_{L+1}$

$q_{L+4}\ B\ L\ q_{L+4}$
$q_{L+4}\ 1\ L\ q_{L+4}$ (terminate)
$q_{L+4}\ \lambda\ 1\ q_{L+5}$.

The effect of Z_4 is to consolidate all of the 1's on the tape into a block, to erase everything else, and finally to add an additional 1. Thus, with respect to Z_4,

$$\begin{aligned} \lambda P q_L Q \rho \rightarrow & \cdots \\ \rightarrow & q_L \lambda P Q \rho \\ \rightarrow & \lambda q_{L+1} P Q \rho \\ \rightarrow & \cdots \\ \rightarrow & \lambda B^s q_{L+1} 1 M \rho \\ \rightarrow & \lambda B^s q_{L+2} B M \rho \\ \rightarrow & q_{L+2} \lambda B^{s+1} M \rho \\ \rightarrow & \lambda q_{L+3} B^{s+1} M \rho \\ \rightarrow & \lambda q_{L+3} 1 B^s M \rho \\ \rightarrow & \lambda 1 q_{L+1} B^s M \rho, \end{aligned}$$

and the process is repeated. If there were originally p 1's present on the tape, we eventually have

$$\begin{aligned} \rightarrow & \cdots \\ \rightarrow & \lambda 1^p B^t q_{L+1} \rho \\ \rightarrow & \lambda 1^p B^t q_{L+4} B \\ \rightarrow & \cdots \\ \rightarrow & q_{L+4} \lambda 1^p \\ \rightarrow & q_{L+5} 1^{p+1}. \end{aligned}$$

Finally, let $Z' = Z_1 \cup Z_2 \cup Z_3 \cup Z_4$. Then, combining (1) to (4) and our present observation, we have

$$\begin{aligned}\text{Res}_{Z'}{}^A [q_1(\overline{m_1, \ldots, m_n})] &= q_{L+5}1^{<\text{Res}_Z{}^A[q_1(\overline{m_1, \ldots, m_n})]>+1} \\ &= q_{\theta(Z')}\overline{\Psi_{Z;A}^{(n)}(m_1, m_2, \ldots, m_n)}.\end{aligned}$$

Lemma 2. *For each n-regular Turing machine Z and each $p > 0$, there is a $(p + n)$-regular Turing machine Z_p such that, whenever*

$$\text{Res}_Z{}^A [q_1(\overline{m_1, \ldots, m_n})] = q_{\theta(Z)}(\overline{r_1, \ldots, r_s}),$$

it is also the case that

$$\text{Res}_{Z_p}{}^A [q_1(\overline{k_1, \ldots, k_p, m_1, \ldots, m_n})] = q_{\theta(Z_p)}(\overline{k_1, \ldots, k_p, r_1, \ldots, r_s}),$$

whereas, whenever $\text{Res}_Z{}^A [q_1(\overline{m_1, \ldots, m_n})]$ *is undefined, so is*

$$\text{Res}_{Z_p}{}^A [q_1(\overline{k_1, \ldots, k_p, m_1, \ldots, m_n})].$$

In other words, Z_p will act on $m_1, \ldots, m_n$ as Z would have, but will leave $k_1, \ldots, k_p$ untouched.

Our proof will be motivated as follows. We will construct Z_p in such a manner that it will rewrite the arguments $k_1, k_2, \ldots, k_p$, replacing 1 by a special symbol ϵ, except that the leftmost 1 will be replaced instead by δ. In addition, one more ϵ will be placed to the right of these transliterated arguments. Then, matters will be so arranged that whenever ϵ is encountered in the course of a computation, the entire block of arguments (now written in terms of ϵ) is moved a single square to the left. Finally, at the termination of the computation, the arguments will be rewritten in their original form.

PROOF. Let δ, ϵ be distinct symbols not in the alphabet of Z. Let U_1 consist of the following quadruples:

$$\begin{array}{l} q_1\ 1\ \delta\ q_1 \\ q_1\ \delta\ R\ q_2 \end{array}$$

$$\left.\begin{array}{l} q_i\ 1\ \epsilon\ q_i \\ q_i\ \epsilon\ R\ q_i \\ q_i\ B\ R\ q_{i+1} \end{array}\right\} \quad \begin{array}{l} 1 < i \leqq p \\ \text{(replace 1 by } \epsilon) \end{array}$$

$$\begin{array}{l} q_{p+1}\ 1\ \epsilon\ q_{p+1} \\ q_{p+1}\ \epsilon\ R\ q_{p+1} \\ q_{p+1}\ B\ \epsilon\ q_{p+2} \\ q_{p+2}\ \epsilon\ R\ q_{p+3}. \end{array}$$

With respect to U_1,

$$
\begin{aligned}
q_1(\overline{k_1, \ldots, k_p, m_1, \ldots, m_n}) \\
&\rightarrow q_1\delta 1^{k_1}B \cdots B1^{k_p+1}B1^{m_1+1}B \cdots B1^{m_n+1} \\
&\rightarrow \delta q_2 1^{k_1}B \cdots B1^{k_p+1}B1^{m_1+1}B \cdots B1^{m_n+1} \\
&\rightarrow \cdots \\
&\rightarrow \cdots \\
&\rightarrow \delta\epsilon^{k_1}B \cdots B\epsilon^{k_p+1}q_{p+1}B1^{m_1+1}B \cdots B1^{m_n+1} \\
&\rightarrow \delta\epsilon^{k_1}B \cdots B\epsilon^{k_p+1}q_{p+2}\epsilon 1^{m_1+1}B \cdots B1^{m_n+1} \\
&\rightarrow \delta\epsilon^{k_1}B \cdots B\epsilon^{k_p+1}\epsilon q_{p+3}(\overline{m_1, \ldots, m_n}),
\end{aligned}
$$

which is final.

Next, let $N = \theta(Z^{(p+2)})$, and let U_2 consist of all of the quadruples of $Z^{(p+2)}$ and, in addition, the following quadruples, where q_i may be any internal configuration of $Z^{(p+2)}$:

$q_i \ \epsilon \ 1 \ q_{N+i}$ (interrupt computation)

$q_{N+i} \ 1 \ L \ q_{N+i}$
$q_{N+i} \ \epsilon \ L \ q_{N+i}$ (search for δ)
$q_{N+i} \ B \ L \ q_{N+i}$
$q_{N+i} \ \delta \ B \ q_{2N+i}$

$q_{2N+i} \ B \ L \ q_{3N+i}$

$q_{3N+i} \ B \ \delta \ q_{3N+i}$ (copy δ)
$q_{3N+i} \ \delta \ R \ q_{4N+i}$

$q_{4N+i} \ \epsilon \ R \ q_{5N+i}$
$q_{4N+i} \ B \ R \ q_{5N+i}$
$q_{4N+i} \ 1 \ B \ q_i$ (resume main computation)

$q_{5N+i} \ \epsilon \ L \ q_{6N+i}$ (observing ϵ, prepare to copy ϵ)
$q_{5N+i} \ B \ L \ q_{7N+i}$ (observing B, perpare to copy B)
$q_{5N+i} \ 1 \ L \ q_{6N+i}$

$q_{6N+i} \ \epsilon \ \epsilon \ q_{8N+i}$ (copy ϵ)
$q_{6N+i} \ B \ \epsilon \ q_{8N+i}$

$q_{7N+i} \ \epsilon \ B \ q_{8N+i}$ (copy B)
$q_{7N+i} \ B \ B \ q_{8N+i}$

$q_{8N+i} \ \epsilon \ R \ q_{4N+i}$ (repeat)
$q_{8N+i} \ B \ R \ q_{4N+i}$.

If P is any tape expression, then, with respect to U_2,

$$
\begin{aligned}
\delta\epsilon^{k_1}B \cdots B\epsilon^{k_p+1}q_i\epsilon P &\rightarrow \delta\epsilon^{k_1}B \cdots B\epsilon^{k_p+1}q_{N+i}1P \\
&\rightarrow \cdots \\
&\rightarrow q_{N+i}\delta\epsilon^{k_1}B \cdots B\epsilon^{k_p+1}1P \\
&\rightarrow q_{2N+i}B\epsilon^{k_1}B \cdots B\epsilon^{k_p+1}1P
\end{aligned}
$$

$$\begin{aligned}
&\to q_{3N+i}BB\epsilon^{k_1}B \cdots B\epsilon^{k_p+1}1P \\
&\to q_{3N+i}\delta B\epsilon^{k_1}B \cdots B\epsilon^{k_p+1}1P \\
&\to \delta q_{4N+i}B\epsilon^{k_1}B \cdots B\epsilon^{k_p+1}1P \\
&\to \delta Bq_{5N+i}\epsilon^{k_1}B \cdots B\epsilon^{k_p+1}1P \\
&\to \cdots \\
&\to \cdots \\
&\to \delta\epsilon^{k_1}B \cdots B\epsilon^{k_p+1}\epsilon q_{5N+i}1P \\
&\to \delta\epsilon^{k_1}B \cdots B\epsilon^{k_p+1}\epsilon q_iBP.
\end{aligned}$$

Thus, under the action of U_2, all of the ϵ's are moved over one square to the left whenever one of them is encountered.

Let $U_3 = U_1 \cup U_2$. Then, with respect to U_3,

$$\begin{aligned}
q_1(\overline{k_1, \ldots, k_p, m_1, \ldots, m_n}) &\to \cdots \\
&\to \delta\epsilon^{k_1}B \cdots B\epsilon^{k_p+1}\epsilon q_{p+3}(\overline{m_1, \ldots, m_n}) \\
&\to \cdots \\
&\to \delta\epsilon^{k_1}B \cdots B\epsilon^{k_p+1}\epsilon q_N(r_1, \ldots, r_s)
\end{aligned}$$

whenever $\text{Res}_Z{}^A\,[q_1(\overline{m_1, \ldots, m_n})]$ is defined; otherwise, there is no A-computation beginning $q_1(\overline{k_1, \ldots, k_p, m_1, \ldots, m_n})$.

Finally, let $L = \theta(U_3)$, and let Z_p consist of all the quadruples of U_3 and, in addition, the following quadruples:

$q_N\ 1\ L\ q_N$
$q_N\ \epsilon\ B\ q_{L+1}$ (erase one ϵ)

$q_{L+1}\ B\ L\ q_{L+1}$
$q_{L+1}\ \epsilon\ 1\ q_{L+1}$ (go left, replacing ϵ and δ by 1)
$q_{L+1}\ 1\ L\ q_{L+1}$
$q_{L+1}\ \delta\ 1\ q_{L+2}$.

Then, using what we have just shown for U_3, we have, with respect to Z_p, whenever $\text{Res}_Z{}^A\,[q_1(\overline{m_1, \ldots, m_n})]$ is defined,

$$\begin{aligned}
q_1(\overline{k_1, \ldots, k_p, m_1, \ldots, m_n}) &\to \cdots \\
&\to \cdots \\
&\to \delta\epsilon^{k}B \cdots B\epsilon^{k_p+1}\epsilon q_N(\overline{r_1, \ldots, r_s}) \\
&\to \delta\epsilon^{k_1}B \cdots B\epsilon^{k_p+1}q_N\epsilon(\overline{r_1, \ldots, r_s}) \\
&\to \delta\epsilon^{k_1}B \cdots B\epsilon^{k_p+1}q_{L+1}B(\overline{r_1, \ldots, r_s}) \\
&\to \cdots \\
&\to q_{L+2}1^{k_1+1}B \cdots B1^{k_p+1}B(\overline{r_1, \ldots, r_s}) \\
&= q_{\theta(Z_p)}(\overline{k_1, \ldots, k_p, r_1, \ldots, r_s}),
\end{aligned}$$

which is final.

In applying and extending Lemmas 1 and 2, it will be important to have available Turing machines that accomplish certain simple transformations on a sequence of arguments.

THE COPYING MACHINES C_p. *For each* $n > 0$ *and* $p \geqq 0$ (note carefully that $p = 0$ is permitted!), *we shall define a* $(p + n)$*-regular Turing machine such that*

$$\text{Res}_{C_p}\,[q_1(\overline{k_1, \ldots, k_p, m_1, \ldots, m_n})] = q_{p+16}(\overline{m_1, \ldots, m_n, k_1, \ldots, k_p, m_1, \ldots . m_n}).$$

That is, the arguments $m_1, \ldots, m_n$ are recopied to the left of the $k_1, \ldots, k_p$.

C_p is to consist of the following quadruples:

$q_1\ 1\ L\ q_1$
$q_1\ B\ L\ q_2$

$q_2\ B\ \lambda\ q_2$
$q_2\ \lambda\ R\ q_3$ (set marker λ)

$q_i\ 1\ R\ q_i$
$q_i\ B\ R\ q_{i+1}$ $3 \leqq i \leqq p + 2$ (move over p blocks of 1's[1])

$q_{p+3}\ 1\ R\ q_{p+3}$
$q_{p+3}\ B\ \delta\ q_{p+3}$ (set marker δ)
$q_{p+3}\ \delta\ R\ q_{p+4}$

$q_{p+4}\ 1\ R\ q_{p+4}$
$q_{p+4}\ B\ R\ q_{p+5}$ (hunt for double blank)
$q_{p+5}\ 1\ R\ q_{p+4}$
$q_{p+5}\ B\ L\ q_{p+6}$

$q_{p+6}\ 1\ L\ q_{p+7}$
$q_{p+6}\ B\ L\ q_{p+7}$

$q_{p+7}\ 1\ \omega\ q_{p+7}$ (seeing 1, replace it by ω, and prepare to copy 1)
$q_{p+7}\ \omega\ L\ q_{p+8}$
$q_{p+7}\ B\ \beta\ q_{p+7}$ (seeing B, replace it by β, and prepare to copy B)
$q_{p+7}\ \beta\ L\ q_{p+11}$
$q_{p+7}\ \delta\ B\ q_{p+15}$ (seeing δ, prepare to terminate)

$q_{p+8}\ 1\ L\ q_{p+8}$
$q_{p+8}\ B\ L\ q_{p+8}$ (go left)
$q_{p+8}\ \delta\ L\ q_{p+8}$
$q_{p+8}\ \lambda\ \omega\ q_{p+10}$ (finding λ, replace it by ω)
$q_{p+8}\ \omega\ 1\ q_{p+9}$ (replace ω by 1)
$q_{p+8}\ \beta\ B\ q_{p+9}$ (replace β by B)

$q_{p+9}\ 1\ L\ q_{p+10}$ (go left one square)
$q_{p+9}\ B\ L\ q_{p+10}$

$q_{p+10}\ B\ \omega\ q_{p+10}$ (copy 1, temporarily using ω)
$q_{p+10}\ \omega\ R\ q_{p+14}$

[1] Note that, if $p = 0$, none of these quadruples occurs.

$q_{p+11}\ 1\ L\ q_{p+11}$
$q_{p+11}\ B\ L\ q_{p+11}$ (go left)
$q_{p+11}\ \delta\ L\ q_{p+11}$
$q_{p+11}\ \omega\ 1\ q_{p+12}$ (replace ω by 1)
$q_{p+11}\ \beta\ B\ q_{p+12}$ (replace β by B)

$q_{p+12}\ 1\ L\ q_{p+13}$ (go left one square)
$q_{p+12}\ B\ L\ q_{p+13}$

$q_{p+13}\ B\ \beta\ q_{p+13}$ (copy B, temporarily using β)
$q_{p+13}\ \beta\ R\ q_{p+14}$

$q_{p+14}\ 1\ R\ q_{p+14}$
$q_{p+14}\ B\ R\ q_{p+14}$ (go right)
$q_{p+14}\ \delta\ R\ q_{p+14}$
$q_{p+14}\ \omega\ 1\ q_{p+6}$ (restore 1, and repeat)
$q_{p+14}\ \beta\ B\ q_{p+6}$ (restore B, and repeat)

$q_{p+15}\ 1\ L\ q_{p+15}$ (go left)
$q_{p+15}\ B\ L\ q_{p+15}$
$q_{p+15}\ \omega\ 1\ q_{p+16}$ (replace ω by 1 and terminate).

With respect to C_0,

$$
\begin{aligned}
q_1(\overline{m_1, \ldots, m_n}) &\rightarrow \cdots \\
&\rightarrow q_2\lambda B(\overline{m_1, \ldots, m_n}) \\
&\rightarrow \lambda q_3 B(\overline{m_1, \ldots, m_n}) \\
&\rightarrow \lambda q_3 \delta(\overline{m_1, \ldots, m_n}) \\
&\rightarrow \cdots \\
&\rightarrow \lambda\delta(\overline{m_1, \ldots, m_{n-1}})B1^{m_n}q_7 1 \\
&\rightarrow \cdots \\
&\rightarrow q_8\lambda\delta(\overline{m_1, \ldots, m_{n-1}})B1^{m_n}\omega \\
&\rightarrow q_{10}\omega\delta(\overline{m_1, \ldots, m_{n-1}})B1^{m_n}\omega \\
&\rightarrow \cdots \\
&\rightarrow \omega\delta(\overline{m_1, \ldots, m_{n-1}})B1^{m_n}q_{14}\omega \\
&\rightarrow \cdots \\
&\rightarrow \cdots \\
&\rightarrow q_{11}\omega 1^{m_n}\delta(\overline{m_1, \ldots, m_{n-1}})\beta 1^{m_n+1} \\
&\rightarrow q_{12}1^{m_n+1}\delta(\overline{m_1, \ldots, m_{n-1}})\beta 1^{m_n+1} \\
&\rightarrow \cdots \\
&\rightarrow q_{13}\beta 1^{m_n+1}\delta(\overline{m_1, \ldots, m_{n-1}})\beta 1^{m_n+1} \\
&\rightarrow \cdots \\
&\rightarrow \cdots \\
&\rightarrow \omega 1^{m_1}B(\overline{m_2, \ldots, m_n})q_7\delta(\overline{m_1, \ldots, m_n}) \\
&\rightarrow \cdots \\
&\rightarrow q_{15}\omega 1^{m_1}B(\overline{m_2, \ldots, m_n})B(\overline{m_1, \ldots, m_n}) \\
&\rightarrow q_{16}(\overline{m_1, \ldots, m_n, m_1, \ldots, m_n}).
\end{aligned}
$$

For $p > 0$,

$$q_1(\overline{k_1, \ldots, k_p, m_1, \ldots, m_n}) \to \cdots$$
$$\to q_2\lambda B(\overline{k_1, \ldots, k_p, m_1, \ldots, m_n})$$
$$\to \cdots$$
$$\to \lambda B(\overline{k_1, \ldots, k_p}) q_{p+3}\delta(\overline{m_1, \ldots, m_n}).$$

The computation now proceeds like the case $p = 0$. The final instantaneous description is

$$q_{p+16}(\overline{m_1, \ldots, m_n, k_1, \ldots, k_p, m_1, \ldots, m_n}).$$

THE TRANSFER MACHINES R_p. *For each $n > 0$ and $p > 0$, we shall define a $(p + n)$-regular Turing machine R_p such that*

$$\text{Res}_{R_p}[q_1(\overline{k_1, \ldots, k_p, m_1, \ldots, m_n})]$$
$$= q_{p+16}(\overline{m_1, \ldots, m_n, k_1, \ldots, k_p}).$$

That is, the arguments $(k_1, \ldots, k_p)$ are interchanged on the tape with the arguments $(m_1, \ldots, m_n)$.

We begin by noting that, in the copying operation of C_p, each 1 that occurs in the tape expression $(\overline{m_1, \ldots, m_n})$ is replaced by ω, which in turn is again replaced by 1. Hence, all we have to do in order to define our Turing machine R_p is *to erase these ω's instead of replacing them by* 1's. Hence, we may define R_p to consist of precisely the quadruples of C_p except that the quadruple

$$q_{p+14}\ \omega\ 1\ q_{p+6}$$

is replaced by

$$q_{p+14}\ \omega\ B\ q_{p+6}.$$

LEMMA 3. *For each n-regular Turing machine Z, there is an n-regular Turing machine Z' such that, whenever*

$$\text{Res}_Z{}^A[q_1(\overline{m_1, \ldots, m_n})] = q_{\theta(Z)}(\overline{r_1, \ldots, r_s}),$$

it is also the case that

$$\text{Res}_{Z'}{}^A[q_1(\overline{m_1, \ldots, m_n})] = q_{\theta(Z')}(\overline{r_1, \ldots, r_s, m_1, \ldots, m_n}),$$

whereas, whenever $\text{Res}_Z{}^A[q_1(\overline{m_1, \ldots, m_n})]$ *is undefined, so is*

$$\text{Res}_{Z'}{}^A[q_1(\overline{m_1, \ldots, m_n})].$$

That is, the original arguments $m_1, \ldots, m_n$ are "preserved" by Z'.

PROOF. By Lemma 2, there is a $2n$-regular Turing machine U such that

$$\text{Res}_U{}^A[q_1(\overline{m_1, \ldots, m_n, m_1, \ldots, m_n})]$$
$$= q_{\theta(U)}(\overline{m_1, \ldots, m_n, r_1, \ldots, r_s}).$$

Then, we need only take

$$Z' = C_0 \cup U^{(15)} \cup R_n^{(14+\theta(U))}.$$

For, with respect to C_0,

$$q_1(\overline{m_1, \ldots, m_n}) \to \cdots \to q_{16}(\overline{m_1, \ldots, m_n, m_1, \ldots, m_n}).$$

With respect to $U^{(15)}$,

$$q_{16}(\overline{m_1, \ldots, m_n, m_1, \ldots, m_n}) \to \cdots \to q_{\theta(U^{(15)})}(\overline{m_1, \ldots, m_n, r_1, \ldots, r_s}).$$

Finally, with respect to $R_n^{(14+\theta(U))}$,

$$q_{\theta(U^{(15)})}(\overline{m_1, \ldots, m_n, r_1, \ldots, r_s}) \to \cdots \to q_{\theta(Z')}(\overline{r_1, \ldots, r_s, m_1, \ldots, m_n}).$$

LEMMA 4. *Let $Z_1, \ldots, Z_p$ be Turing machines. Let $n > 0$. Then, there exists an n-regular Turing machine Z' such that*

$$\mathrm{Res}_{Z'}{}^A\,[q_1(\overline{m_1, \ldots, m_n})] = q_{\theta(Z')}(\overline{\Psi_{Z_1;A}^{(n)}(m_1, \ldots, m_n), \ldots, \Psi_{Z_p;A}^{(n)}(m_1, \ldots, m_n)}).$$

PROOF. Our proof is by induction on p. For $p = 1$, the result is precisely Lemma 1. Let the result be known for $p = k$; we show that it must then also hold for $p = k + 1$.

Let $Z_1, Z_2, \ldots, Z_{k+1}$ be given Turing machines, and let

$$r_i = \Psi_{Z_i;A}^{(n)}(m_1, \ldots, m_n)$$

for $1 \leqq i \leqq k + 1$. By induction hypothesis, there is an n-regular Turing machine Y_1 such that

$$\mathrm{Res}_{Y_1}{}^A\,[q_1(\overline{m_1, \ldots, m_n})] = q_{\theta(Y_1)}(\overline{r_1, \ldots, r_k}).$$

Hence, by Lemma 3, there is an n-regular Turing machine Y_2 such that

$$\mathrm{Res}_{Y_2}{}^A\,[q_1(\overline{m_1, \ldots, m_n})] = q_{\theta(Y_2)}(\overline{r_1, \ldots, r_k, m_1, \ldots, m_n}).$$

By Lemma 1, there is an n-regular Turing machine Y_3 such that

$$\mathrm{Res}_{Y_3}{}^A[q_1(\overline{m_1, \ldots, m_n})] = q_{\theta(Y_3)}\overline{r_{k+1}}$$

Hence, by Lemma 2, there is a $(k + n)$-regular Turing machine Y_4 such that

$$\mathrm{Res}_{Y_4}{}^A\,[q_1(\overline{r_1, \ldots, r_k, m_1, \ldots, m_n})] = q_{\theta(Y_4)}(\overline{r_1, \ldots, r_k, r_{k+1}}).$$

Thus, we need only take $Z' = Y_2 \cup Y_4{}^{(\theta(Y_2)-1)}$ to obtain the result for $p = k + 1$.

2. Composition and Minimalization. The lemmas of the previous section have dealt directly with Turing machines. In the present section, we employ these lemmas in deriving two important results concerning computability.

The first of these concerns the operation of *composition*. That is, we wish to assert, for example, that, if $f(y)$, $g(x)$ are partially computable, then so is $f(g(x))$. This requires, to begin with, agreement as to the domain of this function. We agree that $f(g(x))$ is defined for precisely those numbers a for which a is in the domain of $g(x)$ and $g(a)$ is in the domain of $f(y)$. The value of $f(g(x))$ at a is, of course, the number $f(g(a))$. More generally, we may write

DEFINITION 2.1. *The operation of* **composition** *associates with the functions*[1] $f(\mathfrak{y}^{(m)})$, $g_1(\mathfrak{x}^{(n)})$, $g_2(\mathfrak{x}^{(n)})$, . . . , $g_m(\mathfrak{x}^{(n)})$, *the function*

$$h(\mathfrak{x}^{(n)}) = f(g_1(\mathfrak{x}^{(n)}), g_2(\mathfrak{x}^{(n)}), \ldots, g_m(\mathfrak{x}^{(n)})). \tag{1}$$

This function is defined for precisely those n-tuples $(a_1, \ldots, a_n)$ *for which* $(a_1, \ldots, a_n)$ *is in the domain of each of the functions* $g_i(\mathfrak{x}^{(n)})$, $i = 1, 2, \ldots, m$, *and for which the m-tuple*

$$(g_1(a_1, \ldots, a_n), g_2(a_1, \ldots, a_n), \ldots, g_m(a_1, \ldots, a_n))$$

is in the domain of $f(\mathfrak{y}^{(m)})$. *Its value at* $(a_1, \ldots, a_n)$ *is*

$$f(g_1(a_1, \ldots, a_n), g_2(a_1, \ldots, a_n), \ldots, g_m(a_1, \ldots, a_n)).$$

We now prove

THEOREM 2.1. *Let* $f(\mathfrak{y}^{(m)})$, $g_1(\mathfrak{x}^{(n)})$, $g_2(\mathfrak{x}^{(n)})$, . . . , $g_m(\mathfrak{x}^{(n)})$ *be (partially) A-computable. Let* $h(\mathfrak{x}^{(n)})$ *be given by* (1). *Then* $h(\mathfrak{x}^{(n)})$ *is (partially) A-computable.*

PROOF. By Lemma 4, there is an n-regular Turing machine Z such that

$$\mathrm{Res}_Z{}^A\,[q_1(\overline{\mathfrak{x}^{(n)}})] = q_{\theta(Z)}(\overline{g_1(\mathfrak{x}^{(n)}), g_2(\mathfrak{x}^{(n)}), \ldots, g_m(\mathfrak{x}^{(n)})}).$$

Let Z_1 be chosen so that

$$\Psi_{Z_1;A}^{(m)}(\mathfrak{y}^{(m)}) = f(\mathfrak{y}^{(m)}),$$

and let $Z' = Z \cup Z_1{}^{(\theta(Z)-1)}$. Then, with respect to Z',

$$\begin{aligned} q_1(\overline{\mathfrak{x}^{(n)}}) \to \cdots \to q_{\theta(Z)}(\overline{g_1(\mathfrak{x}^{(n)}), g_2(\mathfrak{x}^{(n)}), \ldots, g_m(\mathfrak{x}^{(n)})}) \\ \to \cdots \to \alpha, \end{aligned}$$

where

$$\langle \alpha \rangle = f(g_1(\mathfrak{x}^{(n)}), g_2(\mathfrak{x}^{(n)}), \ldots, g_m(\mathfrak{x}^{(n)})),$$

[1] Here, for example, $f(\mathfrak{y}^{(m)})$ is an abbreviation of $f(y_1, y_2, \ldots, y_m)$. Cf. Introduction, Sec. 3.1.

when each $g_i(\mathfrak{x}^{(n)})$ and $f(g_1(\mathfrak{x}^{(n)}), \ldots, g_m(\mathfrak{x}^{(n)}))$ is defined; otherwise $\text{Res}_{Z'}{}^A\,[q_1(\mathfrak{x}^{(n)})]$ is undefined. Hence,

$$\begin{aligned}\Psi^{(n)}_{Z';A}(\overline{\mathfrak{x}^{(n)}}) &= f(g_1(\mathfrak{x}^{(n)}), g_2(\mathfrak{x}^{(n)}), \ldots, g_m(\mathfrak{x}^{(n)})) \\ &= h(\mathfrak{x}^{(n)}).\end{aligned}$$

This proves the theorem.

Theorem 2.1 may be restated as follows:

COROLLARY 2.2. *The class of (partially) A-computable functions is closed under the operation of composition.*

Theorem 2.1 enables us to enlarge considerably the class of functions that we know to be computable.

Thus, let

$$f(x_1, x_2) = x_1 \dot{-} x_2;\ g_1(x, y) = (x + 1)(y + 1);\ g_2(x, y) = y + 1.$$

Then, by Examples 1-3.4 and 1-3.7, f and g_1 are computable.

$$g_2(x, y) = S(U_2{}^2(x, y)),$$

where $S(z) = z + 1$, and so is computable, by Examples 1-3.2 and 1-3.6 and Theorem 2.1. Hence, by Theorem 2.1, we can infer the computability of

$$\begin{aligned}h(x, y) &= f(g_1(x, y), g_2(x, y)) \\ &= (x + 1)(y + 1) \dot{-} (y + 1) \\ &= (x + 1)(y + 1) - (y + 1) \\ &= xy + x.\end{aligned}$$

Again, taking $f(x_1, x_2) = x_1 \dot{-} x_2$ and taking

$$g_1(x, y) = xy + x,\ g_2(x, y) = U_1{}^2(x, y) = x,$$

we have the computability of

$$\begin{aligned}k(x, y) &= (xy + x) \dot{-} x \\ &= xy.\end{aligned}$$

Thus, we have proved

COROLLARY 2.3. *xy is computable.*

It is not difficult to see that x^k is computable for each fixed k. We have the result for $k = 1$; to infer it for $k + 1$, given the result for k, we set $f(x_1, x_2) = x_1x_2$, $g_1(x) = x^k$, $g_2(x) = U_1{}^1(x) = x$, and we apply Theorem 2.1. Then, it follows easily that every polynomial with integral coefficients is computable.

The function $|x - y|$† is computable since $|x - y| = (x \dot{-} y) + (y \dot{-} x)$.

† $|x - y| = x - y$ when $x \geq y$; $|x - y| = y - x$ when $y \geq x$.

Functions whose computability we are yet unable to obtain (except. of course, by direct construction of Turing machines) are 2^x and $[\sqrt{x}]$, this last being the largest integer $\leq \sqrt{x}$ (thus, $[\sqrt{5}] = 2$, $[\sqrt{9}] = 3$).

DEFINITION 2.2. *The operation of* **minimalization** *associates with each total function* $f(y, \mathfrak{x}^{(n)})$ *the function* $h(\mathfrak{x}^{(n)})$, *whose value for given* $\mathfrak{x}^{(n)}$ *is the least value of* y, *if one such exists, for which* $f(y, \mathfrak{x}^{(n)}) = 0$, *and which is undefined if no such* y *exists. We write*

$$h(\mathfrak{x}^{(n)}) = \min_y [f(y, \mathfrak{x}^{(n)}) = 0].$$

Thus, for example, consider

$$x/2 = \min_y [|(y + y) - x| = 0].$$

$x/2$ is a partial function, defined only when x is even. It will follow at once from Theorem 2.4 below that $x/2$ is partially computable.

DEFINITION 2.3. *The total function* $f(y, \mathfrak{x}^{(n)})$ *is called* **regular** *if*

$$\min_y [f(y, \mathfrak{x}^{(n)}) = 0]$$

is total.

THEOREM 2.4. *If* $f(y, \mathfrak{x}^{(n)})$ *is A-computable, then*

$$h(\mathfrak{x}^{(n)}) = \min_y [f(y, \mathfrak{x}^{(n)}) = 0]$$

is partially A-computable. Moreover, if $f(y, \mathfrak{x}^{(n)})$ *is regular,* $h(\mathfrak{x}^{(n)})$ *is A-computable.*

Our proof will proceed by constructing a Turing machine that will successively compute $f(0, \mathfrak{x}^{(n)})$, $f(1, \mathfrak{x}^{(n)})$, . . . , until a zero is arrived at. If no zero is ever obtained, the Turing machine will compute "forever."

PROOF. Let U consist of the quadruples

$$\begin{array}{l} q_1\ 1\ L\ q_1 \\ q_1\ B\ L\ q_2 \\ q_2\ B\ 1\ q_3. \end{array}$$

Then, with respect to U,

$$\begin{aligned} q_1(\overline{\mathfrak{x}^{(n)}}) &\rightarrow \cdots \\ &\rightarrow q_3(\overline{0, \mathfrak{x}^{(n)}}), \end{aligned}$$

which is final.

By Lemmas 1 and 3, there is an $(n + 1)$-regular Turing machine Y such that

$$\mathrm{Res}_Y{}^A\ [q_1(\overline{y, \mathfrak{x}^{(n)}})] = q_{\theta(Y)}(\overline{f(y, \mathfrak{x}^{(n)}), y, \mathfrak{x}^{(n)}}).$$

Hence, if we let $N = \theta(Y^{(2)})$, then, with respect to $Y^{(2)}$,

$$\begin{aligned} q_3(\overline{y, \mathfrak{x}^{(n)}}) &\rightarrow \cdots \\ &\rightarrow q_N(\overline{f(y, \mathfrak{x}^{(n)}), y, \mathfrak{x}^{(n)}}). \end{aligned}$$

Let M consist of the quadruples

$$\begin{array}{l} q_N \; 1 \; B \; q_N \\ q_N \; B \; R \; q_{N+1} \\ q_{N+1} \; 1 \; 1 \; q_{N+2} \\ q_{N+1} \; B \; R \; q_{N+4}. \end{array}$$

Thus, if $f(y, \mathfrak{x}^{(n)}) = k > 0$, then, with respect to M,

$$\begin{aligned} q_N(\overline{f(y, \mathfrak{x}^{(n)}), y, \mathfrak{x}^{(n)}}) &= q_N 1 1^k B(\overline{y, \mathfrak{x}^{(n)}}) \\ &\rightarrow \cdots \\ &\rightarrow q_{N+2} 1^k B(\overline{y, \mathfrak{x}^{(n)}}), \end{aligned}$$

which is final.[1]

On the other hand, if $f(y, \mathfrak{x}^{(n)}) = 0$, then, with respect to M,

$$\begin{aligned} q_N(\overline{f(y, \mathfrak{x}^{(n)}), y, \mathfrak{x}^{(n)}}) &= q_N 1 B(\overline{y, \mathfrak{x}^{(n)}}) \\ &\rightarrow \cdots \\ &\rightarrow q_{N+4}(\overline{y, \mathfrak{x}^{(n)}}). \end{aligned}$$

Let Q consist of the quadruples

$$\begin{array}{l} q_{N+2} \; 1 \; B \; q_{N+3} \\ q_{N+2} \; B \; 1 \; q_3 \\ q_{N+3} \; B \; R \; q_{N+2}. \end{array}$$

Then, with respect to Q,

$$\begin{aligned} q_{N+2} 1^k B(\overline{y, \mathfrak{x}^{(n)}}) &\rightarrow \cdots \\ &\rightarrow q_3(\overline{y + 1, \mathfrak{x}^{(n)}}). \end{aligned}$$

By Example 1-3.6 and Lemma 1, there is an $(n + 1)$-regular Turing machine Z_1 such that

$$\begin{aligned} \operatorname{Res}_{Z_1} [q_1(\overline{y, \mathfrak{x}^{(n)}})] &= q_{\theta(Z_1)} \overline{U_1^{n+1}(y, \mathfrak{x}^{(n)})} \\ &= q_{\theta(Z_1)} 1^{y+1}. \end{aligned}$$

Let E consist of all of the quadruples of Z_1 and, in addition, the quadruple

$$q_{\theta(Z_1)} \; 1 \; B \; q_{\theta(Z_1)}.$$

Then, with respect to $E^{(N+3)}$, letting $K = \theta(E^{(N+3)})$,

$$\begin{aligned} q_{N+4}(\overline{y, \mathfrak{x}^{(n)}}) &\rightarrow \cdots \\ &\rightarrow q_K B 1^y. \end{aligned}$$

Now, let

$$Z = U \cup Y^{(2)} \cup M \cup Q \cup E^{(N+3)}.$$

[1] Here, and elsewhere in this proof, we shall omit initial as well as final occurrences of B in instantaneous descriptions, although, strictly speaking, this violates the convention laid down at the beginning of the present chapter.

We shall see that

$$\Psi_{Z;A}^{(n)}(\mathfrak{x}^{(n)}) = h(\mathfrak{x}^{(n)}).$$

To see this, let the numbers $\mathfrak{x}^{(n)}$ be fixed; let $f(i, \mathfrak{x}^{(n)}) = r_i$; and suppose that $r_0 \neq 0, r_1 \neq 0, \ldots, r_{k-1} \neq 0, r_k = 0$. Then, with respect to Z,

$$
\begin{aligned}
q_1(\overline{\mathfrak{x}^{(n)}}) &\to \cdots \\
&\to q_3(\overline{0, \mathfrak{x}^{(n)}}) && \text{(using } U\text{)} \\
&\to \cdots \\
&\to q_N(\overline{r_0, 0, \mathfrak{x}^{(n)}}) && \text{(using } Y^{(2)}\text{)} \\
&\to \cdots \\
&\to q_{N+2}(\overline{r_0 - 1, 0, \mathfrak{x}^{(n)}}) && \text{(using } M\text{)} \\
&\to \cdots \\
&\to q_3(\overline{1, \mathfrak{x}^{(n)}}) && \text{(using } Q\text{)} \\
&\to \cdots \\
&\to \cdots \\
&\to q_3(\overline{k, \mathfrak{x}^{(n)}}) \\
&\to \cdots \\
&\to q_N(\overline{r_k, k, \mathfrak{x}^{(n)}}) && \text{(using } Y^{(2)}\text{)} \\
&= q_N 1 B(\overline{k, \mathfrak{x}^{(n)}}) \\
&\to \cdots \\
&\to q_{N+4}(\overline{k, \mathfrak{x}^{(n)}}) && \text{(using } M\text{)} \\
&\to \cdots \\
&\to q_K B 1^k && \text{(using } E^{(N+3)}\text{).}
\end{aligned}
$$

Hence,

$$
\begin{aligned}
\Psi_{Z;A}^{(n)}(\mathfrak{x}^{(n)}) &= \langle q_K B 1^k \rangle \\
&= k \\
&= \min_y [f(y, \mathfrak{x}^{(n)}) = 0] \\
&= h(\mathfrak{x}^{(n)}).
\end{aligned}
$$

If $r_i \neq 0$ for all i, then Z is never in internal configuration q_{N+4}, and no final instantaneous description is ever reached (that is, the machine computes "forever"). In this case both $\Psi_{Z;A}^{(n)}(\mathfrak{x}^{(n)})$ and $h(\mathfrak{x}^{(n)})$ are undefined.

This completes the proof.

CHAPTER 3

RECURSIVE FUNCTIONS

1. Some Classes of Functions. The results of the previous chapter have made it possible to demonstrate the computability of quite complex functions *without referring back to the original definition of computability in terms of Turing machines.* In the present chapter, we shall develop this systematically. We begin by defining certain classes of functions, employing the operations of *composition* and *minimalization* as defined in the previous chapter.

DEFINITION 1.1. *A function is A-***partial recursive** *or* **partial recursive in** *A if it can be obtained by a finite number of applications of composition and minimalization beginning with the functions of the following list:*

(1) $C_A(x)$, *the characteristic function of A.*
(2) $S(x) = x + 1$.
(3) $U_i^n(x_1, \ldots, x_n) = x_i,\ 1 \leqq i \leqq n$.
(4) $x + y$.
(5) $x \dot{-} y$.
(6) xy.

A function is **partial recursive** *if it is ϕ-partial recursive.*

DEFINITION 1.2. *A function is A-***recursive,** *or* **recursive in** *A if it can be obtained by a finite number of applications of composition and minimalization of regular*[1] *functions, beginning with the functions of the list of Definition* 1.1.

A function is **recursive**[2] *if it is ϕ-recursive.*

COROLLARY 1.1. *Every A-recursive function is total and is A-partial recursive.*

As will appear later, the converse[3] of this is also true.

[1] Cf. Definition 2-2.3.

[2] This is not the way in which recursiveness is usually defined. That a characterization of recursiveness based on applying composition and minimalization beginning with a finite set of initial functions was possible was first proved in Kleene [3]. The present development follows J. Robinson [1].

[3] This converse is not quite obvious, since it seems possible for composition of partial recursive functions to produce a total function that is not recursive.

As thus presented, the concepts of recursive function and partial recursive function seem completely artificial. And indeed, their interest stems entirely from their relation to computability.

Thus, we have at once

THEOREM 1.2. *If a function is recursive, A-recursive, partial recursive, or A-partial recursive, then it is computable, A-computable, partially computable, or partially A-computable, respectively.*

We shall see later that the converse of this is also true; so recursiveness is equivalent to computability.

PROOF. That the initial functions 2 to 5 are computable (and hence partially computable, A-computable, and partially A-computable) was seen in Examples 1-3.1, 1-3.2, 1-3.4, and 1-3.6. For (6), that is, xy, the computability was proved as Corollary 2-2.3. That $C_A(x)$ is A-computable (and hence partially A-computable) was proved as Theorem 1-4.7 (cf. Definition 1-4.6). In particular, $C_\emptyset(x)$ is $\emptyset$-computable, and hence computable (and partially computable); cf. Theorem 1-4.6.

The theorem now follows from Theorems 2-2.1 and 2-2.4.

THEOREM 1.3. *If a function is (partial) recursive, it is A-(partial) recursive for any choice of A.*

PROOF. The proof follows from the observation that

$$C_\emptyset(x) = S(U_1^1(x) \dot{-} U_1^1(x)).$$

We now list some recursive functions. By the above, they are also computable and, hence, A-computable for all A.

(7) $N(x) = 0$. $N(x) = U_1^1(x) \dot{-} U_1^1(x)$.

(8) $\alpha(x) = 1 \dot{-} x$; that is,

$$\begin{aligned} \alpha(0) &= 1, \\ \alpha(x) &= 0 \text{ for } x > 0. \\ \alpha(x) &= S(N(x)) \dot{-} U_1^1(x). \end{aligned}$$

(9) $x^2 = U_1^1(x) \cdot U_1^1(x)$.

(10) $[\sqrt{x}]$, the largest integer $\leqq \sqrt{x}$.

$$\begin{aligned} [\sqrt{x}] &= \min_y [(y+1)^2 \dot{-} x \neq 0] \\ &= \min_y [\alpha((S(U_2^2(x, y)))^2 \dot{-} U_1^2(x, y)) = 0]. \end{aligned}$$

(11) $|x - y| = (x \dot{-} y) + (y \dot{-} x)$.

(12) $[x/y]$. If $y \neq 0$, $[x/y]$ is the greatest integer $\leqq \dfrac{x}{y}$.† If $y = 0$,

† Note that in this context $\frac{3}{2}$, for example, means a definite rational number. As we have understood the symbol x/y (cf. Introduction, Sec. 3.3), we should have to regard $3/2$ as undefined.

$[x/y] = 0$. We have

$$\begin{aligned}[x/y] &= \min_z [y = 0 \vee y(z+1) > x] \\ &= \min_z [y = 0 \vee y(z+1) \dot{-} x \neq 0] \\ &= \min_z [y = 0 \vee \alpha(y(z+1) \dot{-} x) = 0] \\ &= \min_z [y \cdot \alpha(y(z+1) \dot{-} x) = 0].\end{aligned}$$

Here, the symbol $\vee$ (which we have borrowed from symbolic logic) means "or" (cf. Introduction, Secs. 3.4 and 3.5). The difficulty is that minimalization involves a single function set equal to 0. Our device depends on the fact that to say that either one or the other (or both) of a pair of numbers is 0 is equivalent to saying that the product of the numbers is 0.

(13) $R(x, y)$. If $y \neq 0$, $R(x, y)$ is the remainder on dividing x by y. That is,

$$\frac{x}{y} = [x/y] + \frac{R(x, y)}{y};$$

so

$$R(x, y) = x - y[x/y].$$

We take

$$R(x, y) = x \dot{-} y[x/y];$$

so $R(x, 0) = x$.

Thus we see that our present techniques enable us to prove the computability of a large class of functions. However, such functions as x^y, $x!$, the xth prime,[1] are still beyond our powers. The concept of primitive recursion will enable us to deal with them also.

2. Finite Sequences of Natural Numbers. It is a well-known fact that there exist one-one correspondences between the set of natural numbers and the set of ordered pairs of natural numbers and, indeed, that such a correspondence can be set up in an "effective" manner. We show how to set up such a correspondence by recursive (hence computable) functions.

Consider the function

$$J(x, y) = \tfrac{1}{2}[(x + y)^2 + 3x + y].$$

Now, $(x + y)^2 + 3x + y = (x + y)(x + y + 1) + 2x$, which is always even. Hence, $J(x, y)$ is always an integer.[2] $J(x, y)$ is recursive, since

$$J(x, y) = [\{(x + y)^2 + U_1^1(x) + U_1^1(x) + U_1^1(x) + U_1^1(y)\}/S(S(N(x)))]$$

Clearly, with each ordered pair of natural numbers (x, y), the function $J(x, y)$ associates a single natural number z. We shall show that, for

[1] For the meaning of the term "prime," see the Appendix.

[2] In fact,

$$J(x, y) = [1 + 2 + \cdots + (x + y)] + x$$

every natural number z, there is one and only one ordered pair of numbers (x, y) such that $z = J(x, y)$.

Suppose z, x, y are natural numbers such that

$$2z = (x + y)^2 + 3x + y. \tag{1}$$

Then

$$8z + 1 = (2x + 2y + 1)^2 + 8x.$$

Therefore,

$$(2x + 2y + 1)^2 \leqq 8z + 1 < (2x + 2y + 3)^2.$$
$$2x + 2y + 1 \leqq \sqrt{8z + 1} < 2x + 2y + 3.$$

Hence, $[\sqrt{8z + 1}]$ is either $2x + 2y + 1$ or $2x + 2y + 2$. $[\sqrt{8z + 1}] + 1$ is either $2x + 2y + 2$ or $2x + 2y + 3$. Therefore,

$$[([\sqrt{8z + 1}] + 1)/2] = x + y + 1.$$
$$x + y = [([\sqrt{8z + 1}] + 1)/2] - 1. \tag{2}$$

By (1),

$$3x + y = 2z - ([([\sqrt{8z + 1} + 1])/2] - 1)^2. \tag{3}$$

But Eqs. (2) and (3) clearly prove our contention that, for each z, at most one x, y exists satisfying (1).† Such x and y, if they do exist, can be calculated by recursive functions; for, if we write

$$Q_1(z) = [([\sqrt{8z + 1}] + 1)/2] \dot{-} 1,$$
$$Q_2(z) = 2z \dot{-} (Q_1(z))^2,$$

then, clearly, $Q_1(z)$ and $Q_2(z)$ are recursive functions, and (2) and (3) may be written in the form

$$x + y = Q_1(z),$$
$$3x + y = Q_2(z).$$

These equations yield

$$x = [(Q_2(z) \dot{-} Q_1(z))/2] = K(z),$$
$$y = Q_1(z) \dot{-} [(Q_2(z) \dot{-} Q_1(z))/2] = L(z),$$

where $K(z)$ and $L(z)$ are recursive functions. Thus, if x, y, z satisfy (1), then $x = K(z)$, $y = L(z)$. Now, if x and y are chosen arbitrarily, $z = J(x, y)$ satisfies (1). Hence $x = K(J(x, y))$, $y = L(J(x, y))$.

Next, let z be any number. Let r be the largest number such that

$$1 + 2 + \cdots + r \leqq z.$$

Let

$$x = z - (1 + 2 + \cdots + r).$$

† For $\begin{vmatrix} 1 & 1 \\ 3 & 1 \end{vmatrix} \neq 0.$

Then, $x \leqq r$. [For, if $x \geqq r + 1$, then $1 + 2 + \cdots + r + (r + 1) \leqq z$, contradicting the choice of r.] Let $y = r - x$. Then,

$$\begin{aligned} z &= [1 + 2 + \cdots + (x + y)] + x \\ &= \tfrac{1}{2}(x + y)(x + y + 1) + x \\ &= J(x, y). \end{aligned}$$

Thus,

$$\begin{aligned} x &= K(J(x, y)) = K(z), \\ y &= L(J(x, y)) = L(z); \end{aligned}$$

that is,

$$z = J(K(z), L(z)).$$

We have thus proved

THEOREM 2.1. *There exist recursive functions $J(x, y)$, $K(z)$, $L(z)$ such that*

$$\begin{aligned} J(K(z), L(z)) &= z, \\ K(J(x, y)) &= x, \\ L(J(x, y)) &= y. \end{aligned}$$

COROLLARY 2.2. *If $K(z) = K(z')$, $L(z) = L(z')$, then $z = z'$.*

PROOF. $z = J(K(z), L(z)) = J(K(z'), L(z')) = z'$.

This solves the problem of "effectively" obtaining the ordered pairs of integers. We next consider the problem of doing the same for all the finite sequences (of whatever length) of integers.

We have first the

LEMMA. *Let v be divisible*[1] *by the numbers $1, 2, \ldots, n$. Then the numbers $1 + v(i + 1)$, $i = 0, 1, 2, \ldots, n$, are relatively prime*[1] *in pairs.*

PROOF. Let $m_i = 1 + v(i + 1)$. Since v is divisible by $1, 2, \ldots, n$, any divisor of m_i, other than 1, must be greater than n.

Now suppose that $d|m_i$, $d|m_j$, $i > j$. Then $d|(i + 1)m_j - (j + 1)m_i$; that is, $d|i - j$. But $i - j \leqq n$. Hence $d = 1$.

THEOREM 2.3. *Let $a_0, a_1, a_2, \ldots, a_n$ be a finite sequence of integers. Then there are integers u and v such that*

$$R(u, 1 + v(i + 1)) = a_i \qquad i = 0, 1, \ldots, n.$$

PROOF. Let A be the largest of the integers $a_0, a_1, \ldots, a_n$, and let $v = 2A \cdot n!$ Let $m_i = 1 + v(i + 1)$. Then, by the lemma, the m_i are relatively prime in pairs. Also,

$$a_i < v < m_i.$$

Now, by the Chinese remainder theorem (Theorem 12 of the Appendix),

[1] For the meaning of such number-theoretic terms as "divisible" and "relatively prime," see the Appendix.

there exists a number u such that

$$u \equiv a_i \pmod{m_i} \qquad i = 0, 1, \ldots, n.$$

That is,

$$R(u, m_i) = R(a_i, m_i) \qquad i = 0, 1, \ldots, n.$$

But $a_i < m_i$. Hence, $R(a_i, m_i) = a_i$. Thus,

$$R(u, 1 + v(i + 1)) = R(u, m_i) = R(a_i, m_i) = a_i.$$

THEOREM 2.4. *There is a recursive function $T_i(w)$ such that, if $a_0, a_1, \ldots, a_n$ are any integers whatever, there exists a number w_0 such that $T_i(w_0) = a_i$, $i = 0, 1, \ldots, n$.*

PROOF. We define $T_i(w)$ by the equation

$$T_i(w) = R(K(w), 1 + [L(w)(i + 1)]).$$

Clearly, $T_i(w)$ is recursive.

Let us be given the integers $a_0, a_1, \ldots, a_n$. Then, by Theorem 2.3, there exist numbers u and v such that

$$R(u, 1 + v(i + 1)) = a_i \qquad i = 0, 1, \ldots, n.$$

Let $w_0 = J(u, v)$. Then,

$$\begin{aligned} T_i(w_0) &= R(K(J(u, v)), 1 + L(J(u, v)) \cdot (i + 1)) \\ &= R(u, 1 + v(i + 1)) \\ &= a_i \qquad i = 0, 1, \ldots, n. \end{aligned}$$

3. Primitive Recursion. We now consider the question of proving the computability of such functions as x^y and $x!$ Suppose one were asked the value of 7^5. One natural procedure for obtaining the desired result is to obtain successively

$$7^1 = 7, 7^2 = 7^1 \cdot 7 = 7 \cdot 7 = 49, 7^3 = 7^2 \cdot 7 = 49 \cdot 7 = 343,$$

$7^4 = 7^3 \cdot 7 = 343 \cdot 7 = 2{,}401$, $7^5 = 7^4 \cdot 7 = 2{,}401 \cdot 7 = 16{,}807$. This procedure can be represented by the "recursion" equations

$$\begin{aligned} x^1 &= x, \\ x^{y+1} &= x^y \cdot x. \end{aligned}$$

Clearly, the first of these equations could be replaced by $x^0 = 1$. The function $n!$ can be computed similarly from the equations

$$\begin{aligned} 0! &= 1, \\ (n + 1)! &= (n + 1)n! \end{aligned}$$

We shall begin by showing that such a pair of recursion equations always defines one and only one total function.

THEOREM 3.1. *Let $f(\mathfrak{x}^{(n)})$, $g(\mathfrak{x}^{(n+2)})$ be total functions. Then there exists at most one total function $h(\mathfrak{x}^{(n+1)})$ that satisfies the recursion equations*

$$\begin{aligned} h(0, \mathfrak{x}^{(n)}) &= f(\mathfrak{x}^{(n)}), \\ h(z+1, \mathfrak{x}^{(n)}) &= g(z, h(z, \mathfrak{x}^{(n)}), \mathfrak{x}^{(n)}). \end{aligned} \tag{1}$$

PROOF. Suppose $h_1(\mathfrak{x}^{(n+1)})$ and $h_2(\mathfrak{x}^{(n+1)})$ both satisfy (1). We shall see that $h_1(y, \mathfrak{x}^{(n)}) = h_2(y, \mathfrak{x}^{(n)})$ for all y, $\mathfrak{x}^{(n)}$. This is certainly true for $y = 0$ and all $\mathfrak{x}^{(n)}$, since

$$h_1(0, \mathfrak{x}^{(n)}) = f(\mathfrak{x}^{(n)}) = h_2(0, \mathfrak{x}^{(n)}).$$

Suppose it true for $y = z$. Then

$$\begin{aligned} h_1(z+1, \mathfrak{x}^{(n)}) &= g(z, h_1(z, \mathfrak{x}^{(n)}), \mathfrak{x}^{(n)}) \\ &= g(z, h_2(z, \mathfrak{x}^{(n)}), \mathfrak{x}^{(n)}) \\ &= h_2(z+1, \mathfrak{x}^{(n)}). \end{aligned}$$

Thus, $h_1(y, \mathfrak{x}^{(n)}) = h_2(y, \mathfrak{x}^{(n)})$ for all y, $\mathfrak{x}^{(n)}$.

Proving the existence of a function h that satisfies (1) is a peculiar matter. For there are those whose suspicion of nonconstructive arguments is so strong that, although they are happy to accept the existence of such an h, they will insist that our proof of this fact, below, is invalid.

THEOREM 3.2. *Let $f(\mathfrak{x}^{(n)})$, $g(\mathfrak{x}^{(n+2)})$ be total functions. Then, there exists a total function $h(\mathfrak{x}^{(n+1)})$ that satisfies* (1).

PROOF. We consider sets of $(n + 2)$-tuples $(y, \mathfrak{x}^{(n)}, u)$ of numbers. Such a set S of $(n + 2)$-tuples will be called *satisfactory* if

(*a*) For each choice of $\mathfrak{x}^{(n)}$, $(0, \mathfrak{x}^{(n)}, f(\mathfrak{x}^{(n)})) \in S$; and
(*b*) For each choice of $\mathfrak{x}^{(n)}$, if $(z, \mathfrak{x}^{(n)}, u) \in S$, then

$$(z+1, \mathfrak{x}^{(n)}, g(z, u, \mathfrak{x}^{(n)})) \in S.$$

Let Ω be the class of all satisfactory sets S. Then Ω is nonempty, since the set of *all* $(n + 2)$-tuples of numbers is satisfactory. Let S_0 be the intersection of all sets S that are members of Ω. Then it is easy to see that S_0 is satisfactory and that S_0 is contained in every satisfactory set.

Next, suppose that $(0, \mathfrak{x}^{(n)}, u) \in S_0$. Then we have $u = f(\mathfrak{x}^{(n)})$. For otherwise the set obtained on deleting $(0, \mathfrak{x}^{(n)}, u)$ from S_0 would be satisfactory and would not contain S_0.

We next claim that, *for each choice of y and $\mathfrak{x}^{(n)}$, there is one and only one value of u for which $(y, \mathfrak{x}^{(n)}, u) \in S_0$.* By what has just been said, this is true for $y = 0$. Suppose it known for $y = z$. Then for a suitable u, *and for no other,*

$$(z, \mathfrak{x}^{(n)}, u) \in S_0.$$

Since S_0 is satisfactory,

$$(z + 1, \mathfrak{x}^{(n)}, g(z, u, \mathfrak{x}^{(n)})) \in S_0.$$

Suppose that

$$(z + 1, \mathfrak{x}^{(n)}, v) \in S_0 \qquad v \neq g(z, u, \mathfrak{x}^{(n)}).$$

Then the set obtained by deleting $(z + 1, \mathfrak{x}^{(n)}, v)$ from S_0 would clearly be satisfactory and would not contain S_0. This proves our last claim.

This enables us to define the function $h(y, \mathfrak{x}^{(n)})$ by the requirement

$$u = h(y, \mathfrak{x}^{(n)}) \qquad \text{if and only if } (y, \mathfrak{x}^{(n)}, u) \in S_0.$$

The fact that S_0 is satisfactory immediately shows that $h(y, \mathfrak{x}^{(n)})$ satisfies (1).

DEFINITION 3.1. *The operation of* **primitive recursion** *associates with the given total functions* $f(\mathfrak{x}^{(n)})$, $g(\mathfrak{x}^{(n+2)})$ *the function* $h(\mathfrak{x}^{(n+1)})$, *where*

$$\begin{aligned} h(0, \mathfrak{x}^{(n)}) &= f(\mathfrak{x}^{(n)}), \\ h(z + 1, \mathfrak{x}^{(n)}) &= g(z, h(z, \mathfrak{x}^{(n)}), \mathfrak{x}^{(n)}). \end{aligned}$$

THEOREM 3.3. *Let* $h(\mathfrak{x}^{(n+1)})$ *be obtained from* $f(\mathfrak{x}^{(n)})$, $g(\mathfrak{x}^{(n+2)})$ *by primitive recursion. If f and g are A-recursive, then so is h.*

PROOF.[1] By Theorem 2.4, for each choice of $\mathfrak{x}^{(n)}$ and y, there exists at least one number w_0 such that

$$T_i(w_0) = h(i, \mathfrak{x}^{(n)}) \qquad i = 0, 1, \ldots, y.$$

Hence,

$$h(y, \mathfrak{x}^{(n)}) = T_y\left(\min_w \left[(T_0(w) = f(\mathfrak{x}^{(n)})) \wedge \bigwedge_{z=0}^{y-1} (T_{z+1}(w) = g(z, T_z(w), \mathfrak{x}^{(n)}))\right]\right).\dagger$$

Now, to say that a condition holds for all numbers z less than y is equivalent to saying that y is the least number for which it could fail. That is,

$$h(y, \mathfrak{x}^{(n)}) = T_y(\min_w (\{T_0(w) = f(\mathfrak{x}^{(n)})\} \wedge \{y = \min_z [(T_{z+1}(w) \neq g(z, T_z(w), \mathfrak{x}^{(n)})) \vee (z = y)]\})).$$

Let

$$H(y, w, \mathfrak{x}^{(n)}) = \min_z [(T_{z+1}(w) \neq g(z, T_z(w), \mathfrak{x}^{(n)})) \vee (z = y)].$$

[1] This proof is that of J. Robinson [1] following Kleene [3], which in turn follows Gödel [1]. The argument employed could be used to yield another proof of Theorem 3.2.

† The symbol $\bigwedge_{z=0}^{y-1}$ is read, "for all z between 0 and $y - 1$." Cf. Introduction, Sec. 3.6.

Then

$$H(y, w, \mathfrak{x}^{(n)}) = \min_z [|z - y| \cdot \alpha(|T_{z+1}(w) - g(z, T_z(w), \mathfrak{x}^{(n)})|) = 0];$$

so $H(y, w, \mathfrak{x}^{(n)})$ is A-recursive. But

$$\begin{aligned} h(y, \mathfrak{x}^{(n)}) &= T_y(\min_w [(T_0(w) = f(\mathfrak{x}^{(n)})) \wedge (y = H(y, w, \mathfrak{x}^{(n)}))]) \\ &= T_y(\min_w [|T_0(w) - f(\mathfrak{x}^{(n)})| + |y - H(y, w, \mathfrak{x}^{(n)})| = 0]). \end{aligned}$$

Therefore, $h(y, \mathfrak{x}^{(n)})$ is A-recursive. This completes the proof.

Thus, if we take

$$\begin{aligned} f(x) &= S(N(x)) = 1, \\ g(z, u, v) &= uv, \end{aligned}$$

then, clearly, f and g are recursive. Therefore, so is $h(z, x)$, which satisfies

$$\begin{aligned} h(0, x) &= f(x) = 1, \\ h(z + 1, x) &= g(z, h(z, x), x) = xh(z, x). \end{aligned}$$

But $h(y, x) = x^y$ satisfies this pair of equations. Thus, by Theorem 3.1, x^y is recursive and is, therefore, computable.

4. Primitive Recursive Functions. In this section, a certain subclass of the class of recursive functions, the so-called *primitive recursive functions*, will be singled out for study. This class does, in fact, contain all numerical functions ordinarily encountered. Nevertheless, it possesses a certain constructive character[1] which the class of all recursive functions does not possess. Actually, there are many classes[2] which possess these properties and which could be studied instead.

DEFINITION 4.1. *A function is A-***primitive recursive** *(or* **primitive recursive in** *A) if it can be obtained by a finite number of applications of the operations of composition and primitive recursion beginning with functions from the following list:*

(1) $C_A(x)$.
(2) $S(x) = x + 1$.
(3) $N(x) = 0$.
(4) $U_i^n(x_1, \ldots, x_n) = x_i$, $1 \leqq i \leqq n$.

A function is **primitive recursive** *if it is ϕ-primitive recursive.*

THEOREM 4.1. *If a function is A-primitive recursive, then it is A-recursive.*

[1] Thus, for example, it can be proved that there exists a binary recursive function $f(n, x)$ such that, for each singulary primitive recursive function $g(x)$, there exists n for which $f(n, x) = g(x)$.

However, this assertion does not hold with the word "primitive" deleted. For, if such an f could be obtained, $g(x) = f(x, x) + 1$ would yield a contradiction.

[2] Thus, there are the multiply recursive functions and the elementary functions (cf. Péter [1]).

PROOF. The proof is clear from the definitions involved, from the fact [noted in (7) of Sec. 1] that $N(x)$ is recursive, and from Theorem 3.3.

It can be shown that there exist recursive functions that are not primitive[1] recursive.

We now list some primitive recursive functions:

(5) $x + y$.

For

$$x + 0 = U_1^1(x),$$
$$x + (y + 1) = S(x + y).$$

(6) $x \cdot y$.

For

$$x \cdot 0 = N(x),$$
$$x \cdot (y + 1) = (x \cdot y) + U_1^2(x, y).$$

(7) $P(x)$, where $P(0) = 0$, $P(x) = x - 1$ if $x > 0$.

For

$$P(0) = N(x),$$
$$P(x + 1) = U_1^1(x).$$

(8) $x \dot{-} y$.

For

$$x \dot{-} 0 = U_1^1(x),$$
$$x \dot{-} (y + 1) = P(x \dot{-} y).$$

(9) $n!$, where $n! = 1 \cdot 2 \cdot 3 \cdot \cdot \cdot n$.

For

$$0! = S(N(x)),$$
$$(n + 1)! = n! \cdot S(n).$$

(10) x^y.

For

$$x^0 = S(N(x)),$$
$$x^{y+1} = x^y \cdot U_1^2(x, y).$$

(11) $\alpha(x) = 1 \dot{-} x$.

For

$$\alpha(x) = S(N(x)) \dot{-} U_1^1(x).$$

(12) $|x - y|$.

For

$$|x - y| = (x \dot{-} y) + (y \dot{-} x).$$

[1] Cf. Péter [1, pp. 68–72] or R. M. Robinson [2].

In fact, the result follows at once from the statement of the first footnote at the beginning of this section.

The class of primitive recursive functions was first employed in Gödel [1], where Gödel referred to them as "rekursive Funktionen." Kleene [1] formalized and investigated a more general notion proposed by Herbrand and Gödel (Gödel [2]). These functions (our *recursive* functions) Kleene called *general recursive;* and Gödel's rekursive Funktionen Kleene called *primitive recursive* (to distinguish them from the general recursive functions).

We note the following theorems:

THEOREM 4.2. *A function is A-partial recursive if and only if it can be obtained by a finite number of applications of the operations of composition, primitive recursion, and minimalization to the functions in the list of Definition* 4.1.

The class of A-recursive functions will be obtained if minimalization is applied only to regular functions.

PROOF. These results follow at once from the facts that all A-primitive recursive functions are A-recursive and that the functions from the list of Definition 1.1 are A-primitive recursive.

THEOREM 4.3. *If* $f(k, \mathfrak{x}^{(p)})$ *is A-(primitive) recursive, so are*

$$g(n, \mathfrak{x}^{(p)}) = \sum_{k=0}^{n} f(k, \mathfrak{x}^{(p)})$$

and

$$h(n, \mathfrak{x}^{(p)}) = \prod_{k=0}^{n} f(k, \mathfrak{x}^{(p)}).$$

PROOF

$$\begin{aligned} g(0, \mathfrak{x}^{(p)}) &= f(0, \mathfrak{x}^{(p)}), \\ g(n+1, \mathfrak{x}^{(p)}) &= g(n, \mathfrak{x}^{(p)}) + f(n+1, \mathfrak{x}^{(p)}). \end{aligned}$$

$$\begin{aligned} h(0, \mathfrak{x}^{(p)}) &= f(0, \mathfrak{x}^{(p)}), \\ h(n+1, \mathfrak{x}^{(p)}) &= h(n, \mathfrak{x}^{(p)}) \cdot f(n+1, \mathfrak{x}^{(p)}). \end{aligned}$$

THEOREM 4.4. *If a function is primitive recursive, it is A-primitive recursive, for any choice of A.*

PROOF. Like that of Theorem 1.3.

5. Recursive Sets and Predicates. It is convenient to speak of recursive sets and predicates as well as functions. We assume familiarity with the contents of Secs. 3.5 and 3.6 of the Introduction.

DEFINITION 5.1. *Let S be a set of n-tuples. Then, we say that S is* A-(**primitive**) **recursive** *if its characteristic function*[1] $C_S(\mathfrak{x}^{(n)})$ *is.*

We say that S is (**primitive**) **recursive** *if it is* ϕ-*(primitive) recursive.*

THEOREM 5.1. *Let R and S be A-(primitive) recursive sets. Then so are* $R \cup S$, $R \cap S$, *and* $\bar{R}$.

PROOF. The result follows at once from the three identities at the very close of Sec. 3.3 of the Introduction.

DEFINITION 5.2. *The predicate* $P(\mathfrak{x}^{(n)})$ *is called* A-(**primitive**) **recursive** *if its extension* $\{\mathfrak{x}^{(n)} \mid P(\mathfrak{x}^{(n)})\}$ *is.*

$P(\mathfrak{x}^{(n)})$ *is called* (**primitive**) **recursive** *if it is* ϕ-*(primitive) recursive.*

COROLLARY 5.2. $P(\mathfrak{x}^{(n)})$ *is A-(primitive) recursive if and only if its characteristic function is.*

[1] Cf. Sec. 3.3 of the Introduction.

PROOF. The characteristic function of a predicate is precisely that of its extension.

THEOREM 5.3. *Let P and Q be A-(primitive) recursive predicates. Then so are $P \vee Q$, $P \wedge Q$, and $\sim P$.*

PROOF. This follows at once from the identities

$$\begin{aligned} C_{P \vee Q} &= C_P \cdot C_Q, \\ C_{P \wedge Q} &= (C_P + C_Q) \dot{-} (C_P \cdot C_Q), \\ C_{\sim P} &= 1 \dot{-} C_P. \end{aligned}$$

THEOREM 5.4. *If $P(y, \mathfrak{x}^{(n)})$ is an A-(primitive) recursive predicate, then so are*

$$\bigvee_{y=0}^{z} P(y, \mathfrak{x}^{(n)}) \quad \textit{and} \quad \bigwedge_{y=0}^{z} P(y, \mathfrak{x}^{(n)}).$$

PROOF. Let

$$Q(z, \mathfrak{x}^{(n)}) \leftrightarrow \bigvee_{y=0}^{z} P(y, \mathfrak{x}^{(n)}).$$

Then

$$C_Q(z, \mathfrak{x}^{(n)}) = \prod_{y=0}^{z} C_P(y, \mathfrak{x}^{(n)}),$$

from which the first part follows, by Theorem 4.3.

For the second part, we note that

$$\bigwedge_{y=0}^{z} P(y, \mathfrak{x}^{(n)}) \leftrightarrow \sim \bigvee_{y=0}^{z} \sim P(y, \mathfrak{x}^{(n)})$$

and we apply the first part and Theorem 5.3.

The boundedness of the quantification is essential to this result. In fact, as we shall see later, there exists a primitive recursive predicate $R(y, x)$ such that $\bigvee_{y} R(y, x)$ is not recursive.

DEFINITION 5.3. *Let $P(y, \mathfrak{x}^{(n)})$ be an $(n + 1)$-ary predicate. Then, by*

$$f(z, \mathfrak{x}^{(n)}) = \underset{y=0}{\overset{z}{\mathfrak{m}}} P(y, \mathfrak{x}^{(n)})$$

we understand the $(n + 1)$-ary total function that satisfies the equation

$$f(z, \mathfrak{x}^{(n)}) = \min_y [y \leq z \wedge P(y, \mathfrak{x}^{(n)})]$$

where this is defined, and

$$f(z, \mathfrak{x}^{(n)}) = 0$$

elsewhere.

THEOREM 5.5. *If $P(y, \mathfrak{x}^{(n)})$ is A-(primitive) recursive, then so also is* $f(z, \mathfrak{x}^{(n)}) = \underset{y=0}{\overset{z}{\mathfrak{m}}}\, P(y, \mathfrak{x}^{(n)})$.

PROOF. First suppose that there exists $y \leqq z$ such that $P(y, \mathfrak{x}^{(n)})$. Let

$$\varphi(t, \mathfrak{x}^{(n)}) = \prod_{y=0}^{t} C_P(y, \mathfrak{x}^{(n)}).$$

Now, $\varphi(t, \mathfrak{x}^{(n)})$ is the characteristic function of the predicate $\bigvee_{y=0}^{t} P(y, \mathfrak{x}^{(n)})$. Hence, as $t = 0, 1, 2, \ldots$, $\varphi(t, \mathfrak{x}^{(n)})$ remains equal to 1 until a value t_0 of t is encountered that makes $P(t_0, \mathfrak{x}^{(n)})$ true. For this value of t and for all subsequent values, $\varphi(t, \mathfrak{x}^{(n)}) = 0$. Hence,

$$\begin{aligned}\sum_{t=0}^{z} \varphi(t, \mathfrak{x}^{(n)}) &= \sum_{t=0}^{t_0-1} \varphi(t, \mathfrak{x}^{(n)}) \\ &= \sum_{t=0}^{t_0-1} 1 \\ &= t_0 \\ &= \underset{y=0}{\overset{z}{\mathfrak{m}}}\, P(y, \mathfrak{x}^{(n)}).\end{aligned}$$

Thus, assuming that there is a $y \leqq z$ such that $P(y, \mathfrak{x}^{(n)})$,

$$\underset{y=0}{\overset{z}{\mathfrak{m}}}\, P(y, \mathfrak{x}^{(n)}) = \sum_{t=0}^{z} \prod_{y=0}^{t} C_P(y, \mathfrak{x}^{(n)}).$$

Finally, we note that, since $\varphi(z, \mathfrak{x}^{(n)})$ is the characteristic function of the predicate

$$\bigvee_{y=0}^{z} P(y, \mathfrak{x}^{(n)}),$$

the function $\alpha(\varphi(z, \mathfrak{x}^{(n)}))$ is 1 or 0 according as there does or does not exist a $y \leqq z$ such that $P(y, \mathfrak{x}^{(n)})$. Hence, in any case,

$$\begin{aligned}\underset{y=0}{\overset{z}{\mathfrak{m}}}\, P(y, \mathfrak{x}^{(n)}) &= \alpha(\varphi(z, \mathfrak{x}^{(n)})) \cdot \sum_{t=0}^{z} \prod_{y=0}^{t} C_P(y, \mathfrak{x}^{(n)}) \\ &= \alpha\left(\prod_{y=0}^{z} C_P(y, \mathfrak{x}^{(n)})\right) \cdot \sum_{t=0}^{z} \prod_{y=0}^{t} C_P(y, \mathfrak{x}^{(n)}).\end{aligned}$$

The result follows at once from Theorem 4.3.

THEOREM 5.6. *The predicates $x = y$ and $x < y$ are primitive recursive.*

PROOF. Their characteristic functions are $\alpha(\alpha(|x - y|))$ and $\alpha(y \dot{-} x)$, respectively. [Alternatively, the primitive recursiveness of $x = y$ could be established from that of $x < y$ by noting that

$$x = y \leftrightarrow \sim (x < y) \wedge \sim (y < x).]$$

Our present methods enable us to prove the primitive recursiveness of some very complicated functions and predicates. For example:

(1) $y \mid x$. (*y is a divisor*[1] *of x.*) Namely,

$$y \mid x \leftrightarrow \bigvee_{z=0}^{x} x = yz.$$

(2) Prime (x). (*x is a prime*[1] *number.*) Namely,

$$\text{Prime}\ (x) \leftrightarrow (x > 1) \wedge \bigwedge_{z=0}^{x} [(z = 1) \vee (z = x) \vee \sim (z \mid x)].$$

(3) Pr (n). (*The nth prime in order of magnitude, where we arbitrarily take the* 0*th prime equal to* 0.) Namely,

$$\text{Pr}\ (0) = 0,$$

$$\text{Pr}\ (n + 1) = \underset{y=0}{\overset{\text{Pr}(n)!+1}{\mathfrak{M}}} [\text{Prime}\ (y) \wedge y > \text{Pr}\ (n)].$$

In this case it may not be clear either that the recursion equations given do define the function Pr (n) or that the function defined by the equations is primitive recursive.

To see that the equations do define Pr (n) it is necessary only to note that

$$\text{Pr}\ (n) < \text{Pr}\ (n + 1) \leqq \text{Pr}\ (n)! + 1.$$

But this is proved in Theorem 4 of the Appendix.[2]

That the primitive recursiveness of Pr (n) follows from the equations can be seen from the following analysis. Let

$$H(w, z) = \underset{y=0}{\overset{w}{\mathfrak{M}}} [\text{Prime}\ (y) \wedge y > z].$$

Then, $H(w, z)$ is primitive recursive. Next, let

$$\begin{aligned} G(z) &= H(z! + 1, z) \\ &= \underset{y=0}{\overset{z!+1}{\mathfrak{M}}} [\text{Prime}\ (y) \wedge y > z], \end{aligned}$$

[1] Cf. Appendix.

[2] Actually there are far better estimates of Pr $(n + 1)$, for example, Pr $(n + 1) \leqq$ 2 Pr (n).

so that $G(z)$ is primitive recursive. But our second recursion equation now becomes simply

$$\text{Pr}\,(n+1) = G(\text{Pr}\,(n)).$$

(4) $[x/2]$. $\left(\textit{The largest integer} \leqq \frac{x}{2}.\right)$ Namely,

$$[x/2] = \underset{y=0}{\overset{x}{\mathfrak{m}}}\ (2y+2 > x).$$

(5) $J(x, y)$. Namely,

$$J(x, y) = [\{(x+y)^2 + 3x + y\}/2].$$

(6) $K(z)$. Namely,

$$K(z) = \underset{x=0}{\overset{z}{\mathfrak{m}}} \bigvee_{y=0}^{z} [z = J(x, y)].$$

(7) $L(z)$. Namely,

$$L(z) = \underset{y=0}{\overset{z}{\mathfrak{m}}} \bigvee_{x=0}^{z} [z = J(x, y)].$$

CHAPTER 4

TURING MACHINES SELF-APPLIED

1. Arithmetization of the Theory of Turing Machines. In this section we show how the theory of Turing machines and of A-computable functions can be developed in terms of A-primitive recursive functions and *hence by means of Turing machines.* Our method involves using the natural numbers as a code or cipher for expressing the theory of Turing machines.

The basic symbols used in our discussion of Turing machines are

$$R, L$$
$$S_0, S_1, S_2, \ldots$$
$$q_1, q_2, q_3, \ldots.$$

Each of these symbols we associate with an odd number $\geqq 3$, as follows:

$$\begin{array}{cccccccccc} 3, & 5, & 7, & 9, & 11, & 13, & 15, & 17, & 19, & 21, \ldots \\ \uparrow & \uparrow & \uparrow & \uparrow & \uparrow & \uparrow & \uparrow & \uparrow & \uparrow & \uparrow \\ R, & L, & S_0, & q_1, & S_1, & q_2, & S_2, & q_3, & S_3, & q_4, \ldots. \end{array}$$

Thus, for each i, S_i is associated with $4i + 7$, and q_i is associated with $4i + 5$.

Hence, with any expression M there is now associated a finite sequence of odd integers $a_1, a_2, \ldots, a_n$. For example, with the quadruple $q_1\ 1\ R\ q_2$ is associated the sequence 9, 11, 3, 13; with the instantaneous description $q_1 1111$ is associated the sequence 9, 11, 11, 11, 11. We shall now see how to associate a *single* number with each such sequence and hence with each expression.

DEFINITION 1.1. *Let M be an expression consisting of the symbols $\gamma_1, \gamma_2, \ldots, \gamma_n$. Let $a_1, a_2, \ldots, a_n$ be the corresponding integers associated with these symbols. Then the* **Gödel number**[1] **of** *M is the integer*[2]

[1] Actually, the representation of a sequence of integers by a single integer was already accomplished in the previous chapter by means of the function $T_i(w)$. The formal properties of the present representation, however, are slightly simpler.

Both methods for reducing sequences of integers to integers first appear in Gödel [1].

[2] Pr (k) is the kth prime in order of magnitude.

$$r = \prod_{k=1}^{n} \mathrm{Pr}\,(k)^{a_k}.$$

We write gn $(M) = r$. *If* M *is the empty expression, we let* 1 *be the Gödel number of* M, *and we write* gn $(M) = 1$.

Thus, gn $(q_1\, 1\, R\, q_2) = 2^9 \cdot 3^{11} \cdot 5^3 \cdot 7^{13}$.

The following is an immediate consequence of the fundamental theorem of arithmetic (Appendix, Theorem 10).

COROLLARY 1.1. *If* M *and* N *are given expressions such that* gn $(M) =$ gn (N), *then* $M = N$.

DEFINITION 1.2. *If* $n =$ gn (M), *we also write* $M =$ Exp (n).

DEFINITION 1.3. *Let* $M_1, \ldots, M_n$ *be a finite sequence of expressions. Then, by the* **Gödel number of this sequence of expressions** *is understood the number*

$$\prod_{k=1}^{n} \mathrm{Pr}\,(k)^{\mathrm{gn}(M_k)}.$$

Thus, the Gödel number of the sequence $q_1\, 1\, B\, q_1$, $q_1\, B\, R\, q_2$ is $2^{2^9 3^{11} 5^7 7^9} \cdot 3^{2^9 3^7 5^3 7^{13}}$, a rather large number.

COROLLARY 1.2. *No integer is the Gödel number both of an expression and of a sequence of expressions.*

PROOF. A Gödel number either of an expression or of a sequence of expressions is of the form $2^n \cdot m$, $n > 0$, m odd. But, if $2^n \cdot m$ is the Gödel number of an expression, n is odd; whereas, if $2^n \cdot m$ is the Gödel number of a sequence of expressions, n is itself the Gödel number of an expression and, hence, is even.

It is also easy to verify

COROLLARY 1.3. *Two sequences of expressions that have the same Gödel number are identical.*

A computation is a finite *sequence* of expressions; a Turing machine, however, is simply a finite *set*, i.e., order is irrelevant.

DEFINITION 1.4. *Let* Z *be a Turing machine. Let* $M_1, \ldots, M_n$ *be any arrangement of the quadruples of* Z *without repetitions. Then, the Gödel number of the sequence* $M_1, \ldots, M_n$ *is called* **a Gödel number of the Turing machine** Z.

(A Turing machine consisting of n quadruples has $n!$ distinct Gödel numbers.)

DEFINITION 1.5. *For each* $n > 0$ *and for each set of integers* A, *let* $T_n^A(z, x_1, \ldots, x_n, y)$ *be the predicate*[1] *that means, for given* $z, x_1, \ldots, x_n, y$, *that* z *is a Gödel number of a Turing machine* Z, *and that* y *is the*

[1] Predicates having essentially the properties of these "T-predicates" were first considered by Kleene [4].

Gödel number of an A-computation, with respect to Z, beginning with the instantaneous description $q_1(\overline{x_1, \ldots, x_n})$.

$T_n^{\emptyset}(z, x_1, \ldots, x_n, y)$ will be written simply as $T_n(z, x_1, \ldots, x_n, y)$. $T_1^A(z, x, y)$ will be written simply as $T^A(z, x, y)$.

The predicates $T_n^A(z, x_1, \ldots, x_n, y)$ will play an extremely important role in what follows. Their importance stems from the fact that they express, by themselves, the essential elements of the theory of Turing machines. The rest of this section is devoted to a detailed proof that, for each $n > 0$, $T_n^A(z, x_1, \ldots, x_n, y)$ is A-primitive recursive.

The proof proceeds by a detailed list of primitive recursive and A-primitive recursive functions and predicates culminating in the T_n^A predicates. With each function (predicate), a formula is given that proves the function (predicate) to be primitive, or A-primitive, recursive. Most of them will be primitive recursive; those for which we can only assert A-primitive recursiveness will be marked with an "A."

Group I. Functions and Predicates Which Concern Gödel Numbers of Expressions and Sequences of Expressions

$$(1)\ n \text{ Gl } x = \underset{y=0}{\overset{x}{\mathfrak{m}}} [(\Pr(n)^y \mid x) \wedge \sim (\Pr(n)^{y+1} \mid x)].$$

If $x = \text{gn}(M)$, where M consists of the symbols $\gamma_1, \ldots, \gamma_p$, then if $0 < n \leqq p$, n Gl x is the number associated with γ_n, whereas if $n = 0$ or $n > p$, then n Gl $x = 0$.

If x is the Gödel number of the sequence of expressions $M_1, \ldots, M_p$, then if $0 < n \leqq p$, n Gl $x = \text{gn}(M_n)$, whereas if $n = 0$ or $n > p$, then n Gl $x = 0$.

$$(2)\ \mathfrak{L}(x) = \underset{y=0}{\overset{x}{\mathfrak{m}}} \left[(y \text{ Gl } x > 0) \wedge \bigwedge_{i=0}^{x} ((y + i + 1) \text{ Gl } x = 0) \right].$$

If $x = \text{gn}(M)$, then $\mathfrak{L}(x)$ is the number of symbols occurring in M.

If x is the Gödel number of the sequence of expressions $M_1, \ldots, M_n$, then $\mathfrak{L}(x) = n$.

$$(3)\ \text{GN}(x) \leftrightarrow \sim \bigvee_{y=1}^{\mathfrak{L}(x)} [(y \text{ Gl } x = 0) \wedge ((y + 1) \text{ Gl } x \neq 0)].$$

GN (x) holds if and only if there exist *positive* integers a_k, $1 \leqq k \leqq n$, such that

$$x = \prod_{k=1}^{n} \Pr(k)^{a_k}.$$

(4) $\text{Term}\ (x, z) \leftrightarrow \text{GN}\ (z) \wedge \bigvee_{n=0}^{\mathfrak{L}(z)} [(x = n \text{ Gl } z) \wedge (n \neq 0)].$

Term (x, z) holds if and only if $z = \prod_{k=1}^{n} \text{Pr}\ (k)^{a_k}$ for suitable $a_k > 0$, and $x = a_k$ for some k, $1 \leq k \leq n$.

(5) $x * y = x \cdot \prod_{i=0}^{\mathfrak{L}(y) \dot{-} 1} \text{Pr}\ (\mathfrak{L}(x) + i + 1)^{(i+1) \text{ Gl } y}.$

If M and N are expressions, then gn (MN) = gn (M) $*$ gn (N).† If x and y are the Gödel numbers of the sequences of expressions $M_1, \ldots, M_n$, and $N_1, \ldots, N_p$, respectively, then $x * y$ is the Gödel number of the sequence $M_1, \ldots, M_n, N_1, \ldots, N_p$.

Group II. Functions and Predicates Which Concern the Basic Structure of Turing Machines

(6) $\text{IC}\ (x) \leftrightarrow \bigvee_{y=0}^{x} (x = 4y + 9).$

IC (x) holds if and only if x is a number assigned to one of the q_i.

(7) $\text{Al}\ (x) \leftrightarrow \bigvee_{y=0}^{x} (x = 4y + 7).$

Al (x) holds if and only if x is a number assigned to one of the S_i.

(8) $\text{Odd}\ (x) \leftrightarrow \bigvee_{y=0}^{x} (x = 2y + 3).$

Odd (x) holds if and only if x is an odd number ≥ 3.

(9) $\text{Quad}\ (x) \leftrightarrow \text{GN}\ (x) \wedge (\mathfrak{L}(x) = 4) \wedge \text{IC}\ (1 \text{ Gl } x) \wedge \text{Al}\ (2 \text{ Gl } x) \wedge \text{Odd}\ (3 \text{ Gl } x) \wedge \text{IC}\ (4 \text{ Gl } x).$

Quad (x) holds if and only if Exp (x) is a quadruple.

(10) $\text{Inc}\ (x, y) \leftrightarrow \text{Quad}\ (x) \wedge \text{Quad}\ (y) \wedge (1 \text{ Gl } x = 1 \text{ Gl } y) \wedge (2 \text{ Gl } x = 2 \text{ Gl } y) \wedge (x \neq y).$

Inc (x, y) holds if and only if x and y are Gödel numbers of two quadruples beginning with the same two symbols.

† Note that this remains correct if one or both of M, N are empty.

(11) TM $(x) \leftrightarrow$ GN (x)

$$\wedge \bigwedge_{n=1}^{\mathfrak{L}(x)} \left[\text{Quad } (n \text{ Gl } x) \wedge \bigwedge_{m=1}^{\mathfrak{L}(x)} \sim (\text{Inc } (n \text{ Gl } x, m \text{ Gl } x)) \right].$$

TM (x) holds if and only if x is a Gödel number of a Turing machine.

(12) $$\text{MR } (0) = 2^{11}, \quad \text{MR } (n+1) = 2^{11} * \text{MR } (n).$$

That is, MR $(n) = \text{gn } (1^{n+1}) = \text{gn } (\bar{n})$.

(13) $$\text{CU } (n, x) = 0 \text{ if } n \text{ Gl } x \neq 11, \quad \text{CU } (n, x) = 1 \text{ if } n \text{ Gl } x = 11.$$

CU (n, x) is the characteristic function of the predicate n Gl $x \neq 11$ and, as such, is primitive recursive.

(14) $$\text{Corn } (x) = \sum_{n=1}^{\mathfrak{L}(x)} \text{CU } (n, x).$$

If $x = \text{gn } (M)$, then Corn $(x) = \langle M \rangle$.

(15) $U(y) = \text{Corn } (\mathfrak{L}(y) \text{ Gl } y)$.

If y is the Gödel number of the sequence of expressions $M_1, M_2, \ldots, M_n$, then $U(y) = \langle M_n \rangle$.

(16) ID (x)

$$\leftrightarrow \text{GN } (x) \wedge \bigvee_{n=1}^{\mathfrak{L}(x) \dot{-} 1} \left\{ \text{IC } (n \text{ Gl } x) \wedge \bigwedge_{m=1}^{\mathfrak{L}(x)} [(m = n) \vee \text{Al } (m \text{ Gl } x)] \right\}.$$

ID (x) holds if and only if x is the Gödel number of an instantaneous description.

(17) $$\text{Init}_n (x_1, \ldots, x_n) = 2^9 * \text{MR } (x_1) * 2^7 * \text{MR } (x_2) * 2^7 * \cdots * 2^7 * \text{MR } (x_n).$$

[It should be noted that (17) defines an infinite class of functions, one for each value of n.]

$\text{Init}_n (x_1, \ldots, x_n) = \text{gn } (q_1(\overline{x_1, \ldots, x_n}))$.

Group III. The $\rightarrow$ *Relation, etc.*

(18) $\text{Yield}_1 (x, y, z) \leftrightarrow \text{ID } (x) \wedge \text{ID } (y) \wedge \text{TM } (z)$

$$\wedge \bigvee_{F=0}^{x} \bigvee_{G=0}^{x} \bigvee_{r=0}^{x} \bigvee_{s=0}^{x} \bigvee_{t=0}^{y} \bigvee_{u=0}^{y} [(x = F * 2^r * 2^s * G)$$
$$\wedge (y = F * 2^t * 2^u * G) \wedge \text{IC } (r) \wedge \text{IC } (t) \wedge \text{Al } (s) \wedge \text{Al } (u)$$
$$\wedge \text{Term } (2^r \cdot 3^s \cdot 5^u \cdot 7^t, z)].$$

Yield$_1$ (x, y, z) holds if and only if x and y are the Gödel numbers of instantaneous descriptions, z is a Gödel number of a Turing machine Z, and Exp $(x) \rightarrow$ Exp (y) $\quad (Z)$, under Case 1 of Definition 1-1.7.

(19) $$\text{Yield}_2\,(x, y, z) \leftrightarrow \text{ID}\,(x) \wedge \text{ID}\,(y) \wedge \text{TM}\,(z)$$
$$\wedge \bigvee_{F=0}^{x} \bigvee_{G=0}^{x} \bigvee_{r=0}^{x} \bigvee_{s=0}^{x} \bigvee_{t=0}^{x} \bigvee_{u=0}^{y} [(x = F * 2^r * 2^s * 2^t * G)$$
$$\wedge (y = F * 2^s * 2^u * 2^t * G) \wedge \text{IC}\,(r) \wedge \text{IC}\,(u) \wedge \text{Al}\,(s) \wedge \text{Al}\,(t)$$
$$\wedge \text{Term}\,(2^r \cdot 3^s \cdot 5^3 \cdot 7^u, z)].$$

Yield$_2$ is like Yield$_1$, but deals with Case 2 of Definition 1-1.7.

(20) $$\text{Yield}_3\,(x, y, z) \leftrightarrow \text{ID}\,(x) \wedge \text{ID}\,(y) \wedge \text{TM}\,(z)$$
$$\wedge \bigvee_{F=0}^{x} \bigvee_{r=0}^{x} \bigvee_{s=0}^{x} \bigvee_{t=0}^{y} [(x = F * 2^r * 2^s)$$
$$\wedge (y = F * 2^s * 2^t * 2^7) \wedge \text{IC}\,(r) \wedge \text{IC}\,(t) \wedge \text{Al}\,(s)$$
$$\wedge \text{Term}\,(2^r \cdot 3^s \cdot 5^3 \cdot 7^t, z)].$$

Yield$_3$ is like Yield$_1$, but deals with Case 3 of Definition 1-1.7.

(21) $$\text{Yield}_4\,(x, y, z) \leftrightarrow \text{ID}\,(x) \wedge \text{ID}\,(y) \wedge \text{TM}\,(z)$$
$$\wedge \bigvee_{F=0}^{x} \bigvee_{G=0}^{x} \bigvee_{r=0}^{x} \bigvee_{s=0}^{x} \bigvee_{t=0}^{x} \bigvee_{u=0}^{y} [(x = F * 2^r * 2^s * 2^t * G)$$
$$\wedge (y = F * 2^u * 2^r * 2^t * G) \wedge \text{IC}\,(s) \wedge \text{IC}\,(u) \wedge \text{Al}\,(r) \wedge \text{Al}\,(t)$$
$$\wedge \text{Term}\,(2^s \cdot 3^t \cdot 5^5 \cdot 7^u, z)].$$

Yield$_4$ is like Yield$_1$, but deals with Case 4 of Definition 1-1.7.

(22) $$\text{Yield}_5\,(x, y, z) \leftrightarrow \text{ID}\,(x) \wedge \text{ID}\,(y) \wedge \text{TM}\,(z)$$
$$\wedge \bigvee_{G=0}^{x} \bigvee_{r=0}^{x} \bigvee_{s=0}^{x} \bigvee_{t=0}^{y} [(x = 2^r * 2^s * G) \wedge (y = 2^t * 2^7 * 2^s * G)$$
$$\wedge \text{IC}\,(r) \wedge \text{IC}\,(t) \wedge \text{Al}\,(s) \wedge \text{Term}\,(2^r \cdot 3^s \cdot 5^5 \cdot 7^t, z)].$$

Yield$_5$ is like Yield$_1$, but deals with Case 5 of Definition 1-1.7.

(23) $$\text{Yield}\,(x, y, z) \leftrightarrow \text{Yield}_1\,(x, y, z) \vee \text{Yield}_2\,(x, y, z)$$
$$\vee\ \text{Yield}_3\,(x, y, z) \vee \text{Yield}_4\,(x, y, z)$$
$$\vee\ \text{Yield}_5\,(x, y, z).$$

Yield (x, y, z) holds if and only if x and y are the Gödel numbers of instantaneous descriptions, z is a Gödel number of a Turing machine Z, and Exp $(x) \rightarrow$ Exp (y) $\quad (Z)$.

$$(24)\quad \text{Fin}\,(x, z) \leftrightarrow \text{ID}\,(x) \wedge \text{TM}\,(z)$$
$$\wedge \bigvee_{F=0}^{x} \bigvee_{G=0}^{x} \bigvee_{r=0}^{x} \bigvee_{s=0}^{x} \Big\{(x = F * 2^r * 2^s * G) \wedge \text{IC}\,(r) \wedge \text{Al}\,(s)$$
$$\wedge \bigwedge_{n=1}^{\mathfrak{L}(z)} [(1\ \text{Gl}\,(n\ \text{Gl}\ z) \neq r) \vee (2\ \text{Gl}\,(n\ \text{Gl}\ z) \neq s)]\Big\}.$$

Fin (x, z) holds if and only if z is a Gödel number of a Turing machine Z and x is the Gödel number of an instantaneous description final with respect to Z.

$$(25_A)\quad H_A(x, y, z) \leftrightarrow \text{ID}\,(x) \wedge \text{ID}\,(y) \wedge \text{TM}\,(z)$$
$$\wedge \bigvee_{F=0}^{x} \bigvee_{G=0}^{x} \bigvee_{r=0}^{x} \bigvee_{s=0}^{x} \bigvee_{t=0}^{z} \bigvee_{u=0}^{z} \{(x = F * 2^r * 2^s * G)$$
$$\wedge \text{IC}\,(r) \wedge \text{IC}\,(t) \wedge \text{IC}\,(u) \wedge \text{Al}\,(s) \wedge \text{Term}\,(2^r \cdot 3^s \cdot 5^t \cdot 7^u, z)$$
$$\wedge [(C_A(\text{Corn}\,(x)) = 0 \wedge y = F * 2^t * 2^s * G)$$
$$\vee (C_A(\text{Corn}\,(x)) = 1 \wedge y = F * 2^u * 2^s * G)]\}.$$

This formula shows that $H_A(x, y, z)$ is A-primitive recursive. $H_A(x, y, z)$ holds if and only if z is a Gödel number of a Turing machine Z, x and y are Gödel numbers of instantaneous descriptions, and $\text{Exp}\,(x) \underset{A}{\rightarrow} \text{Exp}\,(y) \quad (Z)$.

$$(26_A)\quad \text{Comp}_A\,(y, z) \leftrightarrow \text{TM}\,(z) \wedge \text{GN}\,(y)$$
$$\wedge \bigwedge_{n=1}^{\mathfrak{L}(y) \dot{-} 1} [\text{Yield}\,(n\ \text{Gl}\ y, (n + 1)\ \text{Gl}\ y, z)$$
$$\vee H_A(n\ \text{Gl}\ y, (n + 1)\ \text{Gl}\ y, z)] \wedge \text{Fin}\,(\mathfrak{L}(y)\ \text{Gl}\ y, z).$$

This formula shows that $\text{Comp}_A\,(y, z)$ is A-primitive recursive. $\text{Comp}_A\,(y, z)$ holds if and only if z is a Gödel number of a Turing machine Z and y is the Gödel number of an A-computation of Z.

We have now:

THEOREM 1.4. $T_n{}^A\,(z, x_1, \ldots, x_n, y)$ is A-primitive recursive.

PROOF. By Definition 1.5,

$$T_n{}^A(z, x_1, \ldots, x_n, y) \leftrightarrow \text{Comp}_A\,(y, z) \wedge (1\ \text{Gl}\ y = \text{Init}_n\,(x_1, \ldots, x_n)).$$

2. Computability and Recursiveness. The class of partially A-computable functions was defined in terms of the functions $\Psi^{(n)}_{Z;A}(\mathfrak{x}^{(n)})$ (cf. Definition 1-4.4). We now seek to evaluate these functions in terms of the predicates $T_n{}^A(z, \mathfrak{x}^{(n)}, y)$.

THEOREM 2.1. *Let Z_0 be a Turing machine and let z_0 be a Gödel number of Z_0. Then, the domain of the function $\Psi^{(n)}_{Z_0;A}(\mathfrak{x}^{(n)})$ is equal to the domain* *of* $\min_y T_n{}^A(z_0, \mathfrak{x}^{(n)}, y)$. *Moreover,*

$$\Psi^{(n)}_{Z_0;A}(\mathfrak{x}^{(n)}) = U(\min_y T_n{}^A(z_0, \mathfrak{x}^{(n)}, y)).\dagger$$

Also, if $T_n{}^A(z_0, \mathfrak{x}^{(n)}, y_0)$ *is true for given* $\mathfrak{x}^{(n)}$, *then*

$$y_0 = \min_y T_n{}^A(z_0, \mathfrak{x}^{(n)}, y).$$

PROOF. $\min_y T_n{}^A(z_0, \mathfrak{x}^{(n)}, y)$ is defined for given $\mathfrak{x}^{(n)}$ if and only if there exists an A-computation of Z_0 beginning with $q_1(\overline{\mathfrak{x}^{(n)}})$, that is, if and only if $\Psi^{(n)}_{Z_0;A}(\mathfrak{x}^{(n)})$ is defined. This proves our first contention.

Furthermore, when $y_0 = \min_y T_n{}^A(z_0, \mathfrak{x}^{(n)}, y)$ is defined, y_0 is the Gödel number of an A-computation of Z_0 beginning with $q_1(\overline{\mathfrak{x}^{(n)}})$. (This remark also proves the final assertion of the present theorem.) Hence, $\mathfrak{L}(y_0)$ Gl y_0 is the Gödel number of the *final* instantaneous description α of this A-computation, and $U(y_0) = \text{Corn}\,(\mathfrak{L}(y_0)\text{ Gl } y_0) = \langle\alpha\rangle$. But

$$\langle\alpha\rangle = \Psi^{(n)}_{Z_0;A}(\mathfrak{x}^{(n)}),$$

whence the result follows.

Theorem 2.1 is our most important result thus far. The reader will notice that all of our subsequent results of interest in Parts 1 and 2 are based essentially on this theorem.

COROLLARY 2.2. $f(\mathfrak{x}^{(n)})$ *is partially A-computable if and only if there is a number* z_0 *such that*

$$f(\mathfrak{x}^{(n)}) = U(\min_y T_n{}^A(z_0, \mathfrak{x}^{(n)}, y)).$$

This important result is due to Kleene [4, 6], and is called by him the "normal form theorem." Cf. also Davis [1].

PROOF. That every partially A-computable function can be represented in the desired form is an immediate consequence of our preceding theorem and Definition 1-4.5. The converse results from Theorems 1.4 and 3-1.2.

COROLLARY 2.3. *Every (partially) A-computable function is A-(partial) recursive.*

PROOF. That every partially A-computable function is A-partial recursive follows at once from Corollary 2.2. For an A-computable function, we note that $\min_y T_n{}^A(z_0, \mathfrak{x}^{(n)}, y)$ must be defined for all $\mathfrak{x}^{(n)}$; so the minimalization is of a *regular* function (cf. Definition 2-2.3). Hence, by Definition 3-1.2, the result follows.

Combining Corollary 2.3 with Theorem 3-1.2, we have:

COROLLARY 2.4. *A function is (partially) A-computable if and only if it is A-(partial) recursive.*

† The function $U(y)$ is that defined in (15) of Sec. 1 of the present chapter.

Having proved the equivalence of computability and recursiveness, we shall henceforth use the two terms more or less interchangeably.

The converse of Corollary 3-1.1 now follows trivially.

COROLLARY 2.5. *If a function is total and A-partial recursive, then it is A-recursive.*

PROOF. Such a function is partially A-computable and hence, being total, A-computable. Therefore it is A-recursive.

In Theorem 3-4.2, we saw that the A-(partial) recursive functions could be obtained from the functions $C_A(x)$, $N(x)$, $S(x)$, $U_i^n(\mathfrak{x}^{(n)})$, $1 \leqq i \leqq n$, by employing composition, primitive recursion, and minimalization. Corollary 2.2 shows that minimalization need be employed only *once*.

We take this opportunity to return briefly to our discussion of the adequacy of the identification of the intuitive notion of effective calculability with the precise concept of computability (cf. Chap. 1, Sec. 2). Let us imagine an attempt to describe procedures which are intuitively effective but which cannot be carried out by Turing machines. Such procedures would presumably consist of the iteration of certain atomic operations. Now, our technique of arithmetization of the theory of Turing machines could presumably be modified to apply to such procedures if *the atomic operations correspond to recursive functions.* In other words, a function computed by means of any procedure that consists of the iteration of atomic recursive procedures is itself recursive. In particular, these remarks apply to Turing machines modified, say, to operate on a multidimensional "tape."

3. A Universal Turing Machine.[1] Consider the partial recursive binary function[2] $\varphi(z, x) = U(\min_y T(z, x, y))$. This function is computable; so there is a Turing machine U such that

$$\Psi_U^{(2)}(z, x) = \varphi(z, x).$$

We call this Turing machine *universal.*[3] It can be employed to compute *any partially computable singulary function* as follows: If Z_0 is any Turing machine and if z_0 is a Gödel number of Z_0, then

$$\Psi_U^{(2)}(z_0, x) = \Psi_{Z_0}(x).$$

Thus, if the number z_0 is written on the tape of U, followed by the number x_0, U will compute the number $\Psi_{Z_0}(x_0)$.

[1] The present section is a digression and may be omitted without disturbing continuity.

[2] By the convention immediately following Definition 1.5, $T(z, x, y)$ is $T_1^{\emptyset}(z, x, y)$.

[3] The term *universal* was introduced by Turing [1].

The machine U can also be employed to compute n-ary functions, $n \geqq 1$, by first transforming these into singulary functions. That is, the function $f(x_1, \ldots, x_n)$ can be replaced by

$$g(x) = f(1 \text{ Gl } x, 2 \text{ Gl } x, \ldots, n \text{ Gl } x).$$

Then, to compute $f(x_1, \ldots, x_n)$ for given $x_1, \ldots, x_n$, we compute

$$x = \prod_{k=1}^{n} \text{Pr}\,(k)^{x_k}$$

and then compute $g(x)$.

Thus, from the point of view of recursive-function theory, there is no essential distinction between singulary and n-ary ($n > 1$) functions. It is as though, in analysis, one possessed analytic homeomorphisms between one-dimensional and n-dimensional ($n \geqq 1$) euclidean space.

CHAPTER 5

UNSOLVABLE DECISION PROBLEMS

1. Semicomputable Predicates. In our first four chapters we have developed the theory of computability. In the present chapter we begin our treatment of noncomputability with a discussion of the class of *semicomputable* predicates to be defined below. As we shall see, although all computable predicates are semicomputable, there exist semicomputable predicates that are not computable. In studying semicomputable predicates, we are in effect remaining as close to the domain of the computable as we can. Later, in Part 3, we shall consider predicates that are very badly noncomputable.

DEFINITION 1.1. *A predicate $P(\mathfrak{x}^{(n)})$ is called A-***semicomputable** *if there exists a partially A-computable function whose domain is the set $\{\mathfrak{x}^{(n)} \mid P(\mathfrak{x}^{(n)})\}$.*

$P(\mathfrak{x}^{(n)})$ is called **semicomputable** *if it is ϕ-semicomputable.*

THEOREM 1.1. *Every A-computable predicate is A-semicomputable.*

PROOF. Let $R(\mathfrak{x}^{(n)})$ be A-computable. Then, $\{\mathfrak{x}^{(n)} \mid R(\mathfrak{x}^{(n)})\}$ is the domain of the partially A-computable function $\min_y [C_R(\mathfrak{x}^{(n)}) + y = 0]$.

Next, we shall see that the A-semicomputable predicates are precisely those obtained by prefixing an existential quantifier to an A-computable predicate.

THEOREM 1.2. *Let $R(\mathfrak{x}^{(n)}) \leftrightarrow \bigvee_y P(y, \mathfrak{x}^{(n)})$, where $P(y, \mathfrak{x}^{(n)})$ is A-computable. Then $R(\mathfrak{x}^{(n)})$ is A-semicomputable.*

PROOF. $\{\mathfrak{x}^{(n)} \mid R(\mathfrak{x}^{(n)})\}$ is the domain of the partially A-computable function $\min_y [C_P(y, \mathfrak{x}^{(n)}) = 0]$.

THEOREM 1.3. *Let $R(\mathfrak{x}^{(n)})$ be an A-semicomputable predicate. Then, there exists an A-computable predicate $P(y, \mathfrak{x}^{(n)})$ such that*

$$R(\mathfrak{x}^{(n)}) \leftrightarrow \bigvee_y P(y, \mathfrak{x}^{(n)}).$$

PROOF. By Definition 1.1, $\{\mathfrak{x}^{(n)} \mid R(\mathfrak{x}^{(n)})\}$ is the domain of a partially A-computable function $f(\mathfrak{x}^{(n)})$. But, by Corollary 4-2.2, there is a number z_0 such that

$$f(\mathfrak{x}^{(n)}) = U(\min_y T_n^A(z_0, \mathfrak{x}^{(n)}, y)).$$

Hence, the domain of $f(\mathfrak{x}^{(n)})$ is the set

$$\left\{\mathfrak{x}^{(n)} \mid \bigvee_{y} T_n{}^A(z_0, \mathfrak{x}^{(n)}, y)\right\}.$$

Thus,

$$\{\mathfrak{x}^{(n)} \mid R(\mathfrak{x}^{(n)})\} = \left\{\mathfrak{x}^{(n)} \mid \bigvee_{y} T_n{}^A(z_0, \mathfrak{x}^{(n)}, y)\right\};$$

so

$$R(\mathfrak{x}^{(n)}) \leftrightarrow \bigvee_{y} T_n{}^A(z_0, \mathfrak{x}^{(n)}, y).$$

Actually, we have proved more. Our proof yields the following "enumeration" theorem of Kleene:[1]

THEOREM 1.4. *Let* $R(\mathfrak{x}^{(n)})$ *be any* A*-semicomputable predicate. Then there is some number* z_0 *such that*

$$R(\mathfrak{x}^{(n)}) \leftrightarrow \bigvee_{y} T_n{}^A(z_0, \mathfrak{x}^{(n)}, y).$$

Note that, whereas the proof of Theorem 1.2 was quite direct, the proof of Theorem 1.3 employed the results of our arithmetization. Theorem 1.3 is thus the "deeper" of the two.

Our next result concerns the relationship between A-computability and A-semicomputability.

THEOREM 1.5. $R(\mathfrak{x}^{(n)})$ *is* A*-computable if and only if both* $R(\mathfrak{x}^{(n)})$ *and* $\sim R(\mathfrak{x}^{(n)})$ *are* A*-semicomputable.*[2]

PROOF. If $R(\mathfrak{x}^{(n)})$ is A-computable, then so is $\sim R(\mathfrak{x}^{(n)})$, by Theorem 3-5.3. Hence, by Theorem 1.1, $R(\mathfrak{x}^{(n)})$ and $\sim R(\mathfrak{x}^{(n)})$ are both A-semicomputable.

Next, suppose that both $R(\mathfrak{x}^{(n)})$ and $\sim R(\mathfrak{x}^{(n)})$ are A-semicomputable. Then, by Theorem 1.3, there exist A-computable predicates $P(y, \mathfrak{x}^{(n)})$, $Q(y, \mathfrak{x}^{(n)})$ such that

$$R(\mathfrak{x}^{(n)}) \leftrightarrow \bigvee_{y} P(y, \mathfrak{x}^{(n)}),$$

$$\sim R(\mathfrak{x}^{(n)}) \leftrightarrow \bigvee_{y} Q(y, \mathfrak{x}^{(n)}).$$

Now, for each choice of values for the arguments $\mathfrak{x}^{(n)}$, either $R(\mathfrak{x}^{(n)})$ or $\sim R(\mathfrak{x}^{(n)})$ must hold. Hence, for each such choice of values, there is some value of y for which either $P(y, \mathfrak{x}^{(n)})$ or $Q(y, \mathfrak{x}^{(n)})$ holds. Therefore,

[1] Cf. Kleene [4, 6]. It was Kleene who first recognized the fundamental role partial recursiveness plays in this part of the theory.

[2] This result (with $A = \emptyset$) is due, independently, to Kleene [4], Post [3], and Mostowski [1].

the function

$$h(\mathfrak{x}^{(n)}) = \min_y [P(y, \mathfrak{x}^{(n)}) \vee Q(y, \mathfrak{x}^{(n)})]$$

is total and, therefore, A-computable. But we have

$$R(\mathfrak{x}^{(n)}) \leftrightarrow P(h(\mathfrak{x}^{(n)}), \mathfrak{x}^{(n)}).$$

Hence, $R(\mathfrak{x}^{(n)})$ is A-computable.

Thus we see that, for a predicate to be A-computable, it is necessary and sufficient that both it and its negation be A-semicomputable. Does there exist an A-semicomputable predicate whose negation is not A-semicomputable, that is, an A-semicomputable predicate *that is not A-computable?* Indeed there does!

THEOREM 1.6. *The predicate* $\bigvee_y T^A(x, x, y)$ *is A-semicomputable, but is not A-computable.*

PROOF. Suppose that $\sim \bigvee_y T^A(x, x, y)$ were A-semicomputable. Then, by Theorem 1.4, there would be a number z_0 such that

$$\sim \bigvee_y T^A(x, x, y) \leftrightarrow \bigvee_y T^A(z_0, x, y).$$

But setting $x = z_0$ in this equivalence yields a contradiction.[1]

The special case of this result for which $A = \emptyset$ is of such importance that we shall state it explicitly:

COROLLARY 1.7. *The predicate* $\bigvee_y T(x, x, y)$ *is semicomputable, but not computable.*

We are now in a position to settle the question, raised in connection with Example 1-3.4, about the possibility of "completing" a partially computable function to a computable function.

DEFINITION 1.2. *By the* **completion of a function** $f(x)$ *we mean the function* $g(x)$ *for which*

$$g(x) = f(x), \text{ where } f(x) \text{ is defined,}$$
$$g(x) = 0, \text{ elsewhere.}$$

THEOREM 1.8. *There exists a partially A-computable function whose completion is not A-computable.*

PROOF. Let $\varphi(x) = S(N(x))$; that is, $\varphi(x)$ is the computable function whose value is 1 for all values of x. Let $f(x) = \varphi(\min_y T^A(x, x, y))$.

[1] The reader who is familiar with Cantor's diagonalization method will recognize our proof as an application of it. Theorem 1.4 furnishes the enumeration of all singulary A-semicomputable predicates necessary for the diagonalization.

Then $f(x)$ is partially A-computable. But the completion of $f(x)$ is the characteristic function of the predicate

$$\sim \bigvee_y T^A(x, x, y)$$

and is, therefore, not A-computable.

2. Decision Problems. We are at last in a position to treat, in a precise manner, the question of *decision problems*, discussed in the Introduction. As indicated there, a decision problem inquires "as to the existence of an algorithm for deciding the truth or falsity of a whole class of statements. . . . A *positive solution* to a decision problem consists of giving an algorithm for solving it; a *negative solution* consists of showing that no algorithm for solving the problem exists, or, as we shall say, that the problem is *unsolvable*."

Our identification of effective calculability with computability gives a precise substitute for the intuitive notion of algorithm only in so far as computations on integers are concerned. However, our device of *arithmetization* enables us to replace consideration of finite expressions made up of fixed symbols by consideration of corresponding Gödel numbers. Of course, this replacement is valid only because there exist effective procedures for obtaining the Gödel number of an expression and for obtaining an expression from its Gödel number. Furthermore, an algorithm, in the intuitive sense, can be employed only on a finite expression of this sort. (E.g., an algorithm for adding two numbers given in decimal notation is useless for adding numbers given in Roman notation, except in the presence of auxiliary algorithms for transforming back and forth between the two notations.) Thus, our development should prove adequate for dealing with decision problems of the most varied kinds.

In particular, there is a decision problem associated with every predicate in a very natural way. That is, associated with the predicate $R(x_1, \ldots, x_n)$ is the problem:

To determine, for given numbers $a_1, \ldots, a_n$, *whether or not* $R(a_1, \ldots, a_n)$ *is true.*

This we call *the decision problem for the predicate* $R(x_1, \ldots, x_n)$. Our identification of effective calculability with recursiveness may then be rendered "official" by the following definition.

DEFINITION 2.1. *The* **decision problem** *for a predicate* $R(x_1, \ldots, x_n)$ *is called* **recursively solvable** *if* R *is recursive; otherwise it is called* **recursively unsolvable.**

Thus, Corollary 1.7 may be restated as follows:

COROLLARY 2.1. *The decision problem for the predicate* $\bigvee_y T(x, x, y)$ *is recursively unsolvable.*

Now, let Z be a simple Turing machine. We may associate with Z the following decision problem:

To determine, of a given instantaneous description α, whether or not there exists a computation of Z that begins with α.

That is, we wish to determine whether or not Z, if placed in a given initial state, will eventually halt. We call this problem the *halting problem* for Z. We agree to call the halting problem for a simple Turing machine Z *recursively solvable* or *unsolvable* as the following singulary predicate is or is not computable:

$P_Z(x) \leftrightarrow$ *x is the Gödel number of an instantaneous description α of Z and there exists a computation of Z that begins with α.*

For, intuitively, it is clear that an algorithm for solving the halting problem for Z would be forthcoming if and only if we could be supplied with an algorithm for solving the decision problem for $P_Z(x)$. All that is needed to convert one algorithm into the other is an algorithm for translating between instantaneous descriptions and their Gödel numbers. But we shall show that, for suitable Z, $P_Z(x)$ is not computable. Our method of proof is to show that, if $P_Z(x)$ were computable, so would be $\bigvee_y T(x, x, y)$.

Let Z_0 be such that

$$\Psi_{Z_0}(x) = \min_y T(x, x, y).$$

Then, x belongs to the domain of $\Psi_{Z_0}(x)$ if and only if $\bigvee_y T(x, x, y)$. But x belongs to the domain of $\Psi_{Z_0}(x)$ if and only if $P_{Z_0}(\text{gn}\ (q_1\bar{x}))$. Now $\text{gn}\ (q_1\bar{x})$ is certainly recursive [in fact $\text{gn}\ (q_1\bar{x}) = \text{Init}_1\ (x)$ and so is primitive recursive]. Hence, if $P_{Z_0}(x)$ were computable, so would be the domain of $\Psi_{Z_0}(x)$, and hence also the predicate $\bigvee_y T(x, x, y)$. Since we know the latter to be noncomputable, $P_{Z_0}(x)$ must be noncomputable.

We have proved

THEOREM 2.2. *There exists a Turing machine whose halting problem is recursively unsolvable.*

A related problem is the *printing problem* for a simple Turing machine Z with respect to a symbol S_i. This problem is that of determining, of a given instantaneous description α of Z, whether or not, in the course of its "computation" *beginning with* α, the symbol S_i ever appears on the tape of Z; more precisely,

To determine, of a given instantaneous description α of Z, whether or not there exists a sequence $\alpha_1, \ldots, \alpha_k$ of instantaneous descriptions such that $\alpha = \alpha_1$, $\alpha_{j-1} \rightarrow \alpha_j$ for $1 < j \leqq k$, and α_k contains the symbol S_i.

As before, we interpret the recursive solvability or unsolvability of this problem as meaning the computability or noncomputability, respectively, of a related predicate, in this case $Q_{Z,i}(x)$, true if and only if x is the Gödel number of an instantaneous description α for which the italicized condition, just above, holds.

THEOREM 2.3. *There exists a simple Turing machine Z and a symbol S_k in the alphabet of Z such that the printing problem for Z with respect to S_k is recursively unsolvable.*

PROOF. Let Z_0 be the Turing machine we have obtained whose halting problem is unsolvable. Let S_k be some symbol not in the alphabet of Z_0; q_N some internal configuration not an internal configuration of Z_0. Let Z be obtained from Z_0 by adjoining to it all quadruples of the form

$$q_i \; S_j \; S_k \; q_N,$$

where no quadruple of Z_0 begins with $q_i \; S_i$, where S_j is in the alphabet of Z_0, and where q_i is an internal configuration of Z_0. Then Z duplicates the actions of Z_0, except that, whenever Z_0 would have halted, Z continues one step further, placing the symbol S_k on its tape, and then halts (that is, Z prints S_k when and only when Z_0 halts). Therefore,

$$P_{Z_0}(x) \leftrightarrow Q_{Z,k}(x);$$

so $Q_{Z,k}(x)$ is not computable.

The techniques just applied illustrate the methods usually employed in deriving the unsolvability of decision problems not directly stated in terms of numerical predicates. It might also be mentioned that the unsolvability of essentially these problems was first obtained by Turing [1].

3. Properties of Semicomputable Predicates. Theorems 1.5 and 1.6 show that the class of A-semicomputable predicates is not closed under the operation of negation. In this section, we shall note some operations under which this class is closed.

THEOREM 3.1. *Let $P(y, \mathfrak{x}^{(n)})$ be A-semicomputable, and let*

$$Q(\mathfrak{x}^{(n)}) \leftrightarrow \bigvee_y P(y, \mathfrak{x}^{(n)}).$$

Then $Q(\mathfrak{x}^{(n)})$ is also A-semicomputable.

PROOF. Let

$$P(y, \mathfrak{x}^{(n)}) \leftrightarrow \bigvee_z R(z, y, \mathfrak{x}^{(n)}),$$

where R is A-computable. Then

$$Q(\mathfrak{x}^{(n)}) \leftrightarrow \bigvee_y \bigvee_z R(z, y, \mathfrak{x}^{(n)}) \leftrightarrow \bigvee_t R(K(t), L(t), \mathfrak{x}^{(n)}),$$

which proves the theorem.

THEOREM 3.2. *Let $P(y, \mathfrak{x}^{(n)})$ be A-semicomputable, and let*

$$Q(z, \mathfrak{x}^{(n)}) \leftrightarrow \bigwedge_{y=0}^{z} P(y, \mathfrak{x}^{(n)}).$$

Then $Q(z, \mathfrak{x}^{(n)})$ is also A-semicomputable.[1]

PROOF. Let

$$P(y, \mathfrak{x}^{(n)}) \leftrightarrow \bigvee_{u} R(u, y, \mathfrak{x}^{(n)}),$$

where R is A-computable. Then

$$Q(z, \mathfrak{x}^{(n)}) \leftrightarrow \bigwedge_{y=0}^{z} \bigvee_{u} R(u, y, \mathfrak{x}^{(n)}).$$

Now, to say that, for every y between 0 and z, there is some u such that $R(u, y, \mathfrak{x}^{(n)})$ is true is equivalent to saying that there exists a finite sequence $u_0, u_1, \ldots, u_z$ such that, for each y between 0 and z, $R(u_y, y, \mathfrak{x}^{(n)})$ is true. But, if we set

$$w = \prod_{y=0}^{z} \text{Pr}\,(y+1)^{u_y},$$

this sequence is given by $1 \text{ Gl } w, \ldots, z \text{ Gl } w, (z+1) \text{ Gl } w$. Hence,

$$Q(z, \mathfrak{x}^{(n)}) \leftrightarrow \bigvee_{w} \bigwedge_{y=0}^{z} R((y+1) \text{ Gl } w, y, \mathfrak{x}^{(n)}).$$

By Theorem 3-5.4, this last equivalence yields the desired result.

THEOREM 3.3. *If $P(\mathfrak{x}^{(n)})$ and $Q(\mathfrak{x}^{(n)})$ are A-semicomputable, then so are $P(\mathfrak{x}^{(n)}) \wedge Q(\mathfrak{x}^{(n)})$ and $P(\mathfrak{x}^{(n)}) \vee Q(\mathfrak{x}^{(n)})$.*

PROOF. Let $P(\mathfrak{x}^{(n)}) \leftrightarrow \bigvee_{y} R(y, \mathfrak{x}^{(n)})$ and $Q(\mathfrak{x}^{(n)}) \leftrightarrow \bigvee_{y} S(y, \mathfrak{x}^{(n)})$, where R and S are A-computable. Then

$$P(\mathfrak{x}^{(n)}) \wedge Q(\mathfrak{x}^{(n)}) \leftrightarrow \bigvee_{y} \bigvee_{z} [R(y, \mathfrak{x}^{(n)}) \wedge S(z, \mathfrak{x}^{(n)})],$$

$$P(\mathfrak{x}^{(n)}) \vee Q(\mathfrak{x}^{(n)}) \leftrightarrow \bigvee_{y} [R(y, \mathfrak{x}^{(n)}) \vee S(y, \mathfrak{x}^{(n)})].$$

The result then follows from Theorems 3.1 and 3-5.3.

THEOREM 3.4. *Let $P(y, \mathfrak{x}^{(n)})$ be A-semicomputable, and let $f(\mathfrak{y}^{(m)})$ be A-computable. Then $P(f(\mathfrak{y}^{(m)}), \mathfrak{x}^{(n)})$ is A-semicomputable.*

PROOF. Let

$$P(y, \mathfrak{x}^{(n)}) \leftrightarrow \bigvee_{z} R(z, y, \mathfrak{x}^{(n)}),$$

[1] This theorem is due to Mostowski [2].

where R is A-semicomputable. Then

$$P(f(\mathfrak{y}^{(m)}), \mathfrak{x}^{(n)}) \leftrightarrow \bigvee_z R(z, f(\mathfrak{y}^{(m)}), \mathfrak{x}^{(n)}).$$

4. Recursively Enumerable Sets. The study of A-semicomputable predicates is, in effect, the study of the domains of partially A-computable functions. In this section we shall see that essentially the same results are obtained by studying the ranges of such functions. Moreover, we shall see that, excepting the empty set, it does not matter whether we permit all partially A-computable functions or restrict ourselves to A-computable functions or even to A-primitive recursive functions.

THEOREM 4.1. *Let $\{t \mid P(t)\}$ be the range of a partially A-computable function $f(x)$. Then $P(t)$ is A-semicomputable.*

PROOF. By Corollary 4-2.2, we can find a number z_0 such that

$$f(x) = U(\min_y T^A(z_0, x, y)).$$

Then a number t will be in the range of $f(x)$ if and only if there are numbers x, w such that $t = U(w)$ and $T^A(z_0, x, w)$ is true (cf. the final assertion of Theorem 4-2.1). That is,

$$\{t \mid P(t)\} = \Big\{t \mid \bigvee_x \bigvee_w [t = U(w) \wedge T^A(z_0, x, w)]\Big\}.$$

The result then follows from Theorems 3.1 and 3-5.3.

THEOREM 4.2. *Let $P(x)$ be A-semicomputable, and let $\{x \mid P(x)\} \neq \emptyset$. Then there exists an A-primitive recursive function whose range is $\{x \mid P(x)\}$.*

PROOF. Let $S = \{x \mid P(x)\}$. Since $S \neq \emptyset$, S contains a least element, which we shall call s_0. By Theorem 1.4, there is a number z_0 such that

$$P(x) \leftrightarrow \bigvee_y T^A(z_0, x, y);$$

that is,

$$x \in S \leftrightarrow \bigvee_y T^A(z_0, x, y).$$

Let $\tau^A(x, y)$ be the characteristic function of the A-primitive recursive predicate $T^A(z_0, x, y)$. Let $f(x)$ be defined as follows:

$$\begin{aligned} f(0) &= s_0, \\ f(m+1) &= \tau^A(K(m+1), L(m+1)) \cdot f(m) \\ &\qquad + \alpha(\tau^A(K(m+1), L(m+1))) \cdot K(m+1). \end{aligned}$$

It is easy to see that $f(x)$ is A-primitive recursive. We shall see that the range of f is the set S.

The successive values of f are obtained by searching among the ordered pairs of integers

$$(K(0), L(0)), (K(1), L(1)), (K(2), L(2)), \ldots$$

for pairs $(K(j), L(j))$ for which $T^A(z_0, K(j), L(j))$. As such a pair is found, $K(j)$ is made a value of f. For ordered pairs for which $T^A(z_0, K(j), L(j))$ is false, f simply continues its previous value. Thus, the values taken on by f are precisely the members of the set S.

Combining Theorems 4.1 and 4.2 we have

THEOREM 4.3. *Let* $S = \{x \mid P(x)\}$, *and let* $S \neq \emptyset$. *Then the following statements are all equivalent:*

(1) $P(x)$ *is A-semicomputable.*
(2) S *is the range of an A-primitive recursive function.*
(3) S *is the range of an A-recursive function.*
(4) S *is the range of an A-partial recursive*[1] *function.*

PROOF. By Theorem 4.2, (1) implies (2). Obviously, (2) implies (3), and (3) implies (4). By Theorem 4.1, (4) implies (1). Hence, all four statements are equivalent.

DEFINITION 4.1. *A set S is called A-***recursively enumerable** *either if* $S = \emptyset$ *or if the equivalent conditions* (1) *to* (4) *of Theorem* 4.3 *hold.*

The term "A-recursively enumerable" is motivated by the fact that such a set (if it is nonempty) is, in fact, enumerated by an A-recursive function. Our results regarding semicomputable predicates can, of course, also be stated in terms of recursively enumerable sets.

We begin with

DEFINITION 4.2. We write $\{n\}_A = \left\{x \mid \bigvee_y T^A(n, x, y)\right\}$; $\{n\} = \{n\}_\emptyset$.

In terms of this notation, our enumeration theorem, Theorem 1.4, yields

THEOREM 4.4. *Let S be any A-recursively enumerable set. Then there exists a number n such that* $S = \{n\}_A$. *Moreover, for each n,* $\{n\}_A$ *is an A-recursively enumerable set.*

Theorem 1.5 yields

THEOREM 4.5. *S is A-recursive if and only if S and* $\bar{S}$ *are both A-recursively enumerable.*

[1] $\emptyset$ is also the range of a partial recursive function, namely, the function $f(x)$ which is nowhere defined. $f(x)$ is partial recursive, since $f(x) = min_y\,[x + y + 1 = 0]$. Or $f(x) = \psi_Z(x)$ where Z consists of the quadruples

$$q_1\ 1\ L\ q_1$$
$$q_1\ B\ R\ q_1.$$

DEFINITION 4.3. *We write* $A' = \left\{x \mid \bigvee_y T^A(x, x, y)\right\}$; *we also write* $K = \phi'$.

Then, by Theorem 1.6, we have

THEOREM 4.6. *A' is A-recursively enumerable, but not A-recursive.*

Similarly, Corollary 1.7 yields

COROLLARY 4.7. *K is recursively enumerable, but not recursive.*

By the *decision problem for a set S* of integers is meant the problem of determining, for given n, whether or not $n \in S$.

DEFINITION 4.4. *The decision problem for a set S is called* **recursively solvable** *or* **unsolvable** *according as S is or is not recursive.*

Thus, Corollary 4.7 yields

COROLLARY 4.8. *The recursively enumerable set K has a recursively unsolvable decision problem.*

Theorem 3.3 yields

THEOREM 4.9. *If R and S are A-recursively enumerable, so are* $R \cup S$ *and* $R \cap S$.

Also, by Theorem 3.4, we have

THEOREM 4.10. *If R is A-recursively enumerable and f(x) is A-recursive, then the set* $S = \{x \mid f(x) \in R\}$ *is A-recursively enumerable.*

That the notion of recursive enumerability is capable of so many different formulations suggests that a fundamentally important concept is involved. This does, indeed, appear to be so. Post's original work on the subject[1] was from this point of view, with recursiveness, in effect, defined in terms of recursive enumerability. The relevant intuitive concept is that of *generated set*. A *generated set* of, say, integers is produced by a process that from time to time ejects an integer. Once an integer has been ejected it is placed in the set, and it remains there. We may well be ignorant, however, of the ultimate fate of an integer so far not ejected. In fact, it is not excluded, at any given stage, that no new integers, or even no integers at all, will be subsequently ejected. This circle of ideas, when developed into a formal theory, leads to the *normal systems* of Chap. 6, which, in turn, are shown to represent essentially another formulation of recursive enumerability.

5. Two Recursively Enumerable Sets.[2] If R is a recursively enumerable set that is not recursive, then, as we know, $\bar{R}$ is not recursively enumerable. Hence, to every recursively enumerable set P such that $P \subset \bar{R}$, there corresponds a number x such that $x \notin P$ and $x \in \bar{R}$. We now inquire whether there is an effective procedure by means of which,

[1] Cf. Post [2, 3]. More details appear in his unpublished [8].

[2] The material in this section is not employed in Chaps. 6 and 7, and is employed only incidentally in Chap. 8. It may, therefore, be omitted by the reader who is so minded.

given P, we can obtain x. P, however, cannot be "given." We can, however, give an integer n such that $P = \{n\}$. This leads to

DEFINITION 5.1. *The set R is A-***creative** *if R is A-recursively enumerable and if there exists a recursive function $f(n)$ such that, whenever $\{n\}_A \subset \bar{R}$, then $f(n) \in \bar{R}$, $f(n) \notin \{n\}_A$. R is* **creative** *if it is ϕ-creative.*[1]

Then we have

THEOREM 5.1. *A' is A-creative.*

PROOF. We take $f(n) = n$. We have

$$A' = \left\{n \mid \bigvee_y T^A(n, n, y)\right\}$$
$$= \{n \mid n \in \{n\}_A\}.$$

Hence,

$$\overline{A'} = \{n \mid n \notin \{n\}_A\}.$$

If, for some fixed $n = n_0$,

$$\{n_0\}_A \subset \overline{A'},$$

then, if $x \in \{n_0\}_A$, we have $x \notin \{x\}_A$. Set $x = n_0$. Then, $n_0 \in \{n_0\}_A$ implies $n_0 \notin \{n_0\}_A$. We conclude that $n_0 \notin \{n_0\}_A$, that is, that $f(n_0) \notin \{n_0\}_A$. Moreover, since $n_0 \notin \{n_0\}_A$, $n_0 \in \overline{A'}$; that is, $f(n_0) \in \overline{A'}$. This proves the theorem.

COROLLARY 5.2. K is creative.

PROOF. $K = \phi'$.

It can be shown (cf. Theorem 11-3.1) that the complement of an A-creative set C does in fact contain an infinite recursively enumerable subset. Beginning with $\phi \subset \bar{C}$, we obtain a number $x \in \bar{C}$. Let $\{n\}_A$ be the set whose only element is x. Then we obtain a number $x' \in \bar{C}$, $x' \notin \{n\}_A$, that is, $x' \neq x$. Continuing this process, we generate an infinite sequence of elements of $\bar{C}$. That the set thus obtained is in fact recursively enumerable will be seen later. Thus, we might conjecture that the complements of all recursively enumerable sets contain infinite recursively enumerable subsets. That this is not the case was proved by Post [3].

DEFINITION 5.2. *S is A-***simple** *if*

(1) *S is A-recursively enumerable,*
(2) *$\bar{S}$ is infinite, and*
(3) *$\bar{S}$ contains no infinite A-recursively enumerable subset.*

S is **simple** *if it is ϕ-simple.*[1]

Note that an A-simple set cannot be A-recursive.

We have

THEOREM 5.3. *For every set A, there is an A-simple set.*

[1] This definition, with $A = \phi$, is due to Post [3].

PROOF. Consider the A-partial recursive function

$$\theta_A(i) = K(\min_t [T^A(i, K(t), L(t)) \wedge K(t) > 2i]),$$

and let S^A be the range of $\theta_A(i)$. Then, by Definition 4.1, S^A is A-recursively enumerable. We shall show that S^A is A-simple. But first we pause to indicate the motivation behind the manner in which S^A was defined. In a suitable ordering, $\theta_A(i)$ is the first member of $\{i\}_A$ (if any) greater than $2i$, and S^A is the set of all such first members.

We have the following lemmas:

LEMMA 1. *If $\theta_A(i_0)$ is defined, then $\theta_A(i_0) \in \{i_0\}_A$, $\theta_A(i_0) > 2i_0$.*

PROOF. If $\theta_A(i_0)$ is defined, then there is some value of t such that $T^A(i_0, K(t), L(t))$, $\theta(i_0) = K(t)$, and $\theta_A(i_0) > 2i_0$. Hence,

$$\bigvee_y T^A(i_0, \theta_A(i_0), y); \text{ that is, } \theta_A(i_0) \in \{i_0\}_A.$$

LEMMA 2. *If $\{i\}_A$ is infinite, then $\{i\}_A \cap S^A \neq \phi$.*

PROOF. If $\{i_0\}_A$ is infinite, there is a number m_0 such that $m_0 \in \{i_0\}_A$ and $m_0 > 2i_0$. Hence, $\bigvee_y T^A(i_0, m_0, y)$. Let y_0 be the least such y. (Actually, by the definition of the T-predicates there is at most one such y.) Then $T^A(i_0, m_0, y_0)$. Let $t_0 = J(m_0, y_0)$. Then

$$m_0 = K(J(m_0, y_0)) = K(t_0); \; y_0 = L(J(m_0, y_0)) = L(t_0).$$

Hence, $T^A(i_0, K(t_0), L(t_0))$. Since $K(t_0) = m_0 > 2i_0$, $\theta_A(i_0)$ is defined. Hence, by Lemma 1, $\theta_A(i_0) \in \{i_0\}_A$. But, $\theta_A(i_0) \in S^A$. Hence, $\{i_0\}_A \cap S^A \neq \phi$.

LEMMA 3. *$\overline{S^A}$ contains no infinite A-recursively enumerable subset.*

PROOF. By Lemma 2, S^A has at least one element in common with each infinite A-recursively enumerable set.

LEMMA 4. *$\overline{S^A}$ is infinite.*

PROOF. Our proof is by counting. Let $\sigma_A(x)$ be the number of elements of S^A that are $\leqq x$. We seek to estimate $\sigma_A(2n + 2)$. That is, we seek to estimate how many numbers $\theta_A(i)$ there are, with $\theta_A(i) \leqq 2n + 2$. For this purpose, we may restrict ourselves to values of $i \leqq n$, since, for $i \geqq n + 1$, $\theta_A(i) > 2i \geqq 2n + 2$. Hence, at most, there are the $n + 1$ numbers $\theta_A(0)$, $\theta_A(1)$, . . . , $\theta_A(n)$. Thus, $\sigma_A(2n + 2) \leqq n + 1$.

So we have seen that at most $n + 1$ of the first $2n + 2$ integers belong to S^A. Hence, at least $n + 1$ of them belong to $\overline{S^A}$. Hence, $\overline{S^A}$ is infinite.

Lemmas 3 and 4 give the desired result.

6. A Set Which Is Not Recursively Enumerable. We shall now attempt to formalize the argument of the Introduction, Sec. 1. There we maintained (cf. III of that section) that there was no algorithm for determining whether or not an alleged algorithm was indeed an algorithm. If there were such an algorithm, we could enumerate all algorithms in an effective manner, whereas the existence of such an effective enumeration would lead to a contradiction. This suggests

THEOREM 6.1. *The set of all Gödel numbers of Turing machines Z, for which $\Psi_Z(x)$ is total, is not recursively enumerable.*

PROOF. Let us designate the set of all such Gödel numbers by R, and let us suppose that R is recursively enumerable. Then, since $R \neq \phi$, there would exist a recursive function $f(n)$ whose range is R.

The function $U(\min_y T(f(n), x, y))$ would be total, and hence recursive. Hence, $U(\min_y T(f(x), x, y)) + 1$ would be recursive. Hence, by the very definition of $f(n)$, there would be a number n_0 such that

$$U(\min_y T(f(x), x, y)) + 1 = U(\min_y T(f(n_0), x, y)).$$

Setting $x = n_0$ yields a contradiction.

We may note in passing that this set R is given by

$$R = \left\{z \mid \bigwedge_x \bigvee_y T(z, x, y)\right\}.$$

This expression suggests that $\bar{R}$ is also not recursively enumerable. We shall see later that this is so. (Cf. Theorem 11-1.3.)

PART 2

APPLICATIONS OF THE GENERAL THEORY

CHAPTER 6

COMBINATORIAL PROBLEMS

1. Combinatorial Systems. This chapter is devoted to showing that various decision problems of a combinatorial nature are recursively unsolvable. In this process we shall discover some new formulations of the concept of recursive enumerability.

Combinatorial problems can generally be formulated in terms of finite sequences of fixed objects. We begin by considering an infinite sequence of symbols (or objects) denoted by $a_0, a_1, a_2, a_3, \ldots$. We shall also write 1 for a_0 (the ambiguity resulting because we have also been writing 1 for S_1 should cause no confusion). A finite sequence (possibly of length 0) of these symbols will be called a *word* (or *string*, or *formula*). The *empty word* of length 0 will be written Λ. In this chapter we shall usually employ capital letters, X, Y, etc., as variables ranging over all such words. The result of juxtaposing the pair of words X, Y will be written XY.

We shall also consider *predicates*, the range of whose variables is the set of all words. Such predicates will be represented by capital German letters, $\mathfrak{R}$, $\mathfrak{S}$, etc., and will be called *word predicates.* When there is danger of confusion, we shall refer to predicates whose variables range over the natural numbers (i.e., what we have been calling, simply, *predicates*), as *numerical predicates.*

The technique of Gödel numbers employed in Chap. 4, Sec. 1, can be used to advantage here. With the symbol a_i we associate the odd number $2i + 1$. If $W = a_{i_1}a_{i_2} \cdots a_{i_k}$, we write

$$\text{gn}\,(W) = \prod_{j=1}^{k} \text{Pr}\,(j)^{2i_j+1}.$$

We also set gn $(\Lambda) = 1$. The number gn (W) is called the *Gödel number* of W.

With each word predicate $\mathfrak{R}(X_1, \ldots, X_m)$ is associated a *numerical* predicate $\mathfrak{R}^*(x_1, \ldots, x_m)$, where $\mathfrak{R}^*(x_1, \ldots, x_m)$ is true for given $x_1, \ldots, x_m$ if and only if each x_i, $1 \leq i \leq m$, is the Gödel number of some word W_i such that $\Re(W_1, \ldots, W_m)$ is true. That is,

$$\Re^*(x_1, \ldots, x_m) \leftrightarrow \bigvee_{W_1} \bigvee_{W_2} \cdots \bigvee_{W_m} [(x_1 = \text{gn}(W_1)) \wedge (x_2 = \text{gn}(W_2)) \wedge \cdots \wedge (x_m = \text{gn}(W_m)) \wedge \Re(W_1, \ldots, W_m)].$$

When we speak of a word predicate $\Re$ having attributes that have been formerly defined only for numerical predicates, we always mean that the associated numerical predicate $\Re^*$ has these attributes. Thus, when we say that a word predicate $\Re$ is *recursive,* we mean simply that $\Re^*$ is recursive. The device of Gödel numbering permits us, in effect, to extend our previously developed theory to the present subject matter.

Let $g, h, k; \bar{g}, \bar{h}, \bar{k}$ be six (not necessarily distinct, possibly empty) words. Then we may consider the binary predicate $\Re^{g,h,k}_{\bar{g},\bar{h},\bar{k}}(X, Y)$, which is true for given words X_0, Y_0 if and only if there exist (possibly empty) words P_0, Q_0 such that

$$X_0 = gP_0hQ_0k$$

and

$$Y_0 = \bar{g}P_0\bar{h}Q_0\bar{k};$$

that is,

$$\Re^{g,h,k}_{\bar{g},\bar{h},\bar{k}}(X, Y) \leftrightarrow \bigvee_P \bigvee_Q [(X = gPhQk) \wedge (Y = \bar{g}P\bar{h}Q\bar{k})].$$

This predicate is called the **production** *associated with* $g, h, k, \bar{g}, \bar{h}, \bar{k}$, *and will be symbolized*

$$gPhQk \rightarrow \bar{g}P\bar{h}Q\bar{k}.\dagger$$

COROLLARY 1.1. *The production*

$$gPhQk \rightarrow \bar{g}P\bar{h}Q\bar{k}$$

is a recursive binary predicate.

PROOF. Let the associated numerical predicate be $R(x, y)$. Then,[1]

$$R(x, y) \leftrightarrow \text{GN}(x) \wedge \text{GN}(y) \wedge \bigvee_{p=0}^{x} \bigvee_{q=0}^{x} [(x = \text{gn}(g) * p * \text{gn}(h) * q * \text{gn}(k)) \wedge (y = \text{gn}(\bar{g}) * p * \text{gn}(\bar{h}) * q * \text{gn}(\bar{k}))],$$

which is recursive.

† Actually, we are using the term "production" somewhat ambiguously. Sometimes, we shall mean the ordered sextuple of words $g, h, k, \bar{g}, \bar{h}, \bar{k}$ and, at other times, the associated word predicate $\Re^{g,h,k}_{\bar{g},\bar{h},\bar{k}}$.

The term production was first used by Post [2, 8] in a somewhat wider sense (cf. also, Rosenbloom [1]).

[1] Cf. formulas (3) and (5) of Chap. 4, Sec. 1.

DEFINITION 1.1. *Let* $\mathfrak{R}(X, Y)$ *be a production. And let* X_0, Y_0 *be words such that* $\mathfrak{R}(X_0, Y_0)$ *is true. Then we shall say that* Y_0 *is a* **consequence of X_0 with respect to $\mathfrak{R}$.**

DEFINITION 1.2. *By the* **inverse of the production**

$$gPhQk \rightarrow \bar{g}P\bar{h}Q\bar{k}$$

we mean the production

$$\bar{g}P\bar{h}Q\bar{k} \rightarrow gPhQk.$$

Thus, if the inverse of the production $\mathfrak{R}$ is $\mathfrak{S}$, then Y is a consequence of X with respect to $\mathfrak{R}$ if and only if X is a consequence of Y with respect to $\mathfrak{S}$.

Actually, we are primarily concerned with a few special kinds of production.

DEFINITION 1.3. *Let* g, $\bar{g}$ *be given nonempty words. Then, the production*

$$\Lambda Pg Q\Lambda \rightarrow \Lambda P\bar{g}Q\Lambda$$

is called the **semi-Thue production associated with** g, $\bar{g}$.

This production is written simply

$$PgQ \rightarrow P\bar{g}Q.$$

Y is a consequence of X with respect to this production if and only if Y is obtained from X by replacing g by $\bar{g}$ at some occurrence of g in X.

DEFINITION 1.4. *Let* g, $\bar{g}$ *be given nonempty words. Then, the production*

$$gP\Lambda Q\Lambda \rightarrow \Lambda P\Lambda Q\bar{g}$$

is called the **normal production associated with** g, $\bar{g}$.

This production is written simply

$$gP \rightarrow P\bar{g}.$$

COROLLARY 1.2. *The inverse of a semi-Thue production is a semi-Thue production.*

PROOF. Clearly, the inverse of

$$PgQ \rightarrow P\bar{g}Q$$

is

$$P\bar{g}Q \rightarrow PgQ.$$

DEFINITION 1.5. *A production is called* **antinormal** *if its inverse is normal.*

We indicate the inverse of

$$gP \rightarrow P\bar{g}$$

by

$$P\bar{g} \rightarrow gP.$$

DEFINITION 1.6. *A* **combinatorial system** Γ *consists of a single non-empty word called the* **axiom of** Γ *and a finite set of productions called the* **productions of** Γ.

The **alphabet of** Γ *consists of all letters that occur either in the axiom of* Γ *or in the* $g, h, k, \bar{g}, \bar{h}, \bar{k}$ *that define the productions of* Γ.

By a **word on** Γ *we mean a word in which only letters from the alphabet of* Γ *appear.*

Our interest will center about four special kinds of combinatorial system.

DEFINITION 1.7. *A* **semi-Thue system** *is a combinatorial system all of whose productions are semi-Thue productions.*

A **Thue system** *is a semi-Thue system with the property that the inverse of each of its productions is also one of its productions.*

A **normal system** *is a combinatorial system all of whose productions are normal productions.*

A **Post system**[1] *is a combinatorial system whose productions consist of a finite set of normal productions and their inverses.*

DEFINITION 1.8. *By a* **proof in a combinatorial system** Γ *is meant a finite sequence* $X_1, \ldots, X_m$ *of words such that* X_1 *is the axiom of* Γ *and, for each* i, $1 < i \leqq m$, X_i *is a consequence of* X_{i-1} *with respect to one of the productions of* Γ.

Each of these X_i *is then called a* **step of the proof.**

DEFINITION 1.9. *We say that* W *is a* **theorem of** Γ *and we write*

$$\vdash_\Gamma W$$

if the word W *is the final step of a proof in* Γ.

In this case the proof is called a proof of W *in* Γ.

Note that a theorem of a combinatorial system Γ is necessarily a word on Γ.

DEFINITION 1.10. *A combinatorial system* Γ *is* **monogenic** *if each theorem of* Γ *has at most one immediate consequence with respect to the productions of* Γ.

The set of Gödel numbers of all theorems of the combinatorial system Γ will be written T_Γ.

Thus,

$$\vdash_\Gamma W \leftrightarrow \text{gn}\,(W) \in T_\Gamma.$$

The set T_Γ is a set of integers determined by the system Γ. The association is, however, in a certain sense "unnatural"; that is, the procedure for computing the Gödel number of a theorem of Γ may well be more complicated than the proof of the word in Γ. Moreover, the ele-

[1] This term has been used by Novikoff. Actually, Post systems as such do not occur in Post's work. Cf., however, Markov [1, 3].

ments of T_Γ must have factorizations of a very special kind; e.g., for all Γ, $9 \notin T_\Gamma$. Therefore, with each combinatorial system Γ we shall associate another set of integers, S_Γ, in a more natural way.

We shall write a_i^n for the word $\underbrace{a_i a_i \cdots a_i}_{n}$. Moreover, we shall write $\bar{n} = 1^{n+1} = a_0^{n+1}$.

DEFINITION 1.11. *Let Γ be a combinatorial system. Then, by the* **set of integers generated by** Γ *we shall mean the set*

$$S_\Gamma = \{x \mid \vdash_\Gamma \bar{x}\}.$$

Thus,

$$x \in S_\Gamma \leftrightarrow \vdash_\Gamma \bar{x}.$$

We have at once

COROLLARY 1.3. $S_\Gamma = \{x \mid \text{gn}(\bar{x}) \in T_\Gamma\}$.

We shall see that T_Γ and S_Γ are always recursively enumerable. Moreover, it will turn out that, for every recursively enumerable set R, there is a combinatorial system Γ such that $R = S_\Gamma$.

THEOREM 1.4. *The set T_Γ is recursively enumerable.*

PROOF. We employ the formulas of Group I of Chap. 4, Sec. 1. Let $\mathfrak{R}_1, \ldots, \mathfrak{R}_p$ be the productions of Γ, and let

$$\mathfrak{R}^*(x, y) \leftrightarrow \mathfrak{R}_1^*(x, y) \vee \mathfrak{R}_2^*(x, y) \vee \cdots \vee \mathfrak{R}_p^*(x, y).$$

Then, by Corollary 1.1, $\mathfrak{R}^*(x, y)$ is recursive. Furthermore, let a be the Gödel number of the axiom of Γ. Then

$$T_\Gamma = \Big\{x \mid \text{GN}\,(x) \wedge \bigvee_y \Big[(1 \text{ Gl } y = a) \wedge \bigwedge_{n=1}^{\mathfrak{L}(y) \dot{-} 1} \mathfrak{R}^*(n \text{ Gl } y, (n+1) \text{ Gl } y) \wedge (\mathfrak{L}(y) \text{ Gl } y = x)\Big]\Big\}.$$

Hence, by Theorem 5-4.3, T_Γ is recursively enumerable.

By the *decision problem* for a combinatorial system, we mean the problem of determining, of a given word, whether or not it is a theorem of the system. This leads us to

DEFINITION 1.12. *We say that the decision problem for a combinatorial system Γ is* **recursively solvable** *or* **unsolvable,** *according as T_Γ is or is not a recursive set.*

THEOREM 1.5. *The set S_Γ is recursively enumerable.*

PROOF. By Corollary 1.3,

$$S_\Gamma = \{x \mid \text{gn}\,(\bar{x}) \in T_\Gamma\}.$$

But, by Theorem 1.4, T_Γ is recursively enumerable. Hence, by Theorem 5-4.10, it suffices to show that gn $(\bar{x})$ is recursive.

In fact, it is primitive recursive, since

$$\text{gn}(\bar{0}) = 2,$$
$$\text{gn}(\overline{k+1}) = \text{gn}(\bar{k}) * 2.$$

THEOREM 1.6. *If S_Γ is not recursive, then the decision problem for Γ is recursively unsolvable.*

PROOF. By Corollary 1.3,

$$C_{S_\Gamma}(x) = C_{T_\Gamma}(\text{gn}(\bar{x})).$$

Hence, if T_Γ were recursive, so would be S_Γ.

Let Γ be an arbitrary combinatorial system. We shall show how to construct a new combinatorial system Γ^* whose alphabet consists of only the two letters a_0, a_1 (which we shall write 1, b, respectively) and whose decision problem is recursively solvable if and only if that for Γ is.

To begin with, we set

$$a_i^* = 1b^{i+1}1.$$

If $W = a_{i_1}a_{i_2} \cdots a_{i_p}$, we set $W^* = a_{i_1}^*a_{i_2}^* \cdots a_{i_p}^*$. We also take $\Lambda^* = \Lambda$.

Now, let the axiom of Γ be A, and let its productions be

$$g_iPh_iQk_i \rightarrow \bar{g}_iP\bar{h}_iQ\bar{k}_i \qquad i = 1, 2, \ldots, m.$$

Then, Γ^* is taken to be the combinatorial system whose axiom is A^* and whose productions are

$$g_i^*Ph_i^*Qk_i^* \rightarrow \bar{g}_i^*P\bar{h}_i^*Q\bar{k}_i^* \qquad i = 1, 2, \ldots, m.$$

LEMMA 1. *If $\vdash_\Gamma W$, then $\vdash_{\Gamma^*} W^*$.*

PROOF. Let $W_1, W_2, \ldots, W_n$ be a proof of W in Γ. Then W_1^*, $W_2^*, \ldots, W_n^*$ is clearly a proof of W^* in Γ^*.

We shall say that the word U on 1, b is *regular* if there exists a word W on the a_i such that $U = W^*$.

LEMMA 2. *If $\vdash_{\Gamma^*} W$, then W is regular.*

PROOF. The axiom is regular, and regularity is preserved by the productions.

LEMMA 3. *If $\vdash_{\Gamma^*} W^*$, then $\vdash_\Gamma W$.*

PROOF. Let $U_1, U_2, \ldots, U_n = W^*$ be a proof of W^* in Γ^*. Then, by Lemma 2, $U_1 = W_1^*$, $U_2 = W_2^*, \ldots, U_n = W_n^* = W^*$ for suitable $W_1, \ldots, W_n$. But then $W_1, \ldots, W_n$ is a proof of W in Γ.

Now, let us compare the decision problems of Γ and Γ^*. Intuitively, it is clear from Lemmas 1 and 3 that the decision problems for Γ and Γ^* are solvable or unsolvable, together. However, our definition of *recursive* solvability involves the set T_Γ, and hence some calculations are required.

Let us write

$$\text{Star}(x) = 2^1 * \prod_{i=0}^{[(x-1)/2]} [\text{Pr}(i+1)]^3 * 2^1.$$

Then, if x is the number associated with the symbol a_j (that is, $x = 2j + 1$),

$$\text{Star}(x) = \text{gn}(a_j^*).$$

Next, let

$$\begin{aligned} \text{Ast}(x, 0) &= 0 \\ \text{Ast}(x, k+1) &= \text{Ast}(x, k) * \text{Star}((k+1) \text{ Gl } x) \\ \text{Asterisk}(x) &= \text{Ast}(x, \mathfrak{L}(x)). \end{aligned}$$

Then,

$$\text{Asterisk}(\text{gn}(W)) = \text{gn}(W^*).$$

Now, using Lemmas 1 to 3, we see that

$$x \in T_\Gamma \leftrightarrow \text{Asterisk}(x) \in T_{\Gamma^*},$$

$$x \in T_{\Gamma^*} \leftrightarrow \bigvee_{y=0}^{x} [y \in T_\Gamma \wedge x = \text{Asterisk}(y)].$$

We have proved

THEOREM 1.7. *For every combinatorial system Γ, we can construct a combinatorial system Γ^* whose alphabet consists of two letters and whose decision problem is recursively solvable if and only if that for Γ is. Moreover, if Γ is a semi-Thue system, a Thue system, a normal system, or a Post system, respectively, then so is Γ^*.*

We next turn our attention to the set S_Γ, generated by Γ.

THEOREM 1.8. *For every semi-Thue system Γ, we can construct a semi-Thue system Γ', whose alphabet consists of two letters, such that $S_\Gamma = S_{\Gamma'}$.*

PROOF. We construct Γ^* as above. Then we form Γ' by adjoining to Γ^* the production

$$P1b1Q \rightarrow P1Q.$$

Then we have

$$\begin{aligned} n \in S_\Gamma &\leftrightarrow \vdash_\Gamma 1^{n+1} \\ &\leftrightarrow \vdash_{\Gamma^*} (1b1)^{n+1} \\ &\leftrightarrow \vdash_{\Gamma'} 1^{n+1} \\ &\leftrightarrow n \in S_{\Gamma'}. \end{aligned}$$

THEOREM 1.9. *For every normal system Γ, we can construct a normal system Γ', whose alphabet consists of two letters, such that $S_\Gamma = S_{\Gamma'}$.*†

† Our first effort was to obtain Theorems 1.7 to 1.9 from a single construction. Hartley Rogers showed, by means of a suitable counterexample, that the construction proposed would not work. Various more complex constructions were destroyed by Hilary Putnam, who has also suggested an alternative method of proving the theorems.

PROOF. We construct Γ^* as above. Then, we form Γ' by adjoining to Γ^* the productions

$$1P \to P1,$$
$$b11P \to P1.$$

Then, we have

$$\begin{aligned} n \in S_\Gamma &\leftrightarrow \vdash_\Gamma 1^{n+1} \\ &\leftrightarrow \vdash_{\Gamma^*} (1b1)^{n+1} \\ &\leftrightarrow \vdash_{\Gamma'} (b11)^{n+1} \\ &\leftrightarrow \vdash_{\Gamma'} 1^{n+1} \\ &\leftrightarrow n \in S_{\Gamma'}. \end{aligned}$$

2. Turing Machines and Semi-Thue Systems. We have seen that the set of integers S_Γ, generated by a combinatorial system Γ, is always recursively enumerable. In this section, we shall see that, conversely, for every recursively enumerable set R, there is a combinatorial system Γ such that $R = S_\Gamma$. Moreover, Γ can be chosen to be a semi-Thue system. Later, we shall see that the same is also true for normal systems.

We recall (from Definition 1-1.3 and the discussion following it) that a simple Turing machine is one that contains no quadruple of the form $q_i\, S_j\, q_k\, q_l$, and that a singulary function $f(x)$ is partially computable if and only if there exists a simple Turing machine Z such that $f(x) = \Psi_Z(x)$ (cf. Definitions 1-2.4 and 1-2.5). We now agree to write P_Z for the domain of the function $\Psi_Z(x)$. Then, by Definition 5-1.1 and Theorem 5-4.3, we have

COROLLARY 2.1. *A set S is recursively enumerable if and only if there exists a simple Turing machine Z such that $S = P_Z$.*

Next, we shall show how the theory of simple Turing machines can be interpreted (at least for certain purposes) as a part of the theory of semi-Thue systems. With each simple Turing machine Z and integer m we shall associate a semi-Thue system $\tau_m(Z)$, designed to imitate the behavior of the simple Turing machine Z at the instantaneous description $q_1\overline{m}$. That is, the theorems of $\tau_m(Z)$ are to correspond roughly to the successive instantaneous descriptions of Z. Actually, the operations of a Turing machine are quite suggestive of semi-Thue productions. If we glance at clauses 1, 2, and 4 of Definition 1-1.7, we see that the defining expressions are already in the form of semi-Thue productions. Clauses 3 and 5, however, permit substitutions only at the ends of expressions and, hence, are not in the form of semi-Thue productions. To bring them into this form we introduce a new symbol h, and we place an h at the beginning and end of each instantaneous description of Z. It then becomes possible to express the recalcitrant clauses, also, as semi-Thue productions. Thus, clause 3 becomes

$$Pq_iS_jhQ \to PS_jq_lS_0hQ,$$

since our construction will be such as to make this production effective only when Q is empty. This will be clarified in the formal construction below.

The semi-Thue system $\tau_m(Z)$ is defined[1] as follows:

Alphabet. The alphabet of $\tau_m(Z)$ consists of the alphabet of Z, the internal configurations of Z, and the additional symbols h, q, q'.

Axiom. The axiom of $\tau_m(Z)$ is $hq_1S_1^{m+1}h$, that is, $hq_1\overline{m}h$.

Productions. (1) With each quadruple of Z of the form $q_i\ S_j\ S_k\ q_l$, we include the associated production

$$Pq_iS_jQ \to Pq_lS_kQ.$$

(2) With each quadruple of Z of the form $q_i\ S_j\ R\ q_l$, and each S_k in the alphabet of Z, we include the productions

$$Pq_iS_jS_kQ \to PS_jq_lS_kQ$$

and

$$Pq_iS_jhQ \to PS_jq_lS_0hQ.$$

(3) With each quadruple of Z of the form $q_i\ S_j\ L\ q_l$ and each S_k in the alphabet of Z, we include the productions

$$PS_kq_iS_jQ \to Pq_lS_kS_jQ$$

and

$$Phq_iS_jQ \to Phq_lS_0S_jQ.$$

(4) With each internal configuration q_i of Z and each S_j in the alphabet of Z for which no quadruple of Z begins with $q_i\ S_j$, we include the production

$$Pq_iS_jQ \to PqS_jQ.$$

(5) Finally, we include with each S_i in the alphabet of Z the productions

$$PqS_iQ \to PqQ,$$
$$PqhQ \to Pq'hQ,$$
$$PS_iq'Q \to Pq'Q.$$

In the following lemmas, we shall take Z to be some fixed simple Turing machine and m to be some definite integer. By a *q-symbol*, we shall understand a symbol that is either an internal configuration of Z or one of the symbols q, q'. We shall say that a word is in *standard form* if it can be written hWh, where W contains exactly one occurrence of a q-symbol and no occurrence of h.

LEMMA 1. *Let X be in standard form, and let Y be a consequence of X with respect to one of the productions of $\tau_m(Z)$ or with respect to one of their inverses. Then, Y is in standard form.*

[1] The construction, including the role of h, is that of Post [6].

Each theorem of $\tau_m(Z)$ is in standard form.

PROOF. By direct inspection of (1) to (5) above, it is clear that the property of being in standard form is preserved by each of the productions of $\tau_m(Z)$, as well as by their inverses.

Finally, the axiom $hq_1\bar{m}h$ of $\tau_m(Z)$ is in standard form; hence, so are the theorems of $\tau_m(Z)$.

LEMMA 2. *Let α, β be instantaneous descriptions. Then, $\alpha \to \beta \quad (Z)$*† *if and only if $h\beta h$ is a consequence of $h\alpha h$ with respect to $\tau_m(Z)$.*

PROOF. If $\alpha \to \beta \quad (Z)$, then this is true by virtue of one of the quadruples of Z. But the appropriate corresponding production of $\tau_m(Z)$ will be applicable and will yield $h\beta h$ as a consequence of $h\alpha h$.

Conversely, if $h\beta h$ is a consequence of $h\alpha h$, where α and β are instantaneous descriptions, then it cannot be by virtue of one of the productions of (4) or (5) above, since α contains neither q nor q'. Hence, it must be by virtue of one of the productions of (1), (2), or (3). But, by Definition 1-1.7, we must then have $\alpha \to \beta \quad (Z)$.

LEMMA 3. *Each word in standard form has at most one consequence with respect to the productions of $\tau_m(Z)$.*

$\tau_m(Z)$ is monogenic.

PROOF. Let hWh be a word in standard form. Our proof is by a case analysis.

CASE I. *$W = Pq_iS_jQ$; Z contains a quadruple beginning with $q_i\ S_j$.*

The result follows at once by Lemma 2 and Theorem 1-1.1.

CASE II. *$W = Pq_iS_jQ$; Z does not contain a quadruple beginning with $q_i\ S_j$.*

Then

$$Pq_iS_jQ \to PqS_jQ$$

is the one and only production applicable.

CASE III. *$W = Pq_i$.*

hPq_ih has no consequence.

CASE IV. *The q-symbol of W is q or q'; $W \neq q'$.*

Then, one of the productions of (5) above will be applicable, and no other.

CASE V. *$W = q'$.*

$hq'h$ has no consequence.

Finally, the monogenicity of $\tau_m(Z)$ follows from Definition 1.10.

LEMMA 4. *$\vdash_{\tau_m(Z)} hq'h$ if and only if there are words P, Q such that $\vdash_{\tau_m(Z)} hPqQh$.*

PROOF. If $\vdash_{\tau_m(Z)} hPqQh$, then the productions of (5) eventually yield

$$\vdash_{\tau_m(Z)} hPqh.$$

† For the meaning of $\alpha \to \beta \quad (Z)$, recall Definition 1-1.7.

But this has as a consequence

$$\vdash_{\tau_m(Z)} hPq'h,$$

which, in turn, eventually yields

$$\vdash_{\tau_m(Z)} hq'h.$$

Conversely, suppose that

$$\vdash_{\tau_m(Z)} hq'h.$$

Examining (4) and (5), we see that the only way in which the q_i's with which we begin (recall that the axiom is $hq_1\bar{m}h$) can be replaced by q' is for q to occur in an intermediate step. This proves the lemma.

Lemma 5. *$m \in P_Z$ if and only if $\vdash_{\tau_m(Z)} hq'h$.*

proof. $m \in P_Z$ if and only if m is in the domain of $\Psi_Z(x)$, that is, if and only if there exists a sequence $\alpha_1, \ldots, \alpha_p$ of instantaneous descriptions, where $\alpha_1 = q_1\bar{m}$, α_p is terminal, and where $\alpha_i \rightarrow \alpha_{i+1}$ (Z) for $i = 1, 2, \ldots, p - 1$.

If there exists such a sequence, then, by Lemma 2, $\vdash_{\tau_m(Z)} h\alpha_p h$. But, since α_p is terminal, this last word has a consequence whose q-symbol is q. Then, by Lemma 4, $\vdash_{\tau_m(Z)} hq'h$.

Conversely, suppose that $\vdash_{\tau_m(Z)} hq'h$. Then, by Lemma 4, we can obtain words P_0, Q_0 such that

$$\vdash_{\tau_m(Z)} hP_0qQ_0h.$$

Hence, there is a sequence $W_1, W_2, \ldots, W_r$ of words such that $W_1 = q_1\bar{m}$, $W_r = P_0qQ_0$, and $hW_{i+1}h$ is a consequence of hW_ih for $i = 1, 2, 3, \ldots, r - 1$. Suppose that the q-symbols of $W_1, W_2, \ldots, W_p$, $p < r$, are all internal configurations and that the q-symbol of W_{p+1} is q. Then W_p is a terminal instantaneous description. Furthermore, $W_i \rightarrow W_{i+1}$ (Z) for $i = 1, 2, \ldots, p - 1$. Hence, m is in the domain of $\Psi_Z(x)$.

The results of the preceding lemmas are summarized in the following theorem.

Theorem 2.2. *Associated with each simple Turing machine Z and integer m, there is a semi-Thue system $\tau_m(Z)$ with the following properties:*

(1) *The axiom of $\tau_m(Z)$ is $hq_1\bar{m}h$.*

(2) *The productions and alphabet of $\tau_m(Z)$ depend only on Z and not on the number m.*

(3) *$\tau_m(Z)$ is monogenic.*

(4) *$\vdash_{\tau_m(Z)} hq'h$ if and only if $m \in P_Z$.*

The semi-Thue systems $\tau_m(Z)$, for a given Z, begin with the different axioms $hq_1\bar{m}h$, but, for each $m \in P_Z$, culminate in the same theorem $hq'h$.

This suggests that we invert the systems $\tau_m(Z)$. That is, with each simple Turing machine Z, we associate a semi-Thue system $\sigma(Z)$ as follows:

Alphabet. The alphabet of $\sigma(Z)$ is that of the $\tau_m(Z)$.

Axiom. The axiom of $\sigma(Z)$ is $hq'h$.

Productions. The productions of $\sigma(Z)$ are the *inverses* of those of the $\tau_m(Z)$.

THEOREM 2.3. *$m \in P_Z$ if and only if $\vdash_{\sigma(Z)} hq_1\bar{m}h$.*

PROOF. $m \in P_Z$ if and only if $\vdash_{\tau_m(Z)} hq'h$, which, in turn, is true if and only if $\vdash_{\sigma(Z)} hq_1\bar{m}h$.

Now, let $\sigma'(Z)$ be the semi-Thue system whose alphabet is that of $\sigma(Z)$ with the additional symbol r, whose axiom is the axiom of $\sigma(Z)$ (namely, $hq'h$), and whose productions are those of $\sigma(Z)$ with the following in addition:

(6) $Phq_11Q \to P1rQ$.

(7) $Pr1Q \to P1rQ$.

(8) $P1rhQ \to P1Q$.

We shall show that $\sigma'(Z)$ generates the set P_Z, that is, that $S_{\sigma'(Z)} = P_Z$.

LEMMA 6. *If $\vdash_{\sigma'(Z)} U$, where U is in standard form and is free of occurrences of r, then $\vdash_{\sigma(Z)} U$.*

PROOF. Let $U = hWh$, where W is free of occurrences of h and r.

Let $B_1, B_2, \ldots, B_n$ be a proof in $\sigma'(Z)$, where $B_n = hWh$. We shall show that $B_1, B_2, \ldots, B_n$ is also a proof in $\sigma(Z)$. For, suppose it is not. Then, for some smallest $i = 1, 2, \ldots, n-1$, the transition from B_i to B_{i+1} must require a production of $\sigma'(Z)$ that is not a production of $\sigma(Z)$, that is, one of the productions (6) to (8), above.

Now, by Lemma 1,

$$B_i = hSh,$$

where S contains a single occurrence of some q-symbol and no occurrence of r. A glance at (6), (7), and (8) shows that (6) is the only production that could conceivably be applied to B_i, and then only if $B_i = hq_11Vh$. Then $B_{i+1} = 1rVh$. But, now, neither B_{i+1} nor its consequences can contain a q-symbol. Hence, only (7) and (8) can be applied to them. But this contradicts the fact that $B_n = hWh$, where W is free of r.

LEMMA 7. *$m \in P_Z$ if and only if $\vdash_{\sigma'(Z)} \bar{m}$.*

PROOF. First suppose that $m \in P_Z$. Then, by Theorem 2.3,

$$\vdash_{\sigma(Z)} hq_1\bar{m}h.$$

Hence,

$$\vdash_{\sigma'(Z)} hq_1\bar{m}h.$$

Therefore, by production (6) above,

$$\vdash_{\sigma'(Z)} 1r1^m h.$$

By repeated use of (2),

$$\vdash_{\sigma'(Z)} \bar{m}rh.$$

Finally, by (8),

$$\vdash_{\sigma'(Z)} \bar{m}.$$

Conversely, suppose that

$$\vdash_{\sigma'(Z)} \bar{m}.$$

Let $B_1, B_2, \ldots, B_n$ be a proof in $\sigma'(Z)$, where $B_n = \bar{m}$. Now, B_1 (being $hq'h$) is in standard form. Suppose that $B_1, \ldots, B_s$ are in standard form and that B_{s+1} is not. Then $B_s = hq_1 1Wh$. Thus, $B_{s+1} = 1rWh$. But this can lead to $\bar{m}$ only if $W = 1^m$. Hence, $B_s = hq_1\bar{m}h$. Thus,

$$\vdash_{\sigma'(Z)} hq_1\bar{m}h.$$

By Lemma 6,

$$\vdash_{\sigma(Z)} hq_1\bar{m}h.$$

By Theorem 2.3,

$$m \in P_Z.$$

THEOREM 2.4. *Every recursively enumerable set is generated by a semi-Thue system.*

PROOF. Let S be a recursively enumerable set. Then, by Corollary 2.1, there is a Turing machine Z such that $S = P_Z$. Hence, by Lemma 7, P_Z is generated by $\sigma'(Z)$ (cf. Definition 1.11) if we identify with a_0 the symbol $S_1 = 1$ of $\sigma'(Z)$.

THEOREM 2.5. *Every recursively enumerable set is generated by a semi-Thue system whose alphabet consists of two letters.*

PROOF. Immediate from Theorems 2.4 and 1.8.

THEOREM 2.6. *There exists a semi-Thue system whose alphabet consists of two letters and whose decision problem is recursively unsolvable.*

PROOF. Immediate from Theorems 1.6 and 2.5 and Corollary 5-4.8.

This result gives the recursive unsolvability of a problem which, intuitively, seems quite simple. Moreover, our proofs have shown how an actual semi-Thue system with a recursively unsolvable decision problem could be constructed. The nonrecursive set K is the domain of the partial recursive function $\varphi(x) = \min_y T(x, x, y)$. By the definition of partial recursive function and by the methods of proof of the theorems of Chap. 2, we can obtain a Turing machine Z_0 such that

$$\varphi(x) = \Psi_{Z_0}(x).$$

Next, the methods of the present section enable us to obtain, from the quadruples of Z_0, the productions of $\sigma'(Z_0)$, which, therefore, generates K. Finally, using the method of proof of Theorem 1.8, we obtain a semi-Thue system on two letters which generates K. This last semi-Thue system will have a recursively unsolvable decision problem.

3. Thue Systems. Next, we shall see how a Thue system with an unsolvable decision problem can be obtained.

With each simple Turing machine Z, we associate the Thue system $\rho(Z)$ as follows:

$\rho(Z)$ is obtained from $\sigma(Z)$ by adjoining to the productions of $\sigma(Z)$ the inverses of these productions. That is, the alphabet and axiom of $\rho(Z)$ are those of $\sigma(Z)$, whereas the productions of $\rho(Z)$ consist of those of $\sigma(Z)$ and those of the $\tau_m(Z)$.

THEOREM 3.1. *$\vdash_{\sigma(Z)} W$ if and only if $\vdash_{\rho(Z)} W$.*

PROOF.[1] If $\vdash_{\sigma(Z)} W$, then, clearly, $\vdash_{\rho(Z)} W$.

Conversely, suppose that $\vdash_{\rho(Z)} W$. Let $\mathfrak{R}_1, \ldots, \mathfrak{R}_t$ be the productions of $\sigma(Z)$, and let $\mathfrak{S}_1, \ldots, \mathfrak{S}_t$ be their respective inverses, i.e., the productions of the $\tau_m(Z)$. Let

$$hq'h = W_1, W_2, \ldots, W_p = W$$

be a proof of W in $\rho(Z)$, where no step is repeated. Then each W_j, $1 < j \leqq p$, is a consequence of W_{j-1} with respect to one of the $\mathfrak{R}$'s or $\mathfrak{S}$'s. If only the $\mathfrak{R}$'s were used, we would be through. Hence, we may assume that for some j, $1 < j \leqq p$, and for suitable k, $1 \leqq k \leqq t$, W_j is a consequence of W_{j-1} with respect to $\mathfrak{S}_k$ and that the steps $W_1, W_2, \ldots, W_{j-1}$ do not employ the $\mathfrak{S}$'s. Now $W_{j-1} \neq hq'h$, since none of the $\mathfrak{S}$'s is applicable to $hq'h$. Hence W_{j-1} must be a consequence of W_{j-2} with respect to, say, $\mathfrak{R}_l$. But then W_{j-2} is a consequence of W_{j-1} with respect to $\mathfrak{S}_l$. That is, W_{j-1} has two distinct consequences with respect to the $\mathfrak{S}$'s. Now, by Sec. 2, Lemma 1, W_{j-1} is in standard form, and, therefore, Lemma 3 of Sec. 2 yields a contradiction.

COROLLARY 3.2. *$m \in P_Z$ if and only if $\vdash_{\rho(Z)} hq_1\bar{m}h$.*

PROOF. Immediate from Theorems 2.3 and 3.1.

THEOREM 3.3. *There exists a Thue system ρ whose decision problem is recursively unsolvable.*

PROOF. Let Z_0 be such that P_{Z_0} is not recursive, for instance, $P_{Z_0} = K$. Let $\rho = \rho(Z_0)$. Then

$$\begin{aligned} K &= P_{Z_0} \\ &= \{x \mid \vdash_\rho hq_1\bar{x}h\} \\ &= \{x \mid \text{gn}\,(hq_1\bar{x}h) \in T_\rho\}. \end{aligned}$$

Now, gn $(hq_1\bar{x}h)$ = gn $(hq_1) * $ gn $(\bar{x}) *$ gn (h), which, as in the proof of Theorem 1.5, is recursive. Hence, since K is not recursive, T_ρ is not recursive.

Combining Theorem 3.3 with Theorem 1.7, we have the following:

THEOREM 3.4. *There exists a Thue system ρ whose alphabet consists of two letters and whose decision problem is recursively unsolvable.*

[1] This proof is due to Post [6].

4. The Word Problem for Semigroups. Here we shall see how the considerations of the previous section can be applied to a problem in algebra.

DEFINITION 4.1. *A* **semigroup** *is an arbitrary set of elements, taken together with a binary function f that has this set as the domain of each of its variables, whose range is a subset of this set, and for which the associative relation*

$$f(x, f(y, z)) = f(f(x, y), z)$$

holds for all x, y, z of the set.

We shall follow established mathematical usage and write $x \cdot y$ or xy for $f(x, y)$ wherever no confusion with arithmetic multiplication can arise.

One method for constructing semigroups is to begin with an alphabet. For, clearly, we have

THEOREM 4.1. *The set of all words on some fixed finite alphabet forms a semigroup with respect to the operation of juxtaposition.*

In this case the letters that make up the alphabet are called the *generators* of the semigroup, and, if the alphabet consists of n letters, we speak of the *free semigroup on n generators.*

Note that the empty word Λ serves as an identity; that is,

$$\Lambda W = W\Lambda = W,$$

for all words W.

Now, let $\{g, \bar{g}\}$ be a pair of words (not necessarily nonempty) on some alphabet. Then, the pair $\{g, \bar{g}\}$ is called a *relation* on the generators that make up the alphabet.

Let us consider the free semigroup on n generators with alphabet $a_1, \ldots, a_n$. Let $\{g_i, \bar{g_i}\}$, $i = 1, 2, \ldots, m$, be a finite set of relations on these generators. Then, if A and B are any words on this alphabet, we write $A \sim B$ if there exist words P, Q such that, for some i, $1 \leqq i \leqq m$, either $A = Pg_iQ$ and $B = P\bar{g_i}Q$, or $A = P\bar{g_i}Q$ and $B = Pg_iQ$. Furthermore, we write $A \approx B$ if there exists a sequence of words $A = A_1$, $A_2, \ldots, A_p = B$ such that $A_j \sim A_{j+1}$, $j = 1, 2, \ldots, p - 1$ (we do not exclude $p = 1$). We have

LEMMA 1. *If $A \sim B$, then $B \sim A$.*

LEMMA 2. *$A \approx A$. If $A \approx B$, then $B \approx A$. If $A \approx B$ and $B \approx C$, then $A \approx C$.*

DEFINITION 4.2. *Let $\{g_i, \bar{g_i}\}$, $i = 1, 2, \ldots, m$, be some finite set of relations on some fixed alphabet. Then, with each word A on this alphabet we associate the set $[A]$ of all words B such that $A \approx B$.*

LEMMA 3. *If $A \in [B]$, then $[A] = [B]$. Moreover $[A] = [B]$ if and only if $A \approx B$.*

PROOF. Since $A \in [B]$, $B \approx A$, and also $A \approx B$.

Now, let $C \in [A]$. Then, $A \approx C$, by Definition 4.2. Hence, by Lemma 2, $B \approx C$. Therefore, $C \in [B]$. We conclude that $[A] \subset [B]$.

Also, if $C \in [B]$, then $B \approx C$. Hence $A \approx C$. Therefore $C \in [A]$. We conclude that $[B] \subset [A]$.

Thus, $[A] = [B]$.

Next, suppose $A \approx B$. Then, $B \in [A]$; so, by what we have just proved, $[A] = [B]$.

Finally, suppose $[A] = [B]$. Now, $A \in [A]$. Hence $A \in [B]$. Hence, $A \approx B$.

LEMMA 4. *If $A \sim B$ and $C \sim D$, then $AC \approx BD$.*

PROOF. Let $A = Pg_iQ$, $B = P\bar{g}_iQ$, $C = P'g_jQ'$, $D = P'\bar{g}_jQ'$. Then, $AC \sim AD \sim BD$.

LEMMA 5. *If $A \approx B$ and $C \sim D$, then $AC \approx BD$.*

PROOF. Let $A = A_1 \sim A_2 \sim \cdots \sim A_p = B$. Then, by Lemma 4, $AC = A_1C \approx A_2D \sim A_3D \sim \cdots \sim A_pD = BD$; so $AC \approx BD$.

LEMMA 6. *If $A \approx B$ and $C \approx D$, then $AC \approx BD$.*

PROOF. Let $C = C_1 \sim C_2 \sim \cdots \sim c_q = D$. Then, by Lemma 5, $AC = AC_1 \approx BC_2 \sim BC_3 \sim \cdots \sim BC_q = BD$; so $AC \approx BD$.

LEMMA 7. *If $[A] = [B]$ and $[C] = [D]$, then $[AC] = [BD]$.*

PROOF. This is but a restatement of Lemma 6.

Lemma 7 enables us to define a multiplication operation on the classes $[A]$. If $\alpha = [A]$ and $\beta = [B]$, we write $\alpha\beta$ for the class $[AB]$. Lemma 7 assures us that a unique product is thus obtained.

LEMMA 8. $[A]([B][C]) = ([A][B])[C]$.

PROOF.
$$\begin{aligned} [A]([B][C]) &= [A]([BC]) \\ &= [A(BC)] \\ &= [(AB)C] \\ &= [AB][C] \\ &= ([A][B])[C]. \end{aligned}$$

THEOREM 4.2. *Let $\{g_i, \bar{g}_i\}$ be a set of relations on a given alphabet. Let S be the class of all sets $[A]$, where A is a word on the alphabet. Then, S forms a semigroup, where $[A][B]$ is defined as above.*

PROOF. Immediate from Lemma 8.

DEFINITION 4.3. *The semigroup S of Theorem 4.2 will be called the semigroup with the letters of the given alphabet as* **generators** *and with the $\{g_i, \bar{g}_i\}$ as* **relations.**

Suppose that a definite semigroup is defined for us by means of generators and relations. This by no means implies that we "know" the semigroup algebraically. For example, we may not even know whether it consists of a finite or an infinite number of elements. A decision problem which arises quite naturally in this connection is that of determining, of two given words, whether or not they belong to the same element of the semigroup. That is, one wishes to know, of two words A, B, whether or not $A \approx B$ with respect to a given set of relations. This problem is called the *word problem* for the semigroup in question. That is, if, by being given an alphabet and a finite set of relations on this alphabet, we are presented with a semigroup, the *word problem* for this semigroup is the problem:

To determine, of two words A, B on this alphabet, whether or not $A \approx B$.

If, by means of generators and relations, we are given an arbitrary semigroup, we note that $A \approx B$ is a predicate in the sense of the third paragraph of Sec. 1. Thus the following definition is suggested:

DEFINITION 4.4. *The word problem for a semigroup with a given finite set of generators and relations is* **recursively solvable** *if $A \approx B$ is recursive. Otherwise, it is* **recursively unsolvable.**

It can easily be verified that, in the case of a semigroup on one generator, $A \approx B$ is actually recursive. It is of some interest to examine the associated numerical predicate in the general case. Let the alphabet be $a_{i_1}, a_{i_2}, \ldots, a_{i_n}$. Let the relations be $\{g_i, \bar{g}_i\}$, $i = 1, 2, \ldots, m$. Then, the numerical predicate associated with "A is a word on this alphabet" is

$$W(x) \leftrightarrow \mathrm{GN}\,(x) \wedge \bigwedge_{k=1}^{\mathfrak{L}(x)} [(k \,\mathrm{Gl}\, x = 2i_1 + 1) \vee (k \,\mathrm{Gl}\, x = 2i_2 + 1) \vee \cdots \vee (k \,\mathrm{Gl}\, x = 2i_n + 1)].$$

The numerical predicate associated with $A \sim B$ is

$$\begin{aligned} N(x, y) \leftrightarrow W(x) \wedge W(y) \wedge \bigvee_{p=0}^{x} \bigvee_{q=0}^{x} &\{[(x = p * \mathrm{gn}\,(g_1) * q) \wedge (y = p * \mathrm{gn}\,(\bar{g}_1) * q)] \\ &\vee [(x = p * \mathrm{gn}\,(\bar{g}_1) * q) \wedge (y = p * \mathrm{gn}\,(g_1) * q)] \\ &\vee [(x = p * \mathrm{gn}\,(g_2) * q) \wedge (y = p * \mathrm{gn}\,(\bar{g}_2) * q)] \\ &\vee \cdots \\ &\vee [(x = p * \mathrm{gn}\,(g_m) * q) \wedge (y = p * \mathrm{gn}\,(\bar{g}_m) * q)] \\ &\vee [(x = p * \mathrm{gn}\,(\bar{g}_m) * q) \wedge (y = p * \mathrm{gn}\,(g_m) * q)]\}. \end{aligned}$$

Thus, $A \sim B$ is recursive.

Finally, the numerical predicate associated with $A \approx B$ is

$$\begin{aligned} E(x, y) \leftrightarrow W(x) \wedge W(y) \wedge \bigvee_{z} &\{[1 \,\mathrm{Gl}\, z = x] \\ &\wedge \bigwedge_{k=1}^{\mathfrak{L}(z) \dot{-} 1} [N(k \,\mathrm{Gl}\, z, (k + 1) \,\mathrm{Gl}\, z)] \wedge [\mathfrak{L}(z) \,\mathrm{Gl}\, z = y]\}. \end{aligned}$$

Thus, the numerical predicate associated with $A \approx B$ is *semicomputable.*

Now, let ρ be a Thue system. Let $Pg_iQ \to P\bar{g}_iQ$, $i = 1, 2, \ldots, m$, be the semi-Thue productions that, together with their inverses, make up the productions of ρ. Then, by $\Sigma(\rho)$ we shall understand the semigroup whose alphabet is that of ρ and relations are $\{g_i, \overline{g_i}\}$, $i = 1, 2, \ldots, m$. We have at once

THEOREM 4.3. *Let ρ be a Thue system, and let A_0 be its axiom. Then, $\vdash_\rho B$ if and only if $A_0 \approx B$ with respect to $\Sigma(\rho)$.*

THEOREM 4.4. *Let ρ be a Thue system, and let the word problem for $\Sigma(\rho)$ be recursively solvable. Then, the decision problem for ρ is likewise recursively solvable.*

PROOF. Let A_0 be the axiom of ρ, and let $a_0 = \text{gn}(A_0)$. Let $E(x, y)$ be the numerical predicate associated with $A \approx B$ with respect to $\Sigma(\rho)$. Then

$$T_\rho = \{x \mid E(a_0, x)\},$$

which proves the theorem.

As a result of Theorem 3.3, we now have

THEOREM 4.5. *There is a semigroup Σ, defined by a finite set of generators and relations, whose word problem is recursively unsolvable.*[1]

Moreover, employing Theorem 3.4, we have

THEOREM 4.6. *There exists a semigroup defined by a finite set of relations on two generators whose word problem is recursively unsolvable.*

It has been proved by Turing [4] that the word problem for cancellation semigroups is recursively unsolvable. Boone [1], Markov [1–3, 6–8], Addison, and others have proved the recursive unsolvability of related problems. The word problem for groups presents considerable difficulty. Novikoff [1] has shown that the word problem for groups is recursively unsolvable, using the result of Turing [4]. Boone [1] has, independently of Novikoff, given a direct proof, based on modified Turing machines, of this result.

5. Normal Systems and Post Systems. Let τ be a semi-Thue system whose alphabet (without loss of generality) is $a_1, \ldots, a_n$, whose axiom is A, and whose productions are

$$Pg_iQ \to P\overline{g_i}Q \qquad i = 1, 2, \ldots, m.$$

We construct a *normal system* $\nu(\tau)$ as follows:

Alphabet. $a_1, a_2, \ldots, a_n, a'_1, a'_2, \ldots, a'_n$.

Axiom. A, the axiom of τ.

We agree that, if $W = a_{i_1}a_{i_2} \cdots a_{i_l}$ is any word on τ, then

$$W' = a'_{i_1}a'_{i_2} \cdots a'_{i_l},$$

and that $\Lambda' = \Lambda$.

Productions. (1) $a_iP \to Pa'_i$, $i = 1, 2, \ldots, n$.

(2) $a'_iP \to Pa_i$, $i = 1, 2, \ldots, n$.

(3) $g_iP \to P\overline{g_i}'$, $i = 1, 2, \ldots, m$.

Two words on $\nu(\tau)$ will be called *associates* if they can be obtained from

[1] This result is due to Post [6] and to Markov [1, 3].

each other by using only productions of the forms (1) and (2) above. Thus, the associates of a word

$$a_{i_1}a_{i_2} \cdots a_{i_l}$$

on τ are the $2l$ words

$$\begin{array}{l} a_{i_2}a_{i_3} \cdots a_{i_l}a'_{i_1} \\ a_{i_3}a_{i_4} \cdots a_{i_l}a'_{i_1}a'_{i_2} \\ \cdots\cdots\cdots\cdots \\ a'_{i_1}a'_{i_2} \cdots a'_{i_l} \\ a'_{i_2}a'_{i_3} \cdots a'_{i_l}a_{i_1} \\ \cdots\cdots\cdots\cdots \\ a_{i_1}a_{i_2} \cdots a_{i_l}, \end{array}$$

each being a consequence of the preceding. In particular, note that every word is an associate of itself.

A word on $\nu(\tau)$ will be called *regular* if it can be written in one of the forms PQ' or $P'Q$, where P and Q are words on τ.

We shall show that a word on τ is a theorem of τ if and only if it is a theorem of $\nu(\tau)$. Our proof is presented by means of a series of lemmas, of which the first is obvious.

LEMMA 1. *If* $\vdash_{\nu(\tau)} W$ *and if* V *is an associate of* W, *then* $\vdash_{\nu(\tau)} V$.

LEMMA 2. *If* $\vdash_\tau W$, *then* $\vdash_{\nu(\tau)} W$.

PROOF. The axiom of τ, being also the axiom of $\nu(\tau)$, is certainly a theorem of $\nu(\tau)$. It remains only to show that the property of being a theorem of $\nu(\tau)$ is preserved by the productions of τ.

Thus, suppose it known that

$$\vdash_{\nu(\tau)} Pg_iQ.$$

Then we shall show that

$$\vdash_{\nu(\tau)} P\bar{g}_iQ.$$

For, by Lemma 1,

$$\vdash_{\nu(\tau)} g_iQP'.$$

By applying the appropriate production from (3),

$$\vdash_{\nu(\tau)} QP'\bar{g}_i'.$$

By Lemma 1 again,

$$\vdash_{\nu(\tau)} P\bar{g}_iQ.$$

LEMMA 3. *Every theorem of* $\nu(\tau)$ *is regular and is an associate of a theorem of* τ.

PROOF. Clearly, the axiom of $\nu(\tau)$ is regular and is an associate of a theorem of τ, namely itself. It remains only to show that this property (i.e., the property of being both regular and an associate of a theorem of τ) is preserved by the productions of $\nu(\tau)$.

It is obvious that this property is preserved by the productions included under (1) and (2) above. To see that it is likewise preserved under the productions of (3), suppose that g_iP is regular and has an associate which is a theorem of τ. Since $g_i \neq \Lambda$, the regularity of g_iP implies that we may write $P = P_1P_2'$, where P_1 and P_2 are words on τ. Then, clearly, $g_iP = g_iP_1P_2'$. But the one associate of $g_iP_1P_2'$ that is a word on τ is $P_2g_iP_1$, which must, by our induction hypothesis, be a theorem of τ. Hence, $P_2\overline{g_i}P_1$ is also a theorem of τ. But $P_2\overline{g_i}P_1$ is an associate of $P_1P_2'\overline{g_i}'$, which, moreover, is regular. This completes the proof.

THEOREM 5.1. *Let τ be a semi-Thue system whose alphabet is a_1, a_2, . . . , a_n. Then there is a normal system $\nu(\tau)$ whose alphabet is a_1, a_2, . . . , a_n, a_1', a_2', . . . , a_n', such that the theorems of τ are precisely the theorems of $\nu(\tau)$ that consist entirely of unprimed letters.*

PROOF. We take $\nu(\tau)$ as above. That each theorem of τ is also a theorem of $\nu(\tau)$ is the content of Lemma 2. Now, let W be a theorem of $\nu(\tau)$ free of primed letters. By Lemma 3, W has an associate that is a theorem of τ. But, the one associate of W that is a word on τ is W itself. Hence, $\vdash_\tau W$.

THEOREM 5.2. *Every recursively enumerable set is generated by a normal system.*

PROOF. Immediate from Theorems 2.4 and 5.1.

THEOREM 5.3. *Every recursively enumerable set is generated by a normal system whose alphabet consists of two letters.*[1]

PROOF. Immediate from Theorems 1.9 and 5.2.

THEOREM 5.4. *There exists a normal system whose alphabet consists of two letters and whose decision problem is recursively unsolvable.*

PROOF. Immediate from Theorems 5.3 and 1.6 and Corollary 5-4.8.

Just as we associated a normal system $\nu(\tau)$ with each semi-Thue system τ, we now associate a Post system $\pi(\rho)$ with each Thue system ρ. Let the alphabet of ρ be a_1, . . . , a_n, let its axiom be A, and let its productions be

$$Pg_iQ \leftrightarrow P\overline{g_i}Q \qquad i = 1, 2, \ldots, m,$$

where the double arrow, in an obvious manner, indicates a production and its inverse. We define $\pi(\rho)$ as follows:

Alphabet. $a_1, \ldots, a_n, a_1', \ldots, a_n'$.

Axiom. A, the axiom of ρ.

Productions. (1) $a_iP \leftrightarrow Pa_i'$, $i = 1, 2, \ldots, n$.

(2) $a_i'P \leftrightarrow Pa_i$, $i = 1, 2, \ldots, n$.

(3) $g_iP \leftrightarrow P\overline{g_i}'$, $i = 1, 2, \ldots, m$.

Then we can parallel the argument leading to Theorem 5.1, to prove

[1] This result is due to Post.

THEOREM 5.5. *Let* ρ *be a Thue system whose alphabet is* $a_1, \ldots, a_n$. *Then, there is a Post system* $\pi(\rho)$ *whose alphabet is* $a_1, a_2, \ldots, a_n, a'_1, a'_2, \ldots, a'_n$, *such that the theorems of* ρ *are precisely the theorems of* $\pi(\rho)$ *that consist entirely of unprimed letters.*

Hence,

$$T_\rho = \{x \mid \text{Unprimed}\ (x) \wedge x \in T_{\pi(\rho)}\},$$

where, taking $a'_i = a_{n+i}$,

$$\text{Unprimed}\ (x) \leftrightarrow \bigwedge_{i=1}^{\mathfrak{L}(x)} (i \text{ Gl } x \leqq 2n + 1).$$

Combining this with Theorem 3.3, we have

THEOREM 5.6. *There exists a Post system whose decision problem is recursively unsolvable.*

CHAPTER 7

DIOPHANTINE EQUATIONS

1. Hilbert's Tenth Problem. In his famous address[1] of 1900, Hilbert listed a group of unsolved problems, which were to stand as a challenge to future generations of mathematicians. Included among these was only one *decision problem*, the tenth. This problem is as follows:

To determine, of an arbitrary polynomial equation $P = 0$, with integral (positive, negative, or zero) coefficients, whether or not it has a solution in integers (positive, negative, or zero).

In 1900, such a problem could be imagined as being settled only by actually providing an algorithm for solving it. Hilbert seems to indicate that a belief in the solvability of such problems is an article of faith of the working mathematician. Of course, with our present orientation, it seems quite reasonable to attempt to prove that Hilbert's tenth problem is *unsolvable.* The fact that apparently insurmountable difficulties seem to prevent the development of a general theory of diophantine equations[2] makes it seem likely that Hilbert's tenth problem will indeed prove to be unsolvable.

In this chapter we shall obtain results which yield the recursive unsolvability of a related problem. Moreover, our results will be such that any substantial improvement in their direction will yield the unsolvability of Hilbert's tenth problem. Also, we shall obtain yet one more formulation of the concept of recursive enumerability.

Hilbert's statement of the tenth problem was for solutions in integers, positive, negative, or zero; however, we can easily show that the corresponding problem for nonnegative integral solutions is equivalent to the original problem. For suppose we could solve the problem for nonnegative integral solutions. Then, to determine whether or not $P(x_1, \ldots, x_n) = 0$ has a solution in integers, positive, negative, or zero, it clearly suffices to test for nonnegative integral solutions each of the 2^n equations

[1] Cf. Hilbert [1].

[2] For our purposes, a *diophantine equation* is a polynomial equation $P = 0$ of which only integral solutions are being sought.

$$P(x_1, x_2, \ldots, x_n) = 0, P(-x_1, x_2, \ldots, x_n) = 0, \ldots, P(-x_1, -x_2, \ldots, -x_n) = 0.$$

Conversely, suppose that we could solve the problem for integers, positive, negative, or zero. Then, to determine whether or not $P(x_1, x_2, \ldots, x_n) = 0$ has solutions in nonnegative integers, it suffices to determine whether or not

$$P(p_1^2 + q_1^2 + r_1^2 + s_1^2, p_2^2 + q_2^2 + r_2^2 + s_2^2, \ldots, p_n^2 + q_n^2 + r_n^2 + s_n^2) = 0$$

has solutions in integers. This last is true because of Lagrange's theorem,[1] which states that every nonnegative integer is the sum of four squares.

2. Arithmetical and Diophantine Predicates. By a polynomial, we shall understand a function

$$\sum_{\substack{0 \leqq i_1 \leqq n_1 \\ 0 \leqq i_2 \leqq n_2 \\ \cdots \\ 0 \leqq i_k \leqq n_k}} a_{i_1 i_2 \cdots i_k} x_1^{i_1} x_2^{i_2} \cdots x_k^{i_k}$$

where the numbers $a_{i_1 i_2 \cdots i_k}$ are *integers, positive, negative,* or *zero.* However, the range of the variables $x_1, x_2, \ldots, x_k$ is to be taken as the set of *nonnegative integers.* We continue to use the unmodified word "integer" to mean "nonnegative integer."

DEFINITION 2.1. *Let $P(\mathfrak{x}^{(k)})$ be a polynomial. Then the predicate $P(\mathfrak{x}^{(k)}) = 0$ is called a* **polynomial predicate.**

Examples of polynomial predicates are

$$x + y - 5 = 0, \quad x^3 + y^3 - z^3 = 0, \quad \text{etc.}$$

DEFINITION 2.2. *A predicate $R(\mathfrak{x}^{(n)})$ is* **diophantine**[2] *if there exists a polynomial predicate $S(\mathfrak{x}^{(n)}, \mathfrak{y}^{(m)})$ such that*

$$R(\mathfrak{x}^{(n)}) \leftrightarrow \bigvee_{\mathfrak{y}^{(m)}} S(\mathfrak{x}^{(n)}, \mathfrak{y}^{(m)}).$$

DEFINITION 2.3. *A predicate $R(\mathfrak{x}^{(n)})$ is called* **arithmetical** *if there exists a polynomial predicate $S(\mathfrak{x}^{(n)}, \mathfrak{y}^{(m)})$ such that*

$$R(\mathfrak{x}^{(n)}) \leftrightarrow [M] S(\mathfrak{x}^{(n)}, \mathfrak{y}^{(m)}),$$

where $[M]$ is some sequence of existential and universal quantifiers on $\mathfrak{y}^{(m)}$.

COROLLARY 2.1. *Every polynomial predicate is diophantine; every diophantine predicate is arithmetical.*

COROLLARY 2.2. *Every polynomial predicate is primitive recursive.*

COROLLARY 2.3. *Every diophantine predicate is semicomputable.*

[1] Cf. Hardy and Wright [1, pp. 300, 301].

[2] J. Robinson [2] uses the term *existentially definable.*

PROOF. The result follows at once from Corollary 2.2 and Theorem 5-1.2.

If the converse of Corollary 2.3 were true, we could infer the existence of a diophantine predicate that was not recursive. It is easy to see that this would imply the recursive unsolvability of Hilbert's tenth problem. For, proceeding informally, suppose it known that the predicate

$$\bigvee_{\mathfrak{x}^{(n)}} [P(x, \mathfrak{x}^{(n)}) = 0],$$

where P is a polynomial, is not recursive. Then the problem of determining, for given x_0, whether or not the equation

$$P(x_0, \mathfrak{x}^{(n)}) = 0$$

has a solution in nonnegative integers $\mathfrak{x}^{(n)}$ would be unsolvable. But this would certainly imply (and an arithmetization of the theory of diophantine equations would yield a formal proof of) the recursive unsolvability of the general problem:

To determine, of a given diophantine equation, whether or not it has a solution.

Unfortunately, we shall have to leave open the question of whether or not there exists a diophantine predicate that is not recursive.

COROLLARY 2.4. *If $R(y, \mathfrak{x}^{(n)})$ is a diophantine predicate, then so is* $\bigvee_{y} R(y, \mathfrak{x}^{(n)})$.

THEOREM 2.5. *If R and S are polynomial predicates, so are $R \vee S$ and $R \wedge S$.*

PROOF. $P_1 = 0 \vee P_2 = 0 \leftrightarrow P_1P_2 = 0,$
$P_1 = 0 \wedge P_2 = 0 \leftrightarrow P_1^2 + P_2^2 = 0.$

THEOREM 2.6. *If $R(x, \mathfrak{x}^{(n)})$ is a polynomial predicate and if*

$$S(y, \mathfrak{x}^{(n)}) \leftrightarrow \bigwedge_{x=0}^{y} R(x, \mathfrak{x}^{(n)}),$$

then $S(y, \mathfrak{x}^{(n)})$ is also a polynomial predicate.

PROOF. Let

$$R(x, \mathfrak{x}^{(n)}) \leftrightarrow P(x, \mathfrak{x}^{(n)}) = 0,$$

where P is a polynomial. Then

$$S(y, \mathfrak{x}^{(n)}) \leftrightarrow \bigwedge_{x=0}^{y} [P(x, \mathfrak{x}^{(n)}) = 0]$$

$$\leftrightarrow \sum_{x=0}^{y} P(x, \mathfrak{x}^{(n)})^2 = 0.$$

Let

$$P(x, \mathfrak{x}^{(n)})^2 = \sum_{k=0}^{r} P_k(\mathfrak{x}^{(n)})x^k.$$

Then

$$S(y, \mathfrak{x}^{(n)}) \leftrightarrow \sum_{x=0}^{y} \sum_{k=0}^{r} P_k(\mathfrak{x}^{(n)})x^k = 0$$

$$\leftrightarrow \sum_{k=0}^{r} \left[P_k(\mathfrak{x}^{(n)}) \sum_{x=0}^{y} x^k\right] = 0.$$

But, by a classical theorem of Bernoulli, $\sum_{x=0}^{y} x^k$ is a polynomial in y of degree $k + 1$ with rational coefficients.[1] That is,

$$\sum_{x=0}^{y} x^k = \sum_{j=0}^{k+1} A_{j,k}y^j,$$

where $A_{j,k}$ are rational numbers. Then

$$S(y, \mathfrak{x}^{(n)}) \leftrightarrow \sum_{k=0}^{r} \sum_{j=0}^{k+1} A_{j,k}P_k(\mathfrak{x}^{(n)})y^j = 0.$$

Let D be the least common multiple of the denominators of the numbers $A_{j,k}$, $0 \leqq k \leqq r$, $0 \leqq j \leqq r + 1$. Then $DA_{j,k}$ is an integer for each j,k, and

$$S(y, \mathfrak{x}^{(n)}) \leftrightarrow \sum_{k=0}^{r} \sum_{j=0}^{k+1} DA_{j,k}P_k(\mathfrak{x}^{(n)})y^j = 0.$$

[1] The author is indebted to Professor Lowell Schoenfeld for calling to his attention the following proof of this theorem:

The result for $k = 0$ is obvious. Suppose it known for $k \leqq l$. We note the identity

$$(x + 1)^{l+2} - x^{l+2} = (l + 1)x^{l+1} + \sum_{j=0}^{l} a_j x^j$$

for suitable integers a_j. Hence,

$$x^{l+1} = \frac{1}{l + 1}[(x + 1)^{l+2} - x^{l+2}] + \sum_{j=0}^{l} b_j x^j$$

for suitable rational numbers b_j. Thus,

$$\sum_{x=0}^{y} x^{l+1} = \frac{1}{l + 1}(y + 1)^{l+2} + \sum_{j=0}^{l} \left(b_j \sum_{x=0}^{y} x^j\right),$$

from which the result follows, by induction hypothesis.

THEOREM 2.7. *If $R(x, \mathfrak{x}^{(n)})$ is a polynomial predicate, then so also is* $\bigwedge_x R(x, \mathfrak{x}^{(n)})$.

PROOF. Let

$$R(x, \mathfrak{x}^{(n)}) \leftrightarrow \sum_{k=0}^{r} P_k(\mathfrak{x}^{(n)})x^k = 0.$$

Then

$$\bigwedge_x R(x, \mathfrak{x}^{(n)}) \leftrightarrow \bigwedge_x \Big[\sum_{k=0}^{r} P_k(\mathfrak{x}^{(n)})x^k = 0 \Big]$$

$$\leftrightarrow \bigwedge_{k=0}^{r} [P_k(\mathfrak{x}^{(n)}) = 0]$$

$$\leftrightarrow \sum_{k=0}^{r} [P_k(\mathfrak{x}^{(n)})]^2 = 0.$$

THEOREM 2.8. *If R and S are diophantine predicates, so are $R \vee S$ and $R \wedge S$.*

PROOF

$$\bigvee_{\mathfrak{y}^{(m)}} Q_1 \vee \bigvee_{\mathfrak{z}^{(n)}} Q_2 \leftrightarrow \bigvee_{\mathfrak{y}^{(m)}} \bigvee_{\mathfrak{z}^{(n)}} (Q_1 \vee Q_2).$$

$$\bigvee_{\mathfrak{y}^{(m)}} Q_1 \wedge \bigvee_{\mathfrak{z}^{(n)}} Q_2 \leftrightarrow \bigvee_{\mathfrak{y}^{(m)}} \bigvee_{\mathfrak{z}^{(n)}} (Q_1 \wedge Q_2).$$

The result then follows from Theorem 2.5.

Thus we have seen that the class of diophantine predicates is closed under the operations of conjunction ($\wedge$), alternation ($\vee$), and existential quantification. Later, we shall see that it is *not* closed under negation ($\sim$) and also *not* closed under universal quantification. Whether or not this class is closed under *bounded* universal quantification remains an open question. It will be clear from the results of the following section that an affirmative answer to this question would have as a consequence that every semicomputable predicate is diophantine and, hence, that Hilbert's tenth problem is unsolvable.

We proceed to make a short list of diophantine predicates:

(1) $x \neq 0$.
For

$$x \neq 0 \leftrightarrow \bigvee_y (x - y - 1 = 0).$$

(2) $x < y$.
For

$$x < y \leftrightarrow \bigvee_z [(z \neq 0) \wedge (y - x - z = 0)].$$

but rather in their being *diophantine* (or at least almost diophantine). We find it necessary to represent words by *ordered triples* of integers rather than by the more usual integers themselves.

Let all words on the symbols 1, b be arrayed as follows:

$$\begin{array}{cccccccc} 1 & b \\ 11 & 1b & b1 & bb \\ 111 & 11b & 1b1 & 1bb & b11 & b1b & bb1 & bbb \\ \cdots & \cdots & \cdots & \cdots & \cdots & \cdots & \cdots & \cdots \end{array}$$

Here the nth row contains all words on 1 and b, of length n, arrayed in alphabetical order, with 1 preceding b.

DEFINITION 3.1. *Let W be a word on the letters* 1, b. *Then $L(W)$ is the number of the row in which W appears,* i.e., *the* **length of** W, *and $P(W)$ is the* **position of** W *in this row.* [*We take* $L(\Lambda) = 0$, $P(\Lambda) = 1$.]

By the triple associated with W we mean the ordered triple $(L(W), P(W), 2^{L(W)})$.

Thus, $L(1bb) = 3$ and $P(1bb) = 4$; so the triple associated with $1bb$ is (3, 4, 8). The triple associated with Λ is (0, 1, 1).

LEMMA 1. *(x, y, z) is the triple associated with some word on* 1 *and b if and only if $z = 2^x$ and $0 < y \leqq z$.*

PROOF. This follows from the fact that the xth row of our array consists of 2^x words.

Note that each triple (x, y, z) is associated with at most one word.

The following lemma expresses relations which are evident from our ery definitions of L and P.

LEMMA 2. $L(WW') = L(W) + L(W')$. $P(1W) = P(W)$; $P(bW) =$ $W) + 2^{L(W)}$.

EMMA 3. $P(WW') = 2^{L(W')}(P(W) - 1) + P(W')$.

ROOF. We prove this result by induction on $L(W)$. If $L(W) = 0$, W is the empty word. Hence $P(W) = 1$. In this case $WW' = W'$ hence, $P(WW') = P(W') = 2^{L(W')}(P(W) - 1) + P(W')$.

w, let the result be known for all words whose length is $L(W) - 1$. ow that it must then also hold for W. For we must have either W_1 or $W = bW_1$, where, in either case, $L(W_1) = L(W) - 1$.

suppose that $W = 1W_1$. Then, using Lemma 2 and our induc- othesis, we have

$$\begin{aligned} P(WW') &= P(1W_1W') \\ &= P(W_1W') \\ &= 2^{L(W')}(P(W) - 1) + P(W'). \end{aligned}$$

suppose that $W = bW_1$. Again, using Lemma 2 and our ypothesis, we have

(3) $x \neq y$.
For

$$x \neq y \leftrightarrow (x < y) \vee (y < x).$$

(4) $x \equiv y (\mathrm{mod}\ z)$.†
For

$$x \equiv y(\mathrm{mod}\ z) \leftrightarrow \bigvee_{w} [(x - y - zw = 0) \vee (x - y + zw = 0)].$$

(5) $z = J(x, y)$.
For

$$z = J(x, y) \leftrightarrow 2z - x^2 - y^2 - 2xy - 3x - y = 0.$$

(6) $z = T_i(w)$.
For

$$z = T_i(w) \leftrightarrow \bigvee_{a,b,c} \{[z < b] \wedge [z \equiv a(\mathrm{mod}\ b)] \wedge [b = 1 + (i + 1)c] \wedge [w = J(a, c)$$

(7) $z = T_{i \dot{-} 1}(w)$.
For

$$z = T_{i \dot{-} 1}(w) \leftrightarrow \bigvee_{j} \{z = T_j(w) \wedge [(j = 0 \wedge i = 0) \vee (j + 1$$

(8) NPT (x). *x is not a power of* 2.
For

$$\mathrm{NPT}\ (x) \leftrightarrow \bigvee_{z,w} [x = (2z + 3)w].$$

(9) $\sim$ Prime (x). *x is not a prime number.*

$$\sim \mathrm{Prime}\ (x) \leftrightarrow \bigvee_{y,z} [x = (y + 2)(z + 2)].$$

Later, we shall see that there exists a diophantine
negation is not diophantine. Predicates (8) and (9)
special interest, since it is not known whether or not
a power of 2" and "x is a prime number" are diop

3. Arithmetical Representation of Semicompu
this section, we shall show how to represent sem
in terms of diophantine predicates. We begin w
6-5.3) that every recursively enumerable set can
system on the alphabet 1, b. The extremely
systems suggests that we attempt to accom
of an arithmetization of the theory of nor
principal interest will be not in the *recu*

† Cf. Appendix, Definition 4.

[1] However, Julia Robinson [2] prove that if
are, in a suitable sense, of exponenti order
diophantine. She has also shown th , if z

(3) $x \neq y$.
For

$$x \neq y \leftrightarrow (x < y) \vee (y < x).$$

(4) $x \equiv y (\text{mod } z)$.†
For

$$x \equiv y(\text{mod } z) \leftrightarrow \bigvee_{w} [(x - y - zw = 0) \vee (x - y + zw = 0)].$$

(5) $z = J(x, y)$.
For

$$z = J(x, y) \leftrightarrow 2z - x^2 - y^2 - 2xy - 3x - y = 0.$$

(6) $z = T_i(w)$.
For

$$z = T_i(w) \leftrightarrow \bigvee_{a,b,c} \{[z < b] \wedge [z \equiv a(\text{mod } b)] \wedge [b = 1 + (i + 1)c] \wedge [w = J(a, c)]\}.$$

(7) $z = T_{i \dot{-} 1}(w)$.
For

$$z = T_{i \dot{-} 1}(w) \leftrightarrow \bigvee_{j} \{z = T_j(w) \wedge [(j = 0 \wedge i = 0) \vee (j + 1 = i)]\}.$$

(8) NPT (x). *x is not a power of* 2.
For

$$\text{NPT } (x) \leftrightarrow \bigvee_{z,w} [x = (2z + 3)w].$$

(9) $\sim$ Prime (x). *x is not a prime number.*

$$\sim \text{Prime } (x) \leftrightarrow \bigvee_{y,z} [x = (y + 2)(z + 2)].$$

Later, we shall see that there exists a diophantine predicate whose negation is not diophantine. Predicates (8) and (9) are, therefore, of special interest, since it is not known whether or not the predicates "x is a power of 2" and "x is a prime number" are diophantine.[1]

3. Arithmetical Representation of Semicomputable Predicates. In this section, we shall show how to represent semicomputable predicates in terms of diophantine predicates. We begin with the fact (cf. Theorem 6-5.3) that every recursively enumerable set can be generated by a normal system on the alphabet 1, b. The extremely simple structure of normal systems suggests that we attempt to accomplish our purpose by means of an arithmetization of the theory of normal systems. Of course, our principal interest will be not in the *recursiveness* of various predicates

† Cf. Appendix, Definition 4.

[1] However, Julia Robinson [2] prove that if there exist diophantine predicates that are, in a suitable sense, of exponenti order of growth, then the predicate $z = x^y$ is diophantine. She has also shown th , if $z = x^y$ is diophantine, so is Prime (x).

but rather in their being *diophantine* (or at least almost diophantine). We find it necessary to represent words by *ordered triples* of integers rather than by the more usual integers themselves.

Let all words on the symbols 1, b be arrayed as follows:

$$\begin{array}{llllllll} 1 & b \\ 11 & 1b & b1 & bb \\ 111 & 11b & 1b1 & 1bb & b11 & b1b & bb1 & bbb \\ \cdots & \cdots & \cdots & \cdots & \cdots & \cdots & \cdots & \cdots \end{array}$$

Here the nth row contains all words on 1 and b, of length n, arrayed in alphabetical order, with 1 preceding b.

DEFINITION 3.1. *Let W be a word on the letters* 1, b. *Then $L(W)$ is the number of the row in which W appears,* i.e., *the* **length of** W, *and $P(W)$ is the* **position of** W *in this row.* [*We take* $L(\Lambda) = 0$, $P(\Lambda) = 1$.]

By the triple associated with W we mean the ordered triple $(L(W), P(W), 2^{L(W)})$.

Thus, $L(1bb) = 3$ and $P(1bb) = 4$; so the triple associated with $1bb$ is (3, 4, 8). The triple associated with Λ is (0, 1, 1).

LEMMA 1. *(x, y, z) is the triple associated with some word on* 1 *and b if and only if $z = 2^x$ and $0 < y \leqq z$.*

PROOF. This follows from the fact that the xth row of our array consists of 2^x words.

Note that each triple (x, y, z) is associated with at most one word.

The following lemma expresses relations which are evident from our very definitions of L and P.

LEMMA 2. $L(WW') = L(W) + L(W')$. $P(1W) = P(W)$; $P(bW) = P(W) + 2^{L(W)}$.

LEMMA 3. $P(WW') = 2^{L(W')}(P(W) - 1) + P(W')$.

PROOF. We prove this result by induction on $L(W)$. If $L(W) = 0$, then W is the empty word. Hence $P(W) = 1$. In this case $WW' = W'$ and, hence, $P(WW') = P(W') = 2^{L(W')}(P(W) - 1) + P(W')$.

Now, let the result be known for all words whose length is $L(W) - 1$. We show that it must then also hold for W. For we must have either $W = 1W_1$ or $W = bW_1$, where, in either case, $L(W_1) = L(W) - 1$.

First suppose that $W = 1W_1$. Then, using Lemma 2 and our induction hypothesis, we have

$$\begin{aligned} P(WW') &= P(1W_1W') \\ &= P(W_1W') \\ &= 2^{L(W')}(P(W) - 1) + P(W'). \end{aligned}$$

Finally, suppose that $W = bW_1$. Again, using Lemma 2 and our induction hypothesis, we have

$$\begin{aligned} P(WW') &= P(W_1W') + 2^{L(W_1W')} \\ &= 2^{L(W')}(P(W_1) - 1) + P(W') + 2^{L(W_1W')} \\ &= 2^{L(W')}[P(W_1) + 2^{L(W_1)} - 1] + P(W') \\ &= 2^{L(W')}(P(W) - 1) + P(W'). \end{aligned}$$

LEMMA 4. *If* (x, y, z), (x', y', z') *are the triples associated with the words* W *and* W', *respectively, then the triple associated with* WW' *is* $(x + x', z'(y - 1) + y', zz')$.

PROOF.
$$\begin{aligned} L(WW') &= L(W) + L(W') \\ &= x + x'. \\ P(WW') &= 2^{L(W')}(P(W) - 1) + P(W') \\ &= z'(y - 1) + y'. \\ 2^{L(WW')} &= 2^{L(W)+L(W')} \\ &= 2^{L(W)} \cdot 2^{L(W')} \\ &= zz'. \end{aligned}$$

Now, let us consider some definite normal system ν whose alphabet consists of the letters 1, b. Let the axiom of ν have associated with it the triple (p, q, r). Let the productions of ν be $g_iP \to P\overline{g_i}$, $i = 1, 2, \ldots, n$. Let the triples associated with g_i and $\overline{g_i}$ be (a_i, b_i, c_i) and $(\overline{a_i}, \overline{b_i}, \overline{c_i})$, respectively, $i = 1, 2, \ldots, n$.

Now, consider the predicates

$$\begin{aligned} \text{Prod}_i\,(x, y, z; x', y', z') \leftrightarrow \bigvee_u \bigvee_v \bigvee_w \{&[0 < v \leqq w] \\ &\wedge [(x = a_i + u) \wedge (y = w(b_i - 1) + v) \wedge (z = c_iw)] \\ &\wedge [(x' = u + \overline{a_i}) \wedge (y' = \overline{c_i}(v - 1) + \overline{b_i}) \wedge (z' = w\overline{c_i})]\}, \end{aligned}$$

$i = 1, 2, \ldots, n$. In the first place, it is clear that these predicates are diophantine. Furthermore, suppose that $\text{Prod}_i\,(x, y, z; x', y', z')$ actually holds for six definite numbers $x, y, z; x', y', z'$, and suppose that (x, y, z) is the triple associated with some word X. Then there exist numbers u, v, w satisfying the conditions set forth above. In particular, $0 < v \leqq w$, and $w = z/c_i = 2^x/2^{a_i} = 2^{x-a_i} = 2^u$. Hence, by Lemma 1, the triple (u, v, w) is associated with some word P. Then, by using Lemma 4, we see that $X = g_iP$ and that (x', y', z') is the triple associated with $P\overline{g_i}$. Thus, we see that $\text{Prod}_i\,(x, y, z; x', y', z')$ holds for given $(x, y, z; x', y', z')$, where (x, y, z) is associated with a word X, if and only if (x', y', z') *is the triple associated with a word* Y *that is a consequence of* X *with respect to the ith production of* ν.

Next, let

$$\begin{aligned} \text{Prod}\,(x, y, z; x', y', z') \leftrightarrow\ & \text{Prod}_1\,(x, y, z; x', y', z') \\ & \vee \text{Prod}_2\,(x, y, z; x', y', z') \\ & \vee \cdots \\ & \vee \text{Prod}_n\,(x, y, z; x', y', z'). \end{aligned}$$

Then, we have

LEMMA 5. Prod $(x, y, z; x', y', z')$ *is a diophantine predicate. Moreover, if the triple* (x_0, y_0, z_0) *is associated with some word* X_0, *then* Prod $(x_0, y_0, z_0; x'_0, y'_0, z'_0)$ *is true if and only if* (x'_0, y'_0, z'_0) *is the triple associated with some word* Y_0 *which is a consequence of* X_0 *in* ν.

Let $\mathrm{Th}_\nu\,(x, y, z)$ be true for precisely those triples (x, y, z) associated with some theorem of ν. Then, recalling Theorem 3-2.4,

$$\mathrm{Th}_\nu\,(x, y, z) \leftrightarrow \bigvee_{w,w',w'',l} \{(T_0(w) = p) \wedge (T_0(w') = q) \wedge (T_0(w'') = r)$$

$$\wedge \bigwedge_{k=0}^{l} [\mathrm{Prod}\,(T_{k \dot- 1}(w), T_{k \dot- 1}(w'), T_{k \dot- 1}(w''); T_k(w), T_k(w'), T_k(w''))$$

$$\vee\ (k = 0)] \wedge T_l(w) = x \wedge T_l(w') = y \wedge T_l(w'') = z\}.$$

Now, by (6) and (7) of Sec. 2, $t = T_k(w)$ and $t = T_{k \dot- 1}(w)$ are diophantine. Next, we assert that the predicate

$$\mathrm{Prod}\,(T_{k \dot- 1}(w), T_{k \dot- 1}(w'), T_{k \dot- 1}(w''); T_k(w), T_k(w'), T_k(w''))$$

is diophantine. This follows from Lemma 5 and from the fact that this predicate can be written

$$\bigvee_{t,u,v,t',u',v'} [\mathrm{Prod}\,(t, u, v; t', u', v') \wedge t = T_{k \dot- 1}(w) \wedge u = T_{k \dot- 1}(w')$$

$$\wedge\ v = T_{k \dot- 1}(w'') \wedge t' = T_k(w) \wedge u' = T_k(w') \wedge v' = T_k(w'')].$$

Hence, there is a diophantine predicate $D_1(x, y, z, k, l, w, w', w'')$ such that

$$\mathrm{Th}_\nu\,(x, y, z) \leftrightarrow \bigvee_{w,w',w'',l}\ \bigwedge_{k=0}^{l} D_1(x, y, z, k, l, w, w', w'').$$

Now, recalling Definition 6-1.11,

$$\begin{aligned} n \in S_\nu &\leftrightarrow\ \vdash_\nu 1^{n+1} \\ &\leftrightarrow \mathrm{Th}_\nu\,(n + 1, 1, 2^{n+1}) \\ &\leftrightarrow \bigvee_z \mathrm{Th}_\nu\,(n + 1, 1, z) \\ &\leftrightarrow \bigvee_{z,w,w',w'',l}\ \bigwedge_{k=0}^{l} D_1(n + 1, 1, z, k, l, w, w', w''). \end{aligned}$$

Finally, by Theorem 6-5.3, we have

LEMMA 6. *Let* S *be a recursively enumerable set. Then there exists a diophantine predicate* $D_2(n, z, k, l, w, w', w'')$ *such that*

$$S = \left\{n \mid \bigvee_{z,w,w',w'',l}\ \bigwedge_{k=0}^{l} D_2(n, z, k, l, w, w', w'')\right\}.$$

We next proceed to strip our expression of all but one of its initial existential quantifiers. Thus, for example, we may begin:

$$\bigvee_{z,w,w',w'',l} \bigwedge_{k=0}^{l} D_2(n, z, k, l, w, w', w'')$$
$$\leftrightarrow \bigvee_{u,w',w'',l} \bigwedge_{k=0}^{l} D_2(n, K(u), k, l, L(u), w', w'')$$
$$\leftrightarrow \bigvee_{u,w',w'',l} \bigwedge_{k=0}^{l} \bigvee_{z,w} [D_2(n, z, k, l, w, w', w'') \wedge u = J(z, w)],$$

where the expression in square brackets is diophantine, by (5) of Sec. 2 and Theorem 2.8. Continuing this process, we obtain

$$\bigvee_{v,l} \bigwedge_{k=0}^{l} D_3(n, v, k, l),$$

where D_3 is diophantine. Contraction of this last pair of initial existential quantifiers seems to come up against the obstacle that the variable l occurs not only in D_3 but also as an upper bound of the universal quantifier. Nevertheless, using the fact that $L(y) \leqq y$, we have

$$\bigvee_{v,l} \bigwedge_{k=0}^{l} D_3(n, v, k, l) \leftrightarrow \bigvee_{y} \bigwedge_{k=0}^{L(y)} D_3(n, K(y), k, L(y))$$
$$\leftrightarrow \bigvee_{y} \bigwedge_{k=0}^{y} [D_3(n, K(y), k, L(y)) \vee k > L(y)]$$
$$\leftrightarrow \bigvee_{y} \bigwedge_{k=0}^{y} \bigvee_{v,l} \{[D_3(n, v, k, l) \vee k > l] \wedge y = J(v, l)\}.$$

Thus, we have proved[1]

THEOREM 3.1. *If S is any recursively enumerable set, there exists a diophantine predicate $D(k, x, y)$ such that*

$$S = \left\{x \mid \bigvee_{y} \bigwedge_{k=0}^{y} D(k, x, y)\right\}.$$

THEOREM 3.2. *Let $R(\mathfrak{x}^{(n)})$ be a semicomputable predicate. Then there exists a diophantine predicate $D(k, \mathfrak{x}^{(n)}, y)$ such that*

$$R(\mathfrak{x}^{(n)}) \leftrightarrow \bigvee_{y} \bigwedge_{k=0}^{y} D(k, \mathfrak{x}^{(n)}, y).$$

[1] Cf. Davis [1, 3]. The proof given here is that of [1]. The proof given in [3] (suggested by the referee) is based on a previous proof of Corollary 3.5 below. The present proof yields an independent proof of Corollary 3.5.

PROOF. For $n = 1$, the result is an immediate consequence of Theorems 5-4.3 and 3.1. Suppose the result known for $n = p$, and let us take $n = p + 1$.

We then have the semicomputable predicate $R(\mathfrak{x}^{(p+1)})$. Let $S(\mathfrak{x}^{(p)})$ be defined by

$$S(\mathfrak{x}^{(p)}) \leftrightarrow R(x_1, \ldots, x_{p-1}, K(x_p), L(x_p)).$$

Then S is semicomputable, and, by induction hypothesis, there exists a diophantine predicate $D(k, \mathfrak{x}^{(p)}, y)$ such that

$$S(\mathfrak{x}^{(p)}) \leftrightarrow \bigvee_y \bigwedge_{k=0}^{y} D(k, \mathfrak{x}^{(p)}, y).$$

Now,

$$\begin{aligned} R(\mathfrak{x}^{(p+1)}) &\leftrightarrow S(x_1, \ldots, x_{p-1}, J(x_p, x_{p+1})) \\ &\leftrightarrow \bigvee_y \bigwedge_{k=0}^{y} D(k, x_1, \ldots, x_{p-1}, J(x_p, x_{p+1}), y) \\ &\leftrightarrow \bigvee_y \bigwedge_{k=0}^{y} \bigvee_z [D(k, x_1, \ldots, x_{p-1}, z, y) \wedge z = J(x_p, x_{p+1})]. \end{aligned}$$

This completes the proof.

Employing Corollary 2.3 and Theorem 5-3.2, we see that the converse of our last theorem also holds. That is, we have

THEOREM 3.3. *The predicate $R(\mathfrak{x}^{(n)})$ is semicomputable if and only if there exists a diophantine predicate $D(k, \mathfrak{x}^{(n)}, y)$ such that*

$$R(\mathfrak{x}^{(n)}) \leftrightarrow \bigvee_y \bigwedge_{k=0}^{y} D(k, \mathfrak{x}^{(n)}, y).$$

It is also easy to prove

THEOREM 3.4. *There exists, for each integer n, a diophantine predicate $D_n(k, z, \mathfrak{x}^{(n)}, y)$ such that, for each n-ary semicomputable predicate $R(\mathfrak{x}^{(n)})$, there is some number z_0 for which*

$$R(\mathfrak{x}^{(n)}) \leftrightarrow \bigvee_y \bigwedge_{k=0}^{y} D_n(k, z_0, \mathfrak{x}^{(n)}, y).$$

PROOF. The result is an immediate consequence of the enumeration theorem (Theorem 5-1.4) and Theorem 3.2.

We also have at once

COROLLARY 3.5. *Every semicomputable predicate (and, hence, every recursive predicate) is arithmetical.*[1]

[1] This result is due to Gödel [1].

PROOF. $\bigvee_y \bigwedge_{k=0}^{y} D(k, \mathfrak{x}^{(n)}, y) \leftrightarrow \bigvee_y \bigwedge_k [D(k, \mathfrak{x}^{(n)}, y) \vee (k > y)].$

COROLLARY 3.6. *There exists an arithmetical predicate that is not semicomputable.*

PROOF. By Corollary 3.5, the predicate $\sim T(x, x, y)$ is arithmetical. Hence, so is

$$\sim \bigvee_y T(x, x, y) \leftrightarrow \bigwedge_y \sim T(x, x, y).$$

By Theorem 5-1.6, this yields our result.

THEOREM 3.7. *There exists a diophantine predicate D such that $\sim D$ is not diophantine.*

PROOF. Suppose the result in question did not hold, i.e., that the negation of every diophantine predicate was also diophantine. Then, since

$$\bigwedge_x D \leftrightarrow \sim \bigvee_x \sim D,$$

the class of diophantine predicates would be closed under universal quantification. Hence, by Definition 2.3 and Corollary 2.4, the class of diophantine predicates would be identical with the class of arithmetic predicates. But, by Corollary 3.6, this is a contradiction.

A slight improvement of Theorem 3.1 is easily obtainable; in this the inner existential quantifiers are bounded. This theorem will be useful in Chap. 8. We have[1]

THEOREM 3.8. *If S is any recursively enumerable set, there exists a polynomial $P(k, x, y, x_1, \ldots, x_n)$ such that*

$$S = \left\{x \mid \bigvee_y \bigwedge_{k=0}^{y} \bigvee_{x_1, \ldots, x_n = 0}^{y} (P(k, x, y, x_1, \ldots, x_n) = 0)\right\}.$$

PROOF. By Theorem 3.1, there is a polynomial P such that

$$S = \left\{x \mid \bigvee_y \bigwedge_{k=0}^{y} \bigvee_{x_1, \ldots, x_n} (P(k, x, y, x_1, \ldots, x_n) = 0)\right\}.$$

Now, with each y, we can determine an upper bound for the values of the $x_1, \ldots, x_n$, since only a finite number [that is, $(y + 1)n$] of them occur. Hence,

[1] Cf. Myhill [1].
R. M. Robinson [3] proved that we can take $n = 4$ in this result.

$$S = \left\{x \mid \bigvee_y \bigvee_B \bigwedge_{k=0}^{y} \bigvee_{x_1, \ldots, x_n = 0}^{B} (P(k, x, y, x_1, \ldots, x_n) = 0)\right\}$$

$$= \left\{x \mid \bigvee_t \bigwedge_{k=0}^{K(t)} \bigvee_{x_1, \ldots, x_n = 0}^{L(t)} (P(k, x, K(t), x_1, \ldots, x_n) = 0)\right\}$$

$$= \left\{x \mid \bigvee_t \bigwedge_{k=0}^{t} \bigvee_{x_1, \ldots, x_n = 0}^{L(t)} [(k > K(t)) \vee (P(k, x, K(t), x_1, \ldots, x_n) = 0)]\right\}$$

$$= \left\{x \mid \bigvee_t \bigwedge_{k=0}^{t} \bigvee_{x_1, \ldots, x_n = 0}^{t} \{[(x_1 < L(t)) \wedge (x_2 < L(t)) \wedge \cdots \wedge (x_n < L(t))] \wedge [(k > K(t)) \vee (P(k, x, K(t), x_1, \ldots, x_n) = 0)]\}\right\}$$

$$= \left\{x \mid \bigvee_t \bigwedge_{k=0}^{t} \bigvee_{x_1, \ldots, x_n = 0}^{t} \bigvee_{y,B=0}^{t} \{[(x_1 < B) \wedge (x_2 < B) \wedge \cdots \wedge (x_n < B)] \wedge [(k > y) \vee (P(k, x, y, x_1, \ldots, x_n) = 0)] \wedge (t = J(y, B))\}\right\}$$

$$= \left\{x \mid \bigvee_t \bigwedge_{k=0}^{t} \bigvee_{x_1, \ldots, x_n = 0}^{t} \bigvee_{y,B=0}^{t} \bigvee_{w=0}^{t} \bigvee_{z_1, \ldots, z_n = 0}^{t} \{[(x_1 + z_1 + 1 = B) \wedge (x_2 + z_2 + 1 = B) \wedge \cdots \wedge (x_n + z_n + 1 = B)] \wedge [(k = y + w + 1) \vee (P(k, x, y, x_1, \ldots, x_n) = 0)] \wedge [2t = (x + y)^2 + 3x + y]\}\right\}.$$

DEFINITION 3.2. *A set of integers S is called* **diophantine** *if there exists a singulary diophantine predicate $D(x)$ such that*

$$S = \{x \mid D(x)\}.$$

THEOREM 3.9. *If the sets R, S are diophantine, so are $R \cap S$ and $R \cup S$. There exists, however, a diophantine set S such that $\bar{S}$ is not diophantine.*

PROOF. The first part is an immediate consequence of Theorem 2.8.

For the second part it is required to show that there exists a singulary diophantine predicate $D(x)$ such that $\sim D(x)$ is not diophantine. Now, suppose that no such singulary diophantine predicate exists; i.e., suppose that whenever $D(x)$ is diophantine so is $\sim D(x)$. Then we could show, by induction on n, that if $D(\mathfrak{x}^{(n)})$ is diophantine, so is $\sim D(\mathfrak{x}^{(n)})$. For, if $D(\mathfrak{x}^{(k+1)})$ is diophantine, so is

$$Q(\mathfrak{x}^{(k)}) \leftrightarrow D(x_1, \ldots, x_{k-1}, K(x_k), L(x_{k+1})),$$

since

$$Q(\mathfrak{x}^{(k)}) \leftrightarrow \bigvee_{u,v} [D(x_1, \ldots, x_{k-1}, u, v) \wedge x_k = J(u, v)].$$

By induction hypothesis, so would $\sim Q(\mathfrak{x}^{(k)})$ be diophantine. Hence, so would

$$\sim D(x_1, \ldots, x_{k+1}) \leftrightarrow \sim Q(x_1, \ldots, x_{k-1}, J(x_k, x_{k+1}))$$

be diophantine, yielding a contradiction.

The problem of actually constructing some diophantine set whose complement is not diophantine is open. If such a set could be obtained, and if it were also recursive, it would prove the falsity of the converse of Corollary 2.3.

Let us now consider the special case of Theorem 3.4 for which $n = 1$. If we write

$$S_z = \left\{x \mid \bigvee_y \bigwedge_{k=0}^{y} D_1(k, z, x, y)\right\},$$

then each S_z is a recursively enumerable set, and each recursively enumerable set is equal to some S_z. To see what this implies for the theory of diophantine equations, let us write

$$D_1(k, z, x, y) \leftrightarrow \bigvee_{\mathfrak{x}^{(m)}} [P(k, z, x, \mathfrak{x}^{(m)}, y) = 0],$$

where P is a polynomial. Now, for each choice of x, y, and z, let us write $\Sigma(x, y, z)$ for the system of diophantine equations

$$\begin{aligned} P(0, z, x, \mathfrak{x}^{(m)}, y) &= 0 \\ P(1, z, x, \mathfrak{x}^{(m)}, y) &= 0 \\ \cdots\cdots\cdots\cdots \\ P(y, z, x, \mathfrak{x}^{(m)}, y) &= 0. \end{aligned}$$

Then we see that S_z is simply the set of all numbers x for which there exists a y for which each equation of the system $\Sigma(x, y, z)$ has a solution. Hence, $\overline{S_z}$ is the set of all numbers x such that, no matter what value of y we choose, at least one equation of the system $\Sigma(x, y, z)$ fails to have a solution.

Now, for suitable choice of z_0, we have $S_{z_0} = K$, where K is not recursive. Hence we have

THEOREM 3.10. *There is a number z_0 for which the following problem is recursively unsolvable:*

Given x, to determine whether or not there exists a y for which each equation of the system $\Sigma(x, y, z_0)$ is solvable.

If R is a recursive set, then (and only then) R and $\bar{R}$ are both recursively enumerable. Hence, if R is recursive, there are integers z_1, z_2 such that $R = S_{z_1}$, $\bar{R} = S_{z_2}$. That is, we have

THEOREM 3.11. *Let R be a recursive set. Then there exist integers z_1, z_2 such that*

(1) *R is the set of all numbers x for which there exists a y such that each equation of the system $\Sigma(x, y, z_1)$ has a solution.*

(2) *R is the set of all numbers x such that, no matter what value of y we choose, at least one equation of the system $\Sigma(x, y, z_2)$ fails to have a solution.*

In particular, such numbers z_1, z_2 can be obtained for the set of primes, the set of powers of 2, the set of square-free numbers, etc. That all this can be accomplished with a single polynomial P seems rather surprising. It would be of some interest to obtain explicitly a manageable polynomial having this property.

CHAPTER 8

MATHEMATICAL LOGIC

1. Logics. In this chapter, we shall see how our methods can be applied to systems of symbolic logic. In order that our results may be applicable to a wide class of such systems, we cast our development in abstract form. Our discussion is phrased in terms of word predicates in the sense of the discussion at the beginning of Chap. 6. In addition, we shall deal with sets $\mathfrak{A}$ of words; a set $\mathfrak{A}$ will be called *recursive* if the set $\mathfrak{A}^*$ of all Gödel numbers of words $W \in \mathfrak{A}$ is recursive. Also, a word function $\mathfrak{f}(X)$, one that maps words onto words, is called *recursive* if the function $\mathfrak{f}^*(x)$ is partial recursive, where $\mathfrak{f}^*(x)$ has as its domain the set of all Gödel numbers of words and is such that $\mathfrak{f}^*(\text{gn }(X)) = \text{gn }(\mathfrak{f}(X))$.

DEFINITION 1.1. *By a* **logic** $\mathfrak{L}$ *we understand a recursive set* $\mathfrak{A}$ *of words, called the* **axioms of** $\mathfrak{L}$, *together with a finite set of recursive word predicates, none of which is singulary, called the* **rules of inference of** $\mathfrak{L}$.

When $\mathfrak{R}(Y, X_1, \ldots, X_n)$ *is a rule of inference of* $\mathfrak{L}$, *we shall sometimes say that* Y **is a consequence of** $X_1, \ldots, X_n$ **in** $\mathfrak{L}$ **by** $\mathfrak{R}$.

DEFINITION 1.2. *A finite sequence of words* $X_1, X_2, \ldots, X_n$ *is called a* **proof in a logic** $\mathfrak{L}$ *if, for each* i, $1 \leqq i \leqq n$, *either*

(1) $X_i \in \mathfrak{A}$, *or*

(2) *There exist* $j_1, j_2, \ldots, j_k < i$ *such that* X_i *is a consequence of* $X_{j_1}, X_{j_2}, \ldots, X_{j_k}$ *in* $\mathfrak{L}$ *by one of the rules of inference of* $\mathfrak{L}$.

Each of the X_i, $i = 1, 2, \ldots, n$, *is called a* **step of the proof.**

DEFINITION 1.3. *We say that* W *is a* **theorem of** $\mathfrak{L}$ *or that* W **is provable in** $\mathfrak{L}$ *and we write*

$$\vdash_{\mathfrak{L}} W$$

if there is a proof in $\mathfrak{L}$ *whose final step is* W. *This proof is then called a* **proof of** W **in** $\mathfrak{L}$.

We write $T_{\mathfrak{L}}$ *for the set of all Gödel numbers of theorems of* $\mathfrak{L}$.

Let Γ be a combinatorial system. Then it is natural to associate with Γ the logic $\mathfrak{L}(\Gamma)$ which has a single axiom, namely, the axiom of Γ, and whose rules of inference are the productions of Γ. Now, it is clear that every proof in Γ is also a proof in $\mathfrak{L}(\Gamma)$. It is possible, however, to construct proofs in $\mathfrak{L}(\Gamma)$ that are not proofs in Γ. Thus, for example, if

X_1, X_2, X_3 is a proof in Γ, then X_1, X_2, X_1, X_3 is a proof in $\mathfrak{L}(\Gamma)$ which will not ordinarily be a proof in Γ. However, it is easy to see that

$$\{W \mid \vdash_\Gamma W\} = \{W \mid \vdash_{\mathfrak{L}(\Gamma)} W\}.$$

Hence, for our purposes, it will not be necessary to distinguish between Γ and $\mathfrak{L}(\Gamma)$. When we speak of a combinatorial system Γ as a logic, we are referring to the logic $\mathfrak{L}(\Gamma)$.

Thus, our notion of a logic is so broad as to encompass not only the usual systems of symbolic logic but also combinatorial systems.

THEOREM 1.1. *The set $T_{\mathfrak{L}}$ is recursively enumerable.*

PROOF. Let $\mathfrak{A}^*$ be the (recursive) set of all Gödel numbers of axioms of $\mathfrak{L}$. Let $\mathfrak{R}_1^*(y, \mathfrak{x}^{(n_1)}), \mathfrak{R}_2^*(y, \mathfrak{x}^{(n_2)}), \ldots, \mathfrak{R}_k^*(y, \mathfrak{x}^{(n_k)})$ be the (recursive) associated numerical predicates of the rules of inference of $\mathfrak{L}$. Let $P_{\mathfrak{L}}$ be the class of all numbers x such that

$$x = 2^{\mathrm{gn}(X_1)} 3^{\mathrm{gn}(X_2)} \cdots \Pr(m)^{\mathrm{gn}(X_m)},$$

where $X_1, X_2, \ldots, X_m$ is a proof in $\mathfrak{L}$ (that is, $P_{\mathfrak{L}}$ is the set of all Gödel numbers of proofs in $\mathfrak{L}$). Then

$$\begin{aligned} P_{\mathfrak{L}} = \{x \mid \mathrm{GN}(x) \wedge \bigwedge_{n=1}^{\mathfrak{L}(x)} [(n \mathrm{\,Gl\,} x \in \mathfrak{A}^*) \\ \vee \bigvee_{i_1, i_2, \ldots, i_{n_1}=1}^{n \dot{-} 1} \mathfrak{R}_1^*(n \mathrm{\,Gl\,} x, i_1 \mathrm{\,Gl\,} x, i_2 \mathrm{\,Gl\,} x, \ldots, i_{n_1} \mathrm{\,Gl\,} x) \\ \vee \bigvee_{i_1, i_2, \ldots, i_{n_2}=1}^{n \dot{-} 1} \mathfrak{R}_2^*(n \mathrm{\,Gl\,} x, i_1 \mathrm{\,Gl\,} x, i_2 \mathrm{\,Cl\,} x, \ldots, i_{n_2} \mathrm{\,Gl\,} x) \\ \vee \cdots \\ \vee \bigvee_{i_1, i_2, \ldots, i_{n_k}=1}^{n \dot{-} 1} \mathfrak{R}_k^*(n \mathrm{\,Gl\,} x, i_1 \mathrm{\,Gl\,} x, i_2 \mathrm{\,Gl\,} x, \ldots, i_{n_k} \mathrm{\,Gl\,} x)]\}. \end{aligned}$$

Thus, $P_{\mathfrak{L}}$ is recursive. But

$$T_{\mathfrak{L}} = \left\{x \mid \bigvee_y (y \in P_{\mathfrak{L}} \wedge x = \mathfrak{L}(y) \mathrm{\,Gl\,} y)\right\}.$$

Therefore, $T_{\mathfrak{L}}$ is recursively enumerable.

By the *decision problem for a logic* $\mathfrak{L}$ we mean the problem of determining, of a given word, whether or not it is a theorem of $\mathfrak{L}$.

DEFINITION 1.4. *The decision problem for a logic $\mathfrak{L}$ is* **recursively solvable** *if $T_{\mathfrak{L}}$ is recursive; otherwise it is* **recursively unsolvable.**

COROLLARY 1.2. *There exists a logic $\mathfrak{L}$ whose decision problem is recursively unsolvable.*

PROOF. By the results of Chap. 6 (e.g., Theorem 6-2.6), we know that there exists a combinatorial system whose decision problem is recursively unsolvable.

DEFINITION 1.5. *Let $\mathfrak{L}$ and $\mathfrak{L}'$ be logics. Then we say that $\mathfrak{L}$ is* **translatable into** *$\mathfrak{L}'$ if there exists a recursive word function $\mathfrak{f}(X)$ such that $\vdash_{\mathfrak{L}} X$ if and only if $\vdash_{\mathfrak{L}'} \mathfrak{f}(X)$ and, moreover, if, whenever $X \neq Y$, we also have $\mathfrak{f}(X) \neq \mathfrak{f}(Y)$ (that is, if $\mathfrak{f}$ is one-one).*

THEOREM 1.3. *If $\mathfrak{L}$ is translatable into $\mathfrak{L}'$ and the decision problem for $\mathfrak{L}'$ is recursively solvable, then the decision problem for $\mathfrak{L}$ is recursively solvable.*

Hence, if the decision problem for $\mathfrak{L}$ is recursively unsolvable, then so is that for $\mathfrak{L}'$.

PROOF. Let $T_{\mathfrak{L}'}$ be recursive. Let $\mathfrak{f}(X)$ be a recursive word function such that

$$\vdash_{\mathfrak{L}} X \leftrightarrow \vdash_{\mathfrak{L}'} \mathfrak{f}(X).$$

Then

$$T_{\mathfrak{L}} = \{x \mid \mathfrak{f}^*(x) \in T_{\mathfrak{L}'}\}.$$

Hence,

$$C_{T_{\mathfrak{L}}}(x) = C_{T_{\mathfrak{L}'}}(\mathfrak{f}^*(x)).$$

Now the function on the right is partial recursive, and that on the left is total. Hence, both are in fact recursive. That is, $T_{\mathfrak{L}}$ is recursive.

(Note that the one-one character of $\mathfrak{f}$ was not used in this proof.)

THEOREM 1.4. *For every logic $\mathfrak{L}$ there is a normal system ν, whose alphabet consists of two letters, such that $\mathfrak{L}$ is translatable into ν.*†

PROOF. We define $\mathfrak{f}(X)$ as follows:

$$\mathfrak{f}(X) = 1^{\mathrm{gn}(X)+1}$$

By Theorems 1.1 and 6-5.3, there exists a normal system ν, whose alphabet consists of two letters, which generates the set $T_{\mathfrak{L}}$. For such a normal system, we have

$$\begin{aligned} \vdash_{\mathfrak{L}} X &\leftrightarrow \mathrm{gn}\,(X) \in T_{\mathfrak{L}} \\ &\leftrightarrow \vdash_{\nu} 1^{\mathrm{gn}(X)+1} \\ &\leftrightarrow \vdash_{\nu} \mathfrak{f}(X). \end{aligned}$$

If $\mathfrak{f}(X) = \mathfrak{f}(Y)$, then $1^{\mathrm{gn}(X)+1} = 1^{\mathrm{gn}(Y)+1}$; so $\mathrm{gn}\,(X) = \mathrm{gn}\,(Y)$, and, finally, $X = Y$.

It remains only to show that $\mathfrak{f}(X)$ is recursive. Let $g(x)$ be the function whose domain is the set of Gödel numbers of words on our alphabet and whose value is 1 for all numbers in its domain. Then $g(x)$ is partial

† A somewhat similar theorem is proved by Post [2, 8]. Our result is stronger than Post's in that ours is applicable to more systems than his, and weaker than Post's in that Post supplies a particularly simple translation function.

recursive, since

$$g(x) = \min_y \left[(y = 1) \wedge \text{GN}(x) \wedge \bigwedge_{n=1}^{\mathfrak{L}(x)} \bigvee_{z=0}^{x} (n \text{ Gl } x = 2z + 1) \right].$$

Also, the function gn (1^{x+1}) is recursive, since

$$\text{gn}(1^{0+1}) = 2,$$
$$\text{gn}(1^{(x+1)+1}) = \text{gn}(1^{x+1}) * 2.$$

But

$$\mathfrak{f}^*(x) = \text{gn}(1^{x+1})g(x).$$

Hence $\mathfrak{f}^*(x)$ is partial recursive; so $\mathfrak{f}(X)$ is recursive.

THEOREM 1.5. *If $\mathfrak{L}$ is translatable into $\mathfrak{L}'$ and if $\mathfrak{L}'$ is translatable into $\mathfrak{L}''$, then $\mathfrak{L}$ is translatable into $\mathfrak{L}''$.*

PROOF. There are recursive functions $\mathfrak{f}(X)$, $\mathfrak{g}(X)$ such that

$$\vdash_{\mathfrak{L}} X \leftrightarrow \vdash_{\mathfrak{L}'} \mathfrak{f}(X),$$
$$\vdash_{\mathfrak{L}'} X \leftrightarrow \vdash_{\mathfrak{L}''} \mathfrak{g}(X).$$

Let $\mathfrak{h}(X) = \mathfrak{g}(\mathfrak{f}(X))$. Then, $\mathfrak{h}(X)$ is recursive, and

$$\vdash_{\mathfrak{L}} X \leftrightarrow \vdash_{\mathfrak{L}'} \mathfrak{f}(X)$$
$$\leftrightarrow \vdash_{\mathfrak{L}''} \mathfrak{h}(X).$$

Finally, if $\mathfrak{h}(X) = \mathfrak{h}(Y)$, then $\mathfrak{f}(X) = \mathfrak{f}(Y)$, whence $X = Y$.

2. Incompleteness and Unsolvability Theorems for Logics. We shall begin with some notions which provide a measure of the "deductive power" of a logic without necessitating a discussion of its detailed structure.

DEFINITION 2.1. *A logic $\mathfrak{L}$ is said to be* **semicomplete with respect to a set of integers** *Q if there exists a sequence of words $W_0, W_1, W_2, \ldots$ such that the function $f(n) = \text{gn}(W_n)$ is recursive and such that*

$$Q = \{n \mid \vdash_{\mathfrak{L}} W_n\}.$$

$\mathfrak{L}$ is said to be **complete with respect to** *Q if it is semicomplete with respect to both Q and $\bar{Q}$.*

In terms of the concept of semicompleteness, we can obtain at once a sufficient condition that a logic have a recursively unsolvable decision problem.

THEOREM 2.1. *If $\mathfrak{L}$ is semicomplete with respect to a nonrecursive set Q, the $\mathfrak{L}$ has a recursively unsolvable decision problem.*

PROOF. We have

$$Q = \{n \mid \vdash_{\mathfrak{L}} W_n\}$$
$$= \{n \mid \text{gn}(W_n) \in T_{\mathfrak{L}}\}.$$

Hence, if $T_{\mathfrak{L}}$ were recursive, so would Q be.

COROLLARY 2.2. *If $\mathfrak{L}$ is semicomplete with respect to every recursively enumerable set, then $\mathfrak{L}$ has a recursively unsolvable decision problem.*

Theorem 2.1 and its corollary are actually quite useful in proving the recursive unsolvability of the decision problem for various systems of symbolic logic. The converse of Theorem 2.1 is also true.

THEOREM 2.3. *If $\mathfrak{L}$ has a recursively unsolvable decision problem, then $\mathfrak{L}$ is semicomplete with respect to a nonrecursive set, namely $T_{\mathfrak{L}}$.*

PROOF. Let $\mathfrak{L}$ have a recursively unsolvable decision problem. Then not all words are theorems of $\mathfrak{L}$. Let X_0 be the word of least Gödel number that is not a theorem of $\mathfrak{L}$. We define the sequence W_0, W_1, W_2, . . . as follows:

If n is the Gödel number of a word, $n = \text{gn}\,(W)$, then we set $W_n = W$; otherwise we set $W_n = X_0$.

Then $n \in T_{\mathfrak{L}}$ if and only if $n = \text{gn}\,(W)$, where $\vdash_{\mathfrak{L}} W$. But, by our definition of W_n, this becomes $n \in T_{\mathfrak{L}}$ if and only if $\vdash_{\mathfrak{L}} W_n$. That is,

$$T_{\mathfrak{L}} = \{n \mid \vdash_{\mathfrak{L}} W_n\}.$$

It remains to show that gn (W_n) is recursive. Let

$$W(x) \leftrightarrow \text{GN}\,(x) \wedge \bigwedge_{n=1}^{\mathfrak{L}(x)} \bigvee_{z=0}^{x} (n \text{ Gl } x = 2z + 1).$$

Then $W(x)$ is recursive. Let $w(x)$ be its characteristic function. Then

$$\text{gn}\,(W_n) = n(1 \dot{-} w(n)) + \text{gn}\,(X_0)w(n).$$

Next, we shall see that a logic can be semicomplete only with respect to recursively enumerable sets.

THEOREM 2.4. *If $\mathfrak{L}$ is semicomplete with respect to Q, then Q is recursively enumerable.*

PROOF

$$\begin{aligned} Q &= \{n \mid \vdash_{\mathfrak{L}} W_n\} \\ &= \{n \mid \text{gn}\,(W_n) \in T_{\mathfrak{L}}\}, \end{aligned}$$

and the result follows at once from Theorems 1.1 and 5-4.10.

COROLLARY 2.5. *If Q is not recursively enumerable, then no logic is semicomplete with respect to Q.*

COROLLARY 2.6. *If $\mathfrak{L}$ is complete with respect to Q, then Q is recursive.*

COROLLARY 2.7. *If Q is recursively enumerable but not recursive, then no logic is semicomplete with respect to $\bar{Q}$.*

COROLLARY 2.8. *If Q is recursively enumerable but not recursive, then no logic is complete with respect to Q.*

Theorem 2.4, with its corollaries, represents a decisive limitation on the power of logics. In fact, these results really constitute an abstract form of Gödel's famous incompleteness theorem.[1] Their devastating

[1] Cf. Gödel [1]. Our present formulation is essentially that of Kleene [4, 6].

import stems from the fact that they imply that *an adequate development of the theory of natural numbers, within a logic* $\mathfrak{L}$, *to the point where membership in some given set* Q *of integers can be adequately dealt with within the logic* (i.e., so that a given number n belongs to Q if and only if some corresponding—in an effective manner—word W_n is provable in $\mathfrak{L}$), *is possible only if* Q *happens to be recursively enumerable.* Hence, nonrecursively enumerable sets can, at best, be dealt with in an incomplete manner.

In the next section we shall see how, by assuming more about the basic structure of $\mathfrak{L}$, these results assume a more explicit form.

3. Arithmetical Logics. In this section we shall discuss a large class of logics, which we shall call *arithmetical.* As in the usual systems of symbolic logic, much of the symbolism of these arithmetical logics may be thought of as constituting a "translation" of ordinary mathematical English into a more precise formal language. Although, in principle, these "semantical" ideas can themselves be formalized,[1] we shall not attempt such a formal treatment here, and our development will not depend on any interpretation of the symbols of arithmetical logics. However, solely to aid the reader, we include remarks concerning what we shall call the *intended interpretation* of an arithmetical logic. To emphasize that these remarks are not part of our formal development we place them within curly braces {. . .}.

A logic $\mathfrak{A}$ is to be called an *arithmetical logic* if it has the properties listed below under headings I, II, III, and IV.

I. *Well-formed Formulas.* For each arithmetical logic $\mathfrak{A}$, there is a recursive nonempty set of words called the *well-formed formulas* of $\mathfrak{A}$. We write *w.f.f.* for "well-formed formula." {In the intended interpretation of $\mathfrak{A}$, the w.f.f.'s include the words that represent sentences or predicates.}

All theorems of $\mathfrak{A}$ are w.f.f.'s.

II. *Propositional Connectives.* There are recursive word functions $\mathfrak{J}(A, B)$ and $\mathfrak{N}(A)$. We shall write $[A \supset B]$ for $\mathfrak{J}(A, B)$ and $-A$ for $\mathfrak{N}(A)$. $-A$ is a w.f.f. if and only if A is a w.f.f.; $[A \supset B]$ is a w.f.f. if and only if A and B are w.f.f.'s.† {In the intended interpretation of $\mathfrak{A}$, $\supset$ represents implication and $-$ represents negation.}

If A, B, C are w.f.f.'s of $\mathfrak{A}$, then

$$\vdash_{\mathfrak{A}} [A \supset [B \supset A]], \tag{1}$$

$$\vdash_{\mathfrak{A}} [[A \supset [B \supset C]] \supset [[A \supset B] \supset [A \supset C]]], \tag{2}$$

$$\vdash_{\mathfrak{A}} [[-B \supset -A] \supset [A \supset B]]. \tag{3}$$

[1] Cf. Tarski [1].

† In the theorems about arithmetical logics to be proved below, many of the clauses in our definition will not in fact be employed. The notation and some of the specific formulations used in the remainder of this chapter are taken from Church [5].

Moreover, if

$$\vdash_{\mathfrak{A}} A$$

and

$$\vdash_{\mathfrak{A}} [A \supset B],$$

then

$$\vdash_{\mathfrak{A}} B.$$

This last is called *modus ponens.*

III. *Quantifiers.* There is a denumerable sequence

$$\mathbf{x}_1, \mathbf{x}_2, \mathbf{x}_3, \ldots$$

of distinct words of $\mathfrak{A}$ called *variables.* No variable is a w.f.f. The function $f(n) = \text{gn}(\mathbf{x}_n)$ is recursive. There are no words A, B, C, $A \neq \Lambda$, $B \neq \Lambda$, $C \neq \Lambda$, such that AB and BC are both variables.[1]

There is a recursive binary word function $\mathfrak{G}(M, A)$. We shall write $(M)A$ for $\mathfrak{G}(M, A)$. $(M)A$ is a w.f.f. if and only if A is a w.f.f. and M is a variable. By $(\exists M)A$, we understand $-(M) - A$.

{In the intended interpretation, the words $\mathbf{x}_i$ represent variables whose range is a class that contains the integers.[2] $(\mathbf{x}_i)$ and $(\exists \mathbf{x}_i)$ then represent universal and existential quantification, respectively, over the integers.}

There is a recursive binary word predicate $\mathfrak{B}(M, A)$.

If $\mathfrak{B}(M, A)$, then A is a w.f.f., M is a variable, and M is a part of A (that is, $A = BMC$ for suitable B and C). When $\mathfrak{B}(\mathbf{x}_i, A)$ is true, we say that $\mathbf{x}_i$ is *bound* in A. When $\mathfrak{B}(\mathbf{x}_i, A)$ is false and $\mathbf{x}_i$ is a part of A, we say that $\mathbf{x}_i$ is *free* in A.

If no variable is free in any w.f.f. B that is part of the w.f.f. A, then A is said to be *closed.* We write *c.w.f.f.* for "closed w.f.f."

{In the intended interpretation, it is the c.w.f.f.'s that represent sentences.}

$\mathfrak{B}(\mathbf{x}_i, (\mathbf{x}_i)A)$ is always true.

If $\mathfrak{B}(\mathbf{x}_i, A)$ and A is a part of the w.f.f. B, then $\mathfrak{B}(\mathbf{x}_i, B)$.

There is a recursive word function $\mathfrak{S}(X, M, N)$ such that, if X is a w.f.f. and M is a variable, then $\mathfrak{S}(X, M, N)$ is the word that results on replacing M by N at all of its occurrences in X: $\mathfrak{S}([X \supset Y], M, N)$ is $[\mathfrak{S}(X, M, N) \supset \mathfrak{S}(Y, M, N)]$. $\mathfrak{S}(-X, M, N)$ is $-\mathfrak{S}(X, M, N)$. If $i \neq j$, then $\mathfrak{S}((\mathbf{x}_j)X, \mathbf{x}_i, N)$ is $(\mathbf{x}_j)\mathfrak{S}(X, \mathbf{x}_i, N)$.

There is a recursive word function $\mathfrak{C}(W)$. If W is a w.f.f., then $\mathfrak{C}(W)$ is a c.w.f.f., and $\mathfrak{C}(W)$ is obtained by prefixing zero or more universal quantifiers to W. If W is a c.w.f.f., $\mathfrak{C}(W) = W$. If W is not a w.f.f., $\mathfrak{C}(W)$ is not a w.f.f. $\mathfrak{C}(W)$ is called the *closure* of W.

It follows that the class of c.w.f.f.'s is recursive, since W is a c.w.f.f. if and only if $\mathfrak{C}(W) = W$.

[1] Thus, it is impossible for two variables which occur in the same word to overlap.

[2] Thus, the class may be the set of integers or it may contain other objects as well.

$\vdash_{\mathfrak{A}} A$ if and only if $\vdash_{\mathfrak{A}} (\mathbf{x}_i)A$. From this last it follows at once that $\vdash_{\mathfrak{A}} A$ if and only if $\vdash_{\mathfrak{A}} \mathfrak{C}(A)$.

IV. *Integers.* Associated with each integer n is a word that we write n^* and that we call the *numeral associated with the number n.* The function $f(n) = \text{gn}\ (n^*)$ is recursive. If $n \neq m$, then $n^* \neq m^*$.

If A is a w.f.f., if $\mathbf{x}_i$ is free in A, and if N is a variable or a numeral, then $\mathfrak{S}(A, \mathbf{x}_i, N)$ is a w.f.f.

If $\mathbf{x}_i$ and $\mathbf{x}_j$ are not bound in A, then

$$\vdash_{\mathfrak{A}} [\mathfrak{S}(A, \mathbf{x}_i, \mathbf{x}_j) \supset (\exists \mathbf{x}_i) A], \tag{4}$$
$$\vdash_{\mathfrak{A}} [\mathfrak{S}(A, \mathbf{x}_i, m^*) \supset (\exists \mathbf{x}_i) A]. \tag{5}$$

If $\mathbf{x}_i$ is not free in A, then

$$\vdash_{\mathfrak{A}} [(\mathbf{x}_i)[A \supset B] \supset [A \supset (\mathbf{x}_i)B]]. \tag{6}$$

This completes our definition of what constitutes an *arithmetical logic.*

DEFINITION 3.1. *A w.f.f. W of $\mathfrak{A}$ is called n-***ary** *if the variables $\mathbf{x}_1$, $\mathbf{x}_2, \ldots, \mathbf{x}_n$ are free in W and if no other variables are free in W.*

DEFINITION 3.2. *If W is an n-ary w.f.f. of $\mathfrak{A}$ and if $(x_1, x_2, \ldots, x_n)$ is an n-tuple of integers, we write $W(x_1, x_2, \ldots, x_n)$ to denote the w.f.f. obtained from W on replacing $\mathbf{x}_i$ by x_i^* at all occurrences of $\mathbf{x}_i$ in W, for $i = 1, 2, \ldots, n$.*

DEFINITION 3.3. *An arithmetical logic $\mathfrak{A}$ is called* **consistent** *if, for no word A, do we have $\vdash_{\mathfrak{A}} A$ and $\vdash_{\mathfrak{A}} - A$.*

If $\mathfrak{A}$ is not consistent, it is called **inconsistent.**

COROLLARY 3.1. *$\mathfrak{A}$ is inconsistent if and only if all w.f.f.'s of $\mathfrak{A}$ are theorems of $\mathfrak{A}$.*

PROOF. If all w.f.f.'s are theorems, then $\mathfrak{A}$ is certainly inconsistent.

Conversely, suppose that $\vdash_{\mathfrak{A}} A$ and $\vdash_{\mathfrak{A}} - A$. Let B be any w.f.f. of $\mathfrak{A}$. By II (1),

$$\vdash_{\mathfrak{A}} [- A \supset [- B \supset - A]].$$

Hence, by modus ponens (II),

$$\vdash_{\mathfrak{A}} [- B \supset - A].$$

By II (3),

$$\vdash_{\mathfrak{A}} [[- B \supset - A] \supset [A \supset B]].$$

Hence, by modus ponens,

$$\vdash_{\mathfrak{A}} [A \supset B],$$

and, by modus ponens again,

$$\vdash_{\mathfrak{A}} B.$$

DEFINITION 3.4. *An arithmetical logic* $\mathfrak{A}$ *is called* ω-**consistent** *if there does not exist a* 1*-ary w.f.f.* W *such that, for all integers* m, $\vdash_{\mathfrak{A}} W(m)$ *and also* $\vdash_{\mathfrak{A}} - (x_1)W$.

An arithmetical logic that is not ω-**consistent** *is called* ω-**inconsistent.**

{Taking into account the intended interpretation of A and the meaning of universal quantification, we see that, in an ω-*inconsistent* arithmetical logic $\mathfrak{A}$, there is a c.w.f.f. A of $\mathfrak{A}$ such that $\vdash_{\mathfrak{A}} A$, whereas the sentence A represents is false. Thus, ω-inconsistency is, from the point of view of intended interpretations, as undesirable as inconsistency.}

COROLLARY 3.2. *If* $\mathfrak{A}$ *is* ω*-consistent, it is consistent.*

PROOF. Suppose $\mathfrak{A}$ to be inconsistent. Let W be any 1-ary w.f.f. of $\mathfrak{A}$. Then, by Corollary 3.1, for all m, $\vdash_{\mathfrak{A}} W(m)$ and $\vdash_{\mathfrak{A}} - (x_1)W$. Hence, $\mathfrak{A}$ is ω-inconsistent.

DEFINITION 3.5. *Let* $P(x_1, \ldots, x_n)$ *be a predicate. Then we say that* P *is* **completely representable** *in an arithmetical logic* $\mathfrak{A}$ *if there exists an* n*-ary w.f.f.* W *of* $\mathfrak{A}$ *such that*

(1) *For each* n*-tuple* $(x_1, \ldots, x_n)$ *for which* $P(x_1, \ldots, x_n)$ *is true,* $\vdash_{\mathfrak{A}} W(x_1, \ldots, x_n)$.

(2) *For each* n*-tuple* $(x_1, \ldots, x_n)$ *for which* $P(x_1, \ldots, x_n)$ *is false,* $\vdash_{\mathfrak{A}} - W(x_1, \ldots, x_n)$.

THEOREM 3.3. *If* $R(\mathfrak{x}^{(n)})$ *is completely representable in an arithmetical logic* $\mathfrak{A}$*, then* $R(\mathfrak{x}^{(n)})$ *is recursive.*

PROOF. There is an n-ary w.f.f. W such that

$$\begin{aligned} R(\mathfrak{x}^{(n)}) &\leftrightarrow \vdash_{\mathfrak{A}} W(\mathfrak{x}^{(n)}) \\ &\leftrightarrow \text{gn}\,(W(\mathfrak{x}^{(n)})) \in T_{\mathfrak{A}}, \\ \sim R(\mathfrak{x}^{(n)}) &\leftrightarrow \vdash_{\mathfrak{A}} - W(\mathfrak{x}^{(n)}) \\ &\leftrightarrow \text{gn}\,(- W(\mathfrak{x}^{(n)})) \in T_{\mathfrak{A}}. \end{aligned}$$

We first note that the functions gn $(W(\mathfrak{x}^{(n)}))$, gn $(- W(\mathfrak{x}^{(n)}))$ are recursive. This follows at once from the fact that $\mathfrak{S}$ and $-$ are recursive word functions and from

$$W(\mathfrak{x}^{(n)}) = \mathfrak{S}(\cdots \mathfrak{S}(\mathfrak{S}(\mathfrak{S}(W, x_1, x_1^*), x_2, x_2^*), x_3, x_3^*), \ldots, x_n, x_n^*).$$

Now, by Theorem 1.1, the set $T_{\mathfrak{A}}$ is recursively enumerable. Hence, by Theorem 5-3.4, the predicates $R(\mathfrak{x}^{(n)})$, $\sim R(\mathfrak{x}^{(n)})$ are semicomputable. Therefore, by Theorem 5-1.5, $R(\mathfrak{x}^{(n)})$ is recursive.

DEFINITION 3.6. *A class* Δ *of binary predicates is called a* **basis** *if, for every recursively enumerable set* Q*, there is a predicate* $R(x, y)$ *such that* $R(x, y) \in \Delta$ *and*

$$Q = \left\{x \mid \bigvee_y R(x, y)\right\}.$$

By the results of Chap. 5, the class of binary recursive predicates is a basis. By the enumeration theorem (Theorem 5-1.4), the class of predicates of the form $T(z_0, x, y)$ is a basis. Hence, the class of binary *primitive recursive predicates* is a basis. Moreover, by Theorem 7-3.8, the class of all predicates of the form $\bigwedge_{k=0}^{y} \bigvee_{x_1,\ldots,x_n=0}^{y} P(k, x, y, \mathfrak{x}^{(n)}) = 0$, where P is a polynomial, is a basis.

DEFINITION 3.7. *An arithmetical logic $\mathfrak{A}$ is called* **adequate** *if there is some basis Δ such that, for every predicate $P \in \Delta$, P is completely representable in $\mathfrak{A}$.*

We are now in a position to apply the results of Sec. 2.

THEOREM 3.4. *If $\mathfrak{A}$ is an ω-consistent and adequate arithmetical logic, then $\mathfrak{A}$ is semicomplete with respect to every recursively enumerable set.*

PROOF. Let Q be some recursively enumerable set. Then since $\mathfrak{A}$ is adequate, there is a predicate $R(x, y)$, completely representable in $\mathfrak{A}$, such that

$$Q = \left\{x \mid \bigvee_{y} R(x, y)\right\}.$$

Thus, there exists a 2-ary w.f.f. W such that

When $R(x, y)$ is true, $\vdash_{\mathfrak{A}} W(x, y)$,

When $R(x, y)$ is false, $\vdash_{\mathfrak{A}} - W(x, y)$.

Let U be the 1-ary w.f.f. $(\exists \mathbf{x}_2)W$. We shall see that

$$Q = \{x \mid \vdash_{\mathfrak{A}} U(x)\}. \tag{7}$$

For each integer n, we write $W(n, \mathbf{x}_2)$ for $\mathfrak{S}(W, \mathbf{x}_1, n^*)$. Now, suppose that $n_0 \in Q$. Then, for some number y_0, $R(n_0, y_0)$ is true. Hence, $\vdash_{\mathfrak{A}} W(n_0, y_0)$. Now, by IV(5),

$$\vdash_{\mathfrak{A}} [W(n_0, y_0) \supset (\exists \mathbf{x}_2)W(n_0, \mathbf{x}_2)].$$

Hence, by modus ponens,

$$\vdash_{\mathfrak{A}} (\exists \mathbf{x}_2)W(n_0, \mathbf{x}_2).$$

That is,

$$\vdash_{\mathfrak{A}} U(n_0).$$

Conversely, suppose that $\vdash_{\mathfrak{A}} U(n_0)$. That is,

$$\vdash_{\mathfrak{A}} (\exists \mathbf{x}_2)W(n_0, \mathbf{x}_2)$$

or

$$\vdash_{\mathfrak{A}} - (\mathbf{x}_2) - W(n_0, \mathbf{x}_2).$$

Hence, by the ω-consistency of $\mathfrak{A}$, there is some integer m_0 for which the w.f.f. $- W(n_0, m_0)$ is not a theorem of $\mathfrak{A}$. Hence, $R(n_0, m_0)$ must be true, since, if it were false, we should have $\vdash_{\mathfrak{A}} - W(n_0, m_0)$. Hence, finally, $n_0 \in Q$, and we have proved (7). By Definition 2.1, it remains only to show that the function $f(x) = \text{gn}\,(U(x))$ is recursive. But this is clear, since

$$\text{gn}\,(U(x)) = \text{gn}\,(\mathfrak{S}(U, \mathbf{x}_1, x^*)).$$

THEOREM 3.5. *If $\mathfrak{A}$ is an ω-consistent and adequate arithmetical logic, then $\mathfrak{A}$ has a recursively unsolvable decision problem.*[1]

PROOF. Immediate from Theorem 3.4 and Corollary 2.2.

DEFINITION 3.8. *An arithmetical logic $\mathfrak{A}$ is said to be* **complete** *if, for every c.w.f.f. A of $\mathfrak{A}$, either $\vdash_{\mathfrak{A}} A$ or $\vdash_{\mathfrak{A}} - A$.*

If $\mathfrak{A}$ is not complete, it is called **incomplete.**

THEOREM 3.6. *If an arithmetical logic $\mathfrak{A}$ is complete, then its decision problem is recursively solvable.*

PROOF. Let F be the set of all Gödel numbers of c.w.f.f.'s of $\mathfrak{A}$, so that F is recursive. Let P^+ be the set of all Gödel numbers of theorems of $\mathfrak{A}$ that are c.w.f.f.'s; P^-, the set of all Gödel numbers of nontheorems of $\mathfrak{A}$ that are c.w.f.f.'s.

We first wish to show that P^+ is recursive. Now, $P^+ = F \cap T_{\mathfrak{A}}$; so, by Theorems 1.1 and 5-4.9, P^+ is recursively enumerable. Since $\mathfrak{A}$ is complete, a nonprovable c.w.f.f. is one whose negation is provable. Hence,

$$P^- = F \cap \{x \mid \mathfrak{N}^*(x) \in T_{\mathfrak{A}}\}.$$

[Here $\mathfrak{N}^*(x)$ is the associated numerical function of $-$.] Thus, by Theorems 1.1, 5-4.9, and 5-4.10, P^- is also recursively enumerable. But

$$\overline{P^+} = P^- \cup \bar{F};$$

so, by Theorem 5-4.9, $\overline{P^+}$ is recursively enumerable. Hence, by Theorem 5-4.5, P^+ is recursive.

But a w.f.f. of $\mathfrak{A}$ is a theorem if and only if its closure is a theorem. Hence,

$$T_{\mathfrak{A}} = \{x \mid \mathfrak{C}^*(x) \in P^+\}$$

and, therefore, $T_{\mathfrak{A}}$ is recursive.

THEOREM 3.7 (Gödel's Incompleteness Theorem). *If $\mathfrak{A}$ is an ω-consistent and adequate arithmetical logic, then $\mathfrak{A}$ is incomplete.*

PROOF. Immediate from Theorems 3.5 and 3.6.

Thus, for any arithmetical logic $\mathfrak{A}$ satisfying the hypotheses of Theorem 3.7, there is a c.w.f.f. A of $\mathfrak{A}$ such that neither A nor $-A$ is a theorem of $\mathfrak{A}$. Or, as we may say, A is *undecidable.* However, our proof of Theorem 3.7 gives no indication of what form such an A would take. In order to determine this, we turn to the incompleteness theorem of Sec. 2.

SECOND PROOF OF THEOREM 3.7. Let Q be a recursively enumerable set which is not recursive. As in the proof of Theorem 3.4,

$$\{x \mid \vdash_{\mathfrak{A}} U(x)\} = Q.$$

[1] Essentially this result was first proved by Church. Cf. Church [1].

Suppose that $\vdash_{\mathfrak{A}} - U(n_0)$. Then $n_0 \in \bar{Q}$ [if $n_0 \in Q$, $\vdash_{\mathfrak{A}} U(n_0)$, which would contradict the consistency of $\mathfrak{A}$]. Thus,

$$\{x \mid \vdash_{\mathfrak{A}} - U(x)\} \subset \bar{Q}.$$

But, by Corollary 2.7,

$$\{x \mid \vdash_{\mathfrak{A}} - U(x)\} \neq \bar{Q}.$$

Hence, there is a number n_0 such that $n_0 \in \bar{Q}$, but it is not the case that $\vdash_{\mathfrak{A}} - U(n_0)$. On the other hand, it is also not the case that $\vdash_{\mathfrak{A}} U(n_0)$, since this would imply $n_0 \in Q$. This completes the proof.

{Thus we see that, if $\mathfrak{A}$ satisfies the hypotheses of Theorem 3.7, there is a c.w.f.f. A of $\mathfrak{A}$ that is undecidable and whose interpretation is that $n_0 \in Q$, where n_0 is some integer and where Q is a recursively enumerable set which is not recursive. Changing the logic in various ways can succeed only in changing the value of n_0. By Theorem 7-3.1, such a proposition has the simple arithmetic form $\bigvee_{y} \bigwedge_{k=0}^{y} \bigvee_{\mathfrak{x}^{(n)}} [P(k, n_0, y, \mathfrak{x}^{(n)}) = 0]$, where the polynomial remains unchanged as the logic is varied.

Next, let $Q = S$, where S is the *simple* set constructed in Chap. 5, Sec. 5. Then, since $\{x \mid \vdash_{\mathfrak{A}} - U(x)\}$ is recursively enumerable, it must be *finite*. That is, if the hypotheses of Theorem 3.7 are satisfied by an arithmetical logic $\mathfrak{A}$, of the c.w.f.f.'s $- U(n)$ of $\mathfrak{A}$ chosen as formalizations of the propositions $n \in \bar{S}$, at most a finite number can be theorems of $\mathfrak{A}$.

If we take $Q = K = \left\{x \mid \bigvee_{y} T(x, x, y)\right\}$, we may make use of the fact that, by Corollary 5-5.2, K is creative. Hence, there is a recursive function $f(x)$ such that, if

$$\{m\} = \{n \mid \vdash_{\mathfrak{A}} - U(n)\} \subset \bar{K},$$

then $f(m) \in \bar{K}$, whereas $f(m) \notin \{m\}$, that is, $- U(f(m))$ is not a theorem of $\mathfrak{A}$. If we now construct $\mathfrak{A}'$ by adjoining the w.f.f. $- U(f(m))$ as an additional axiom, we have a new arithmetical logic in which the undecidable proposition $f(m) \in \bar{K}$ has been decided. Of course, the process is capable of iteration.}

In order to verify that the results of the present section actually do hold for some definite arithmetical logic, it suffices to show that the logic in question is ω-consistent and adequate.

The proof of adequacy of a logic is usually not very difficult to supply. Some economy is often made possible by a shrewd choice of a basis. A very popular choice is the class of binary primitive recursive predicates. (E.g., Gödel [1] and Hilbert and Bernays [1] make use of this basis.) Often it is simpler to use recursive predicates directly, making use of a formulation like that of our Definition 3-1.2, which does not involve primitive recursion. For the system Z of Hilbert and Bernays [1], the class of predicates of the form

$$\bigwedge_{k=0}^{y} \bigvee_{x_1, \ldots, x_n=0}^{y} [P(k, x, y, \mathfrak{x}^{(n)}) = 0],$$

where P is a polynomial, is an ideal choice.

The question of ω-consistency is more delicate. {On the one hand, it is ordinarily possible to give a seemingly valid proof of the ω-consistency of many arithmetical logics by explicitly invoking their intended interpretation. That is, one can ordinarily prove:

(1) The axioms of the logic are true.[1]

(2) The rules of inference of the logic preserve truth.

(3) If the logic is ω-inconsistent, at least one of its theorems is false.

Such a proof of ω-consistency is open to the objection of *circularity.* For, in the process of showing that no theorems of the logic are false, one is tacitly using this very assertion on an informal level. This is not a matter of pedantry; at least four logics which were originally proposed quite seriously have, in fact, turned out to be inconsistent. Another difficulty is associated with the fact that in many arithmetical logics $\mathfrak{A}$, within which it is possible to find a c.w.f.f. C whose interpretation is that $\mathfrak{A}$ is consistent, it can be proved that C is not a theorem[2] of $\mathfrak{A}$. Thus a proof of the consistency (and hence certainly the ω-consistency) of a logic requires the use of methods not formalized within the logic.}

These considerations make our results seem somewhat equivocal. However, in their negative aspect there is really no equivocation. Either an adequate arithmetical logic is ω-inconsistent {in which case it is possible to prove false statements within it} or it has an unsolvable decision problem and is subject to the limitation of Gödel's incompleteness theorem.

4. First-order Logics. In this section we deal with yet another class of logics, the *first-order logics.*

The alphabet of a first-order logic $\mathfrak{F}$ contains the symbols

$$- \supset [\ ] \ , \ (\ )$$

Also in the alphabet there is a denumerable infinity of symbols called *individual variables.* We write these

$$\boldsymbol{x}, \boldsymbol{y}, \boldsymbol{z}, \boldsymbol{x}_1, \boldsymbol{y}_1, \boldsymbol{z}_1, \boldsymbol{x}_2, \boldsymbol{y}_2, \boldsymbol{z}_2, \ldots .$$

[1] In the case of w.f.f.'s containing free variables, "true" means true for all values of these variables.

An "abuse of language" is involved here. When we say that an axiom is true, we mean of course that the sentence it represents is true.

[2] Cf. Gödel [1].

The remaining symbols are *individual constants*, *predicate symbols*, or *function symbols*. Each predicate symbol and function symbol has associated with it an integer $\geqq 1$, called its *degree*. By an *individual symbol* we understand either an individual variable or an individual constant.

Of course, our development has been in terms of the fixed alphabet

$$a_0, a_1, a_2, \ldots .$$

Hence, we shall assume that the classes of individual variables, of individual constants, of predicate symbols, and of function symbols are recursive, and, also, that there is a partial recursive function $Q(n)$ such that, if a_n is a predicate symbol or a function symbol, then $Q(n)$ is the degree of a_n.

A word X is called a *term* of a first-order logic $\mathfrak{F}$ if there exists a sequence $X_1, \ldots, X_n$ of words such that X_n is X and, for each i, $1 \leqq i \leqq n$, either

(1) X_i is an individual symbol, or
(2) X_i is $a(X_{j_1}, X_{j_2}, \ldots, X_{j_m})$, where $j_1, j_2, \ldots, j_m < i$ and where a is a function symbol of degree m.

A word W is called a *well-formed formula* (abbreviated *w.f.f.*) of a first-order logic $\mathfrak{F}$ if there exists a sequence $W_1, \ldots, W_n$ of words such that W_n is W and, for each i, $1 \leqq i \leqq n$, either

(1) W_i is $p(X_1, \ldots, X_m)$, where $X_1, \ldots, X_m$ are terms and p is a predicate symbol of degree m, or
(2) W_i is $[W_j \supset W_k]$, where $j, k < i$, or
(3) W_i is $- W_j$, where $j < i$, or
(4) W_i is $(v)W_j$, where $j < i$ and v is an individual variable.

In the case of a function or predicate symbol a of degree 2, we often write (XaY) for $a(X, Y)$.

A first-order logic has two rules of inference, namely:

MODUS PONENS. *This rule of inference is given by the 3-ary predicate $\mathfrak{M}(Y, X_1, X_2)$, which is true when and only when X_2 is $[X_1 \supset Y]$.*

GENERALIZATION. *This rule of inference is given by the predicate $\mathfrak{G}(Y, X)$, which is true when and only when Y is $(v)X$, where v is an individual variable.*

It is left to the reader to verify that these rules of inference are recursive predicates.

An occurrence of an individual variable v in a w.f.f. W is a *bound occurrence* if it is in a w.f. part of W of the form $(v)A$. An occurrence of v in a w.f.f. W which is not bound is called a *free occurrence*. A w.f.f. in which no individual variable has a free occurrence is called *closed*.

We write *c.w.f.f.* for "closed w.f.f." If W is a w.f.f., v is an individual variable, and X is a term, we define $\mathfrak{S}(W, v, X)$ as follows:

If no free occurrence of v in W is in a w.f. part of W of the form $(w)A$, where w occurs in X, then $\mathfrak{S}(W, v, X)$ is the result of replacing v by X at all free occurrences of v in W; otherwise $\mathfrak{S}(W, v, X)$ is W.

The axioms of a first-order logic include all words obtained (subject to the specific restriction mentioned below) by replacing v by an individual variable, X by a term, and A, B, C by w.f.f.'s in the schemata:

(1) $[A \supset [B \supset A]]$.
(2) $[[A \supset [B \supset C]] \supset [[A \supset B] \supset [A \supset C]]]$.
(3) $[[-B \supset -A] \supset [A \supset B]]$.
(4) $[(v)A \supset \mathfrak{S}(A, v, X)]$.
(5) $[(v)[A \supset B] \supset [A \supset (v)B]]$, where v has no free occurrence in A.

In addition to these, a first-order logic contains a *finite* (possibly zero) number of additional axioms, all of which are c.w.f.f.'s; these are called the *special axioms* of the logic.

THEOREM 4.1. *If A is any w.f.f. of a first-order logic $\mathfrak{F}$, then* $\vdash_{\mathfrak{F}} [A \supset A]$.

PROOF. By schema (2),

$$\vdash_{\mathfrak{F}} [[A \supset [[A \supset A] \supset A]] \supset [[A \supset [A \supset A]] \supset [A \supset A]]].$$

By schema (1),

$$\vdash_{\mathfrak{F}} [A \supset [[A \supset A] \supset A]]$$

and

$$\vdash_{\mathfrak{F}} [A \supset [A \supset A]].$$

Applying modus ponens twice,

$$\vdash_{\mathfrak{F}} [A \supset A].$$

DEFINITION 4.1. *Let $\mathfrak{F}$ be a first-order logic and let $A_1, \ldots, A_n$ be c.w.f.f.'s of $\mathfrak{F}$. Then, $\mathfrak{F}(A_1, \ldots, A_n)$ is the first-order logic obtained from $\mathfrak{F}$ by adjoining to its special axioms those of $A_1, \ldots, A_n$ that are not axioms of $\mathfrak{F}$.*

THEOREM 4.2. *If $\mathfrak{F}$ is a first-order logic and A is a c.w.f.f. of $\mathfrak{F}$, then* $\vdash_{\mathfrak{F}(A)} W$ *if and only if* $\vdash_{\mathfrak{F}} [A \supset W]$.

PROOF. If $\vdash_{\mathfrak{F}} [A \supset W]$, then $\vdash_{\mathfrak{F}(A)} [A \supset W]$. But then, by modus ponens, $\vdash_{\mathfrak{F}(A)} W$.

Conversely, suppose that $\vdash_{\mathfrak{F}(A)} W$. Let $W_1, \ldots, W_n$ be a proof of W in $\mathfrak{F}(A)$, so that W_n is W. We shall show that, for each i, $1 \leqq i \leqq n$, $\vdash_{\mathfrak{F}} [A \supset W_i]$. Suppose this result known for all $j < i$. We distinguish several cases.

CASE I. *W_i is A.* Then, $[A \supset W_i]$ is $[A \supset A]$, which, by Theorem 4.1, is a theorem of $\mathfrak{F}$.

CASE II. *W_i is an axiom of $\mathfrak{F}$.* Then, $\vdash_{\mathfrak{F}} W_i$. By schema (1), $\vdash_{\mathfrak{F}} [W_i \supset [A \supset W_i]]$. By modus ponens, $\vdash_{\mathfrak{F}} [A \supset W_i]$.

CASE III. *W_j is $[W_k \supset W_i]$ for $j, k < i$.* Then, by induction hypothesis,

$$\vdash_{\mathfrak{F}} [A \supset W_k],$$
$$\vdash_{\mathfrak{F}} [A \supset [W_k \supset W_i]].$$

By schema (2),

$$\vdash_{\mathfrak{F}} [[A \supset [W_k \supset W_i]] \supset [[A \supset W_k] \supset [A \supset W_i]]].$$

By modus ponens twice,

$$\vdash_{\mathfrak{F}} [A \supset W_i].$$

CASE IV. *W_i is $(v)W_j$ for $j < i$, where v is an individual variable.* By induction hypothesis, $\vdash_{\mathfrak{F}} [A \supset W_j]$. By generalization, $\vdash_{\mathfrak{F}} (v)[A \supset W_j]$. Since A is a c.w.f.f., schema (5) is applicable, and

$$\vdash_{\mathfrak{F}} [(v)[A \supset W_j] \supset [A \supset (v)W_j]].$$

By modus ponens,

$$\vdash_{\mathfrak{F}} [A \supset (v)W_j],$$

that is,

$$\vdash_{\mathfrak{F}} [A \supset W_i].$$

This completes the proof.

COROLLARY 4.3. *If $\mathfrak{F}$ is a first-order logic and $A_1, \ldots, A_n$ are c.w.f.f.'s of $\mathfrak{F}$, then $\vdash_{\mathfrak{F}(A_1,\ldots,A_n)} W$ if and only if*

$$\vdash_{\mathfrak{F}} [A_1 \supset [A_2 \supset [A_3 \supset \cdots \supset [A_n \supset W]\cdots]]].$$

PROOF. It suffices to apply Theorem 4.2 repeatedly.

THEOREM 4.4. *If $\mathfrak{F}$ is a first-order logic and $A_1, \ldots, A_n$ are c.w.f.f.'s of $\mathfrak{F}$, then $\mathfrak{F}(A_1, \ldots, A_n)$ is translatable into $\mathfrak{F}$.*

PROOF. Recalling Definition 1.5, it is necessary only, by Corollary 4.3, to remark that, if

$$\mathfrak{f}(W) = [A_1 \supset [A_2 \supset \cdots \supset [A_n \supset W]\cdots]],$$

then $\mathfrak{f}(W)$ is recursive.

DEFINITION 4.2. *A first-order logic $\mathfrak{F}$ is called* **unspecialized** *if it has no special axioms.*

THEOREM 4.5. *Every first-order logic is translatable into an unspecialized first-order logic.*

PROOF. Let $\mathfrak{F}$ be any first-order logic, and let $\mathfrak{F}'$ be like $\mathfrak{F}$ except that $\mathfrak{F}'$ has no special axioms. Thus, $\mathfrak{F}'$ is unspecialized. Moreover, let the special axioms of $\mathfrak{F}$ be $A_1, \ldots, A_n$. Then, $\mathfrak{F} = \mathfrak{F}'(A_1, \ldots, A_n)$. Hence, by Theorem 4.4, $\mathfrak{F}$ is translatable into $\mathfrak{F}'$.

We now introduce the unspecialized first-order logic $\mathfrak{F}_0$, the *full first-order logic.* $\mathfrak{F}_0$ has a denumerable infinity of individual constants and, for each integer $m \geqq 1$, a denumerable infinity of function and predicate symbols of degree m. In more detail, the seven symbols

$$- \supset [\ ] \ , (\ )$$

are identified with $a_0, a_1, a_2, a_3, a_4, a_5, a_6$, respectively. For $n > 6$, a_n is an *individual variable* if $L(n) = 0$ and $K(n)$ is even, an *individual constant* if $L(n) = 0$ and $K(n)$ is odd. Thus, $x, y, z, x_1, y_1, z_1, x_2, y_2, z_2, \ldots$ are identified with $a_{J(4,0)}, a_{J(6,0)}, a_{J(8,0)}, \ldots$. For $n > 6$, $L(n) > 0$, $L(n)$ even, a_n is a *predicate symbol of degree* $\frac{1}{2}L(n)$. For $n > 6$, $L(n)$ odd, a_n is a *function symbol of degree* $\frac{1}{2}(L(n) + 1)$.

$\mathfrak{F}_0$ is essentially the logic called the *engere Prädikatenkalkül* in Hilbert and Ackermann [1] and called the *pure first-order functional calculus* by Church [4, 5].†

The importance of $\mathfrak{F}_0$ stems from

THEOREM 4.6. *Every unspecialized first-order logic is translatable into* $\mathfrak{F}_0$.

PROOF. Let $\mathfrak{F}$ be some unspecialized first-order logic. We define a one-one recursive function $f(n)$ on the integers as follows:

For each n, we check the role played by a_n in $\mathfrak{F}$. If a_n is one of

$$- \supset [\ ] \ , (\ )$$

in $\mathfrak{F}$, $f(n)$ is 0, 1, 2, 3, 4, 5, or 6, respectively. If $f(x)$ is defined for $x < n$, and if a_n is an *individual variable* in $\mathfrak{F}$, an *individual constant* in $\mathfrak{F}$, a *predicate symbol of degree m* in $\mathfrak{F}$, or a *function symbol of degree m* in $\mathfrak{F}$,

† Actually, these logics, as they were originally formulated, do not have provision for function symbols or individual constants. However, this difference is not an essential one. Cf. Hilbert and Bernays [1, p. 422] and Kleene [6, p. 408]. Although the detailed proof is quite involved, the idea behind it is quite simple. We set aside a predicate symbol $=$ of degree 2, assuming the special axioms (1) to (3), below. Then, to eliminate an individual constant c, we set aside a predicate symbol C of degree 1 and assume the special axioms (cf. the next footnote)

$$(\exists x)C(x),$$
$$[C(x) \supset [C(y) \supset (x = y)]].$$

To eliminate a function symbol f of degree m, we set aside a predicate symbol F of degree $m + 1$ and assume the special axioms

$$(\exists y)F(x_1, x_2, \ldots, x_m, y),$$
$$[F(x_1, x_2, \ldots, x_m, y) \supset [F(x_1, x_2, \ldots, x_m, z) \supset (y = z)]].$$

It is then necessary to replace individual constants and function symbols by these predicates in the appropriate way. For example, $G(c)$ becomes

$$(x)[C(x) \supset G(x)].$$

then $f(n)$ is the least integer p such that $f(x) \neq p$ for $x < n$, and a_p is an *individual variable* in $\mathfrak{F}_0$, an *individual constant* in $\mathfrak{F}_0$, a *predicate symbol of degree m* in $\mathfrak{F}_0$, or a *function symbol of degree m* in $\mathfrak{F}_0$, respectively. Now, we define the function $\mathfrak{f}$ as follows:

If $X = a_{r_1}a_{r_2} \cdots a_{r_k}$, then $\mathfrak{f}(X) = a_{f(r_1)}a_{f(r_2)} \cdots a_{f(r_k)}$. It is easy to see that $\mathfrak{f}(X)$ satisfies the conditions of Definition 1.5. Hence, $\mathfrak{F}$ is translatable into $\mathfrak{F}_0$.

THEOREM 4.7. *If the decision problem for $\mathfrak{F}_0$ is recursively solvable, then so is that for every first-order logic.*

PROOF. Immediate from Theorems 1.3, 1.5, 4.5, and 4.6.

Theorem 4.7 was known (necessarily on the informal level) to Hilbert and his school well before the formal development of recursive-function theory. On the basis of this theorem, Hilbert declared that the decision problem for $\mathfrak{F}_0$ (often referred to simply as the *Entscheidungsproblem*) was the central problem of mathematical logic. Subsequently, there was much research directed toward a positive solution of this problem. In part, this took the form of reductions of the decision problem for $\mathfrak{F}_0$ to that for classes of special w.f.f.'s, for example, the class of w.f.f.'s of the form[1]

$$(\exists x_1)(\exists x_2)(\exists x_3)(y_1)(y_2)\cdots(y_n)A,$$

where A is free of quantifiers. Other work was directed toward showing that, for larger and larger classes of w.f.f.'s, the decision problem could be solved. In particular, it was proved that the decision problem could be solved for the class of all w.f.f.'s of the form[1]

$$(\exists x_1)(\exists x_2)(y_1)(y_2)\cdots(y_n)A,$$

where A is free of quantifiers. Thus, it seemed that but the smallest advance in either direction would suffice to settle the problem.

In fact, it was proved by Church[2] and by Turing[3] (independently of each other) that the decision problem for $\mathfrak{F}_0$ is recursively unsolvable. By Theorem 4.7, this will follow at once if we can produce any first-order logic whose decision problem is recursively unsolvable.[4]

We shall prove

THEOREM 4.8. *Every normal system ν with an alphabet of two letters is translatable into a first-order logic $\mathfrak{F}_\nu$.*

[1] Here, by $(\exists x_i)W$ is understood $-(x_i) - W$. Cf. Church [5].

[2] Cf. Church [2].

[3] Cf. Turing [1].

[4] It might be expected that we would achieve this by constructing a first-order logic which is also an ω-consistent and adequate arithmetical logic. Indeed, the proof of Church [2] follows essentially these lines. A simpler development is to be found in Tarski, Mostowski, and R. M. Robinson [1]. However, because of the difficulties associated with ω-consistency, we choose to proceed otherwise.

Assuming this result for the moment, Theorems 1.3 and 6-5.4 immediately yield[1]

COROLLARY 4.9. *There exists a first-order logic whose decision problem is recursively unsolvable.*

Finally, using Theorem 4.7, we have

COROLLARY 4.10. *The decision problem for $\mathfrak{F}_0$ is recursively unsolvable.*

It remains only to give the proof of Theorem 4.8.

Let ν be a normal system on the alphabet 1, b. Let A_0 be its axiom, and let its productions be $g_iP \to P\bar{g}_i$, $i = 1, 2, \ldots, m$. Then we let the alphabet of $\mathfrak{F}_\nu$ contain the individual constants $\mathbf{1}$ and $\boldsymbol{b}$, the function symbol $\star$ of degree 2, the predicate symbol $\boldsymbol{T}$ of degree 1, and the predicate symbol $=$ of degree 2.

With each word W on the alphabet 1, b, we associate a *term* W' of $\mathfrak{F}_\nu$ as follows:

$$1' = \mathbf{1},\ b' = \boldsymbol{b};$$
$$(W1)' = (W' \star \mathbf{1}),\ (Wb)' = (W' \star \boldsymbol{b}).$$

Finally, we let $\mathfrak{f}(W) = \boldsymbol{T}(W')$. Clearly, $\mathfrak{f}(W)$ is recursive. The special axioms of $\mathfrak{F}_\nu$ will be chosen in such a way that $\vdash_\nu W$ if and only if $\vdash_{\mathfrak{F}_\nu} \mathfrak{f}(W)$.

The special axioms of $\mathfrak{F}_\nu$ are as follows:

(1) $(x_1)(x_1 = x_1)$.
(2) $(x_1)(x_2)[(x_1 = x_2) \supset (x_2 = x_1)]$.
(3) $(x_1)(x_2)(x_3)[(x_1 = x_2) \supset [(x_2 = x_3) \supset (x_1 = x_3)]]$.
(4) $(x_1)(x_2)[(x_1 = x_2) \supset [\boldsymbol{T}(x_1) \supset \boldsymbol{T}(x_2)]]$.
(5) $(x_1)(x_2)(x_3)[(x_1 = x_2) \supset ((x_1 \star x_3) = (x_2 \star x_3))]$.
(6) $(x_1)(x_2)(x_3)[(x_1 = x_2) \supset ((x_3 \star x_1) = (x_3 \star x_2))]$.
(7) $(x_1)(x_2)(x_3)((x_1 \star (x_2 \star x_3)) = ((x_1 \star x_2) \star x_3))$.
(8) $\boldsymbol{T}(A_0')$.
(9) $(x_1)[\boldsymbol{T}((g_i' \star x_1)) \supset \boldsymbol{T}((x_1 \star \bar{g}_i'))]$, $i = 1, 2, \ldots, m$.

It is required to prove that $\vdash_\nu W$ if and only if $\vdash_{\mathfrak{F}_\nu} \boldsymbol{T}(W')$.

A term of $\mathfrak{F}_\nu$ is called a *constant* if it contains no individual variables. If X is a constant, we write $\{X\}$ to denote the word on 1, b obtained from X by replacing $\mathbf{1}$ by 1, $\boldsymbol{b}$ by b, and by removing all parentheses and all stars. For example, $\{W'\} = W$. Two constants X, Y are called *associates* if $\{X\} = \{Y\}$.

Then, we have

LEMMA 1. *If X and Y are constants, then $\vdash_{\mathfrak{F}_\nu} (X = Y)$ if and only if X and Y are associates.*

[1] Alternatively, Corollary 4.9 can be deduced from Theorem 4.8 by using Theorem 6-5.2 and Corollary 2.2.

PROOF. If X and Y are associates, then $\{X\} = \{Y\}$ so $\{X\}$ and $\{Y\}$ certainly are of the same length. If this length is 1 or 2, the result is quite obvious. If the length is 3, then if, say, $\{X\} = \{Y\} = 1b1$, the only possibilities for X, Y are $(\mathbf{1} \star (b \star \mathbf{1}))$, $((\mathbf{1} \star b) \star \mathbf{1})$. Now, by axiom (7),

$$\vdash_{\mathfrak{F}_\nu} (x_1)(x_2)(x_3)((x_1 \star (x_2 \star x_3)) = ((x_1 \star x_2) \star x_3)).$$

By axiom schema (4) for first-order logics,

$$\vdash_{\mathfrak{F}_\nu} [(x_1)(x_2)(x_3)((x_1 \star (x_2 \star x_3)) = ((x_1 \star x_2) \star x_3)) \supset (x_2)(x_3)((\mathbf{1} \star (x_2 \star x_3)) = ((\mathbf{1} \star x_2) \star x_3))],$$

$$\vdash_{\mathfrak{F}_\nu} [(x_2)(x_3)((\mathbf{1} \star (x_2 \star x_3)) = ((\mathbf{1} \star x_2) \star x_3)) \supset (x_3)((\mathbf{1} \star (b \star x_3)) = ((\mathbf{1} \star b) \star x_3))],$$

$$\vdash_{\mathfrak{F}_\nu} [(x_3)((\mathbf{1} \star (b \star x_3)) = ((\mathbf{1} \star b) \star x_3)) \supset ((\mathbf{1} \star (b \star \mathbf{1})) = ((\mathbf{1} \star b) \star \mathbf{1}))].$$

By modus ponens three times,

$$\vdash_{\mathfrak{F}_\nu} ((\mathbf{1} \star (b \star \mathbf{1})) = ((\mathbf{1} \star b) \star \mathbf{1}));$$

that is,

$$\vdash_{\mathfrak{F}_\nu} (X = Y).$$

A similar argument works for any other word of length 3. The result for lengths > 3 is proved by mathematical induction.

The converse follows by noting that the only equalities that can be obtained outright are those coming from axioms (1) and (7), which certainly cannot give the equality of nonassociates. Axioms (2) to (6), however, can go from equalities of associates only to equalities of associates.

LEMMA 2. *If* $\vdash_\nu W$, *then* $\vdash_{\mathfrak{F}_\nu} \mathbf{T}(W')$.

PROOF. Let $W_1, W_2, \ldots, W_n$ be a proof in ν, where W_n is W. We shall show that, for each i, $1 \leqq i \leqq n$, $\vdash_{\mathfrak{F}_\nu} T(W'_i)$. This is clearly true for $i = 1$, since W_1 is A_0. Suppose it true for $i = k$. Now, for suitable j, $W_k = g_jP$, $W_{k+1} = P\bar{g}_j$.

By axiom schema (4) for first-order logics,

$$\vdash_{\mathfrak{F}_\nu} [(x_1)[T((g'_j \star x_1)) \supset T((x_1 \star \overline{g}'_j))] \supset [T((g'_j \star P')) \supset T((P' \star \overline{g}'_j))]].$$

This, with axiom (9) and modus ponens, yields

$$\vdash_{\mathfrak{F}_\nu} [T((g'_j \star P')) \supset T((P' \star \overline{g}'_j))].$$

By Lemma 1, we have

$$\vdash_{\mathfrak{F}_\nu} ((g_jP)' = (g'_j \star P')).$$

By the use of axiom (4), this will give

$$\vdash_{\mathfrak{F}_\nu} [T((g_jP)') \supset T((g'_j \star P'))].$$

Now, using our induction hypothesis and modus ponens twice, we have

$$\vdash_{\mathfrak{F}_\nu} T((P' \star \overline{g_j'})).$$

And then, by Lemma 1 and axiom (4),

$$\vdash_{\mathfrak{F}_\nu} T((P\overline{g_j})');$$

that is,

$$\vdash_{\mathfrak{F}_\nu} T(W'_{k+1}).$$

LEMMA 3. *If X is a constant and $\vdash_\nu \{X\}$, then $\vdash_{\mathfrak{F}_\nu} \mathrm{T}(X)$.*

PROOF. If $X = \{X\}'$, we are through, by Lemma 2. If not, X is an associate of $Y = \{X\}'$. Then, by Lemmas 1 and 2,

$$\vdash_{\mathfrak{F}_\nu} (X = Y),$$
$$\vdash_{\mathfrak{F}_\nu} T(Y).$$

By axiom (2),

$$\vdash_{\mathfrak{F}_\nu} (Y = X).$$

By axiom (4),

$$\vdash_{\mathfrak{F}_\nu} [T(Y) \supset T(X)].$$

Hence,

$$\vdash_{\mathfrak{F}_\nu} T(X).$$

LEMMA 4. *If X is a constant and $\vdash_{\mathfrak{F}_\nu} T(X)$, then $\vdash_\nu \{X\}$.*

PROOF. The only axiom of $\mathfrak{F}_\nu$ of the form $T(X)$, where X is constant, has the desired property.

To derive $T(X)$, X constant, in $\mathfrak{F}_\nu$, axiom (4) or (9) must be used.[1] But, by Lemma 1, axiom (4) can produce $T(X)$ as a theorem only if we already know that $\vdash_{\mathfrak{F}_\nu} T(Y)$, where X and Y are associates. And axiom (9) can be used to produce $T(X)$ as a theorem only if $\{X\}$ is a consequence of $\{Y\}$ by one of the productions of ν and if $\vdash_{\mathfrak{F}_\nu} T(Y)$.

LEMMA 5. *$\vdash_\nu W$ if and only if $\vdash_{\mathfrak{F}_\nu} T(W')$.*

PROOF. Clear from Lemmas 2 and 4.

This completes the proof of Theorem 4.8.

5. Partial Propositional Calculi. The alphabet of a *partial propositional calculus* $\mathfrak{P}$ consists of the symbols

$$- \quad \supset \quad [\quad]$$

and of the denumerable infinitude of symbols

$$p_1, q_1, r_1, p_2, q_2, r_2, p_3, q_3, r_3, \ldots$$

called *propositional variables.*

A word W is called a *well-formed formula* (abbreviated w.f.f.) of a partial propositional calculus if there exists a sequence $W_1, \ldots, W_n$

[1] This fact follows from Gentzen's Hauptsatz. Cf. Kleene [6].

of words such that W_n is W and for each i, $1 \leqq i \leqq n$, either

(1) W_i is a propositional variable,
(2) W_i is $[W_j \supset W_k]$, $j, k < i$, or
(3) W_i is $- W_j$, $j < i$.

A partial propositional calculus has two rules of inference, namely,

MODUS PONENS. *This rule of inference is given by the 3-ary predicate* $\mathfrak{M}(Y, X_1, X_2)$, *which is true when and only when* X_2 *is* $[X_1 \supset Y]$.

SUBSTITUTION. *This rule of inference is given by the predicate* $\mathfrak{S}(Y, X)$, *which is true when and only when* Y *results from* X *on replacing some propositional variable at all of its occurrences in* X *by a w.f.f.* W.

Let U be some set which consists of two distinct objects (e.g., the numbers 0 and 1); these objects we shall write T and F and we shall call *truth values.* A function whose domain is the set of n-tuples of truth values, and whose values are always T or F, is called a *truth function.* With each w.f.f. A of a partial propositional calculus we associate a truth function $\bar{A}$, as follows:

(1) If A is a propositional variable, then $\bar{A}$ is the identity function.

(2) If A is $[B \supset C]$, then $\bar{A}$ is F only for arguments that make $\bar{B} = T$ and $\bar{C} = F$; otherwise $\bar{A}$ is T.

(3) If A is $- B$, then $\bar{A}$ is T when and only when $\bar{B}$ is F.

The w.f.f. A is called a *tautology* if $\bar{A}$ takes on only the value T. We leave it to the reader to demonstrate

THEOREM 5.1. *The class of tautologies is recursive.*

Our final stipulation concerning partial propositional calculi is that they shall have only a finite number of axioms, *all of which are tautologies.*

THEOREM 5.2. *Every theorem of a partial propositional calculus is a tautology.*

PROOF. The axioms are tautologies, and, as one can readily verify, the rules of inference preserve the property of being a tautology.

In particular, we may consider $\mathfrak{P}_0$, the *full partial propositional calculus*, or, more simply, the *propositional calculus*, whose axioms are the three w.f.f.'s:

(1) $[p_1 \supset [q_1 \supset p_1]]$.
(2) $[[p_1 \supset [q_1 \supset r_1]] \supset [[p_1 \supset q_1] \supset [p_1 \supset r_1]]]$.
(3) $[[- q_1 \supset - p_1] \supset [p_1 \supset q_1]]$.

DEFINITION 5.1. *A partial propositional calculus* $\mathfrak{P}$ *is called* **complete** *if every tautology is a theorem of* $\mathfrak{P}$.

COROLLARY 5.3. *If* $\mathfrak{P}$ *is complete, then the decision problem for* $\mathfrak{P}$ *is recursively solvable.*

PROOF. Immediate from Theorems 5.1 and 5.2.

One of the fundamental results of modern logic (due to Post) is

THEOREM 5.4. *$\mathfrak{P}_0$ is complete.*[1]

COROLLARY 5.5. *The decision problem for $\mathfrak{P}_0$ is recursively solvable.*

These results led to the belief that, in this domain at least, no unsolvable problems were to be found. However, Linial and Post were able[2] to show that there exists a partial propositional calculus whose decision problem is recursively unsolvable. We obtain their result by proving:

THEOREM 5.6. *Every semi-Thue system σ with an alphabet of two letters is translatable into a partial propositional calculus.*

Let σ have the alphabet 1, b, the axiom A_0, and the productions $Pg_iQ \to P\bar{g}_iQ$, $i = 1, 2, \ldots, m$. We set

$$\begin{aligned} 1' &= - - [- p_1 \supset - p_1] \\ b' &= - - - - - [- p_1 \supset - p_1] \\ (W1)' &= [W' \,\&\, 1'] \\ (Wb)' &= [W' \,\&\, b'], \end{aligned}$$

where by $[A \,\&\, B]$ is understood $-[A \supset - B]$. Then if W is any word on 1, b, the w.f.f. W' is a tautology. We define the partial propositional calculus $\mathfrak{P}_\sigma$ to have the axioms:

(1) $[[p_1 \,\&\, [q_1 \,\&\, r_1]] \supset [[p_1 \,\&\, q_1] \,\&\, r_1]]$.
(2) $[[[p_1 \,\&\, q_1] \,\&\, r_1] \supset [p_1 \,\&\, [q_1 \,\&\, r_1]]]$.
(3) $[[p_1 \supset q_1] \supset [[q_1 \supset p_1] \supset [[r_1 \,\&\, p_1] \supset [r_1 \,\&\, q_1]]]]$.
(4) $[[p_1 \supset q_1] \supset [[q_1 \supset p_1] \supset [[p_1 \,\&\, r_1] \supset [q_1 \,\&\, r_1]]]]$.
(5) A_0'; $[p_1 \supset p_1]$.
(6) $[[[p_1 \,\&\, g_i'] \,\&\, q_1] \supset [[p_1 \,\&\, \bar{g}_i'] \,\&\, q_1]]$, $i = 1, 2, \ldots, m$.†

It is easy to verify that these axioms are all tautologies.

A word X is called *regular* if there exists a finite sequence $X_1, X_2, \ldots, X_n$, where $X_n = X$, such that, for each $i = 1, 2, \ldots, n$, either X_i is $1'$ or b', or X_i is $[X_j \,\&\, X_k]$, $j, k < i$. Thus, W' is always regular, as is, for example, $[[1' \,\&\, b'] \,\&\, [b' \,\&\, 1']]$.

If X is a regular word, we let $\langle X \rangle$ be the word on 1, b obtained by first replacing all occurrences of b' in X by b, then replacing all remaining occurrences of $1'$ by 1, and finally removing all occurrences of [,], and &. Two regular words X, Y are called *associates* if $\langle X \rangle = \langle Y \rangle$.

LEMMA 1. *If X and Y are associates, then $\vdash_{\mathfrak{P}_\sigma} [X \supset Y]$ and also $\vdash_{\mathfrak{P}_\sigma} [Y \supset X]$.*

PROOF. By mathematical induction, employing axioms (1) to (4).

[1] For a short and elegant proof of this result (due to Kalmár [1]), see Church [5].

[2] Cf. Linial and Post [1].

† Three additional tautalogies giving the effect of the semi-Thue productions with empty P or Q or both should be added here.

LEMMA 2. *If* W_2 *is a consequence of* W_1 *by one of the productions of* σ *and* $\vdash_{\mathfrak{P}_\sigma} W_1'$, *then* $\vdash_{\mathfrak{P}_\sigma} W_2'$.

PROOF. Let $W_1 = Pg_iQ$, $W_2 = P\bar{g}_iQ$. Then we have

$$\vdash_{\mathfrak{P}_\sigma} (Pg_iQ)'.$$

By Lemma 1,

$$\vdash_{\mathfrak{P}_\sigma} [(Pg_iQ)' \supset [[P' \,\&\, g_i'] \,\&\, Q']].$$

By Axiom (6),

$$\vdash_{\mathfrak{P}_\sigma} [[[P' \,\&\, g_i'] \,\&\, Q'] \supset [[P' \,\&\, \bar{g}_i'] \,\&\, Q']].$$

By Lemma 1,

$$\vdash_{\mathfrak{P}_\sigma} [[[P' \,\&\, \bar{g}_i'] \,\&\, Q'] \supset (P\bar{g}_iQ)'].$$

By modus ponens, three times,

$$\vdash_{\mathfrak{P}_\sigma} (P\bar{g}_iQ)',$$

that is,

$$\vdash_{\mathfrak{P}_\sigma} W_2'.$$

LEMMA 3. *If* $\vdash_\sigma W$, *then* $\vdash_{\mathfrak{P}_\sigma} W'$.

PROOF. A_0' is an axiom and hence a theorem. The result then follows from Lemma 2.

LEMMA 4. *If* X *is regular and* $\vdash_{\mathfrak{P}_\sigma} X$, *then* $\vdash_\sigma \langle X\rangle$.

PROOF. The only axiom of $\mathfrak{P}_\sigma$ that is regular is A_0'.

Substitution in axioms (1) to (4) and (6) and subsequent modus ponens can yield only regular words from regular words. Axioms (1) and (2) can go only from a word to one of its associates. Axiom (6) can yield regular Y as a theorem of $\mathfrak{P}_\sigma$ only if $\langle Y\rangle$ is a consequence of $\langle X\rangle$ by one of the productions of σ, and we already have $\vdash_{\mathfrak{P}_\sigma} X$. Finally, axioms (3) and (4) can do no harm since, in a semi-Thue system, when Q is a consequence of P, then RQ is a consequence of RP, and QR is a consequence of PR.

LEMMA 5. $\vdash_\sigma W$ *if and only if* $\vdash_{\mathfrak{P}_\sigma} W'$.

PROOF. Clear from Lemmas 3 and 4.

Lemma 5, coupled with the fact that the function $\mathfrak{f}(W) = W'$ is recursive, completes the proof of Theorem 5.6.

THEOREM 5.7. *There exists a partial propositional calculus with a recursively unsolvable decision problem.*

PROOF. The result follows at once from Theorems 5.6, 1.3, and 6-2.6.

Let σ_0 be some definite semi-Thue system on 1, b with a recursively unsolvable decision problem. Then, $\mathfrak{P}_{\sigma_0}$ also has a recursively unsolvable decision problem. Let W be some word on 1, b, and let the partial propositional calculus $\mathfrak{R}(W)$ have as axioms the axioms of $\mathfrak{P}_{\sigma_0}$ and, in addition,

(7) $[W' \supset [p_1 \supset [q_1 \supset p_1]]]$.
(8) $[W' \supset [[p_1 \supset [q_1 \supset r_1]] \supset [[p_1 \supset q_1] \supset [p_1 \supset r_1]]]]$.
(9) $[W' \supset [[- q_1 \supset - p_1] \supset [p_1 \supset q_1]]]$.

Then we have

THEOREM 5.8. $\vdash_{\mathfrak{P}_{\sigma_0}} W'$ *if and only if* $\mathfrak{R}(W)$ *is complete.*

PROOF. If $\vdash_{\mathfrak{P}_{\sigma_0}} W'$, then $\vdash_{\mathfrak{R}(W)} W'$; so, by applying modus ponens to (7) to (9), the axioms of $\mathfrak{P}_0$ are theorems of $\mathfrak{R}(W)$. Hence, all tautologies, being theorems of $\mathfrak{P}_0$, are theorems of $\mathfrak{R}(W)$.

Conversely, suppose that $\mathfrak{R}(W)$ is complete. Then, in particular, $\vdash_{\mathfrak{R}(W)} W'$. Now, axioms (7) to (9) are useless as the antecedent of a modus ponens operation in a proof of W'. If we are to apply modus ponens to these axioms to derive a shorter formula, either W' or the result of substituting in W' must already be available. But axioms (1) to (6) will not yield a result of substituting in W', except via a proof of W' itself. Hence, W' can be proved on the basis of axioms (1) to (6) alone; that is, $\vdash_{\mathfrak{P}_{\sigma_0}} W'$.

By a suitable use of Gödel numbers, we can define what we mean by the recursive solvability or unsolvability of the decision problem:

To determine of a given partial propositional calculus whether or not it is complete.

Theorem 5.8 will then immediately give the recursive unsolvability of this problem; that is,

THEOREM 5.9. *The problem of determining, of a given partial propositional calculus, whether or not it is complete, is recursively unsolvable.*[1]

DEFINITION 5.2. *A partial propositional calculus* $\mathfrak{P}$ *is called* **independent** *if for no axiom A of* $\mathfrak{P}$ *is A a theorem of the partial propositional calculus* $\mathfrak{P}'$, *obtained by deleting A from the axioms of* $\mathfrak{P}$.

Let σ_0 be as above, let W be some word on 1, b, and let the partial propositional calculus $\mathfrak{P}_W$ have as axioms the axioms of $\mathfrak{P}_{\sigma_0}$, and, in addition:

(7′) $- - [p_1 \supset p_1]$.
(8′) $[W' \supset - - [p_1 \supset p_1]]$.

Then, we have

THEOREM 5.10. $\vdash_{\mathfrak{P}_{\sigma_0}} W'$ *if and only if* $\mathfrak{P}_W$ *is not independent.*

PROOF. Suppose $\vdash_{\mathfrak{P}_{\sigma_0}} W'$. Let $\mathfrak{S}_W$ be obtained by deleting axiom (7′) from $\mathfrak{P}_W$. Then, $\vdash_{\mathfrak{S}_W} W'$. Hence, by axiom (8′) and modus ponens, $\vdash_{\mathfrak{S}_W} - - [p_1 \supset p_1]$; so $\mathfrak{P}_W$ is not independent.

Conversely, suppose that $\mathfrak{P}_W$ is not independent. Now, it is easy to see that $\mathfrak{P}_{\sigma_0}$ is independent. Moreover, adjoining axiom (7′) to $\mathfrak{P}_{\sigma_0}$ changes nothing essential; only words obtained by direct substitutions in axiom

[1] This result is due to Linial and Post [1].

(7′) are obtained as new theorems. Hence, $\vdash_{\mathfrak{S}_W} - - [p_1 \supset p_1]$, since none of axioms (1) to (6) is derivable when axiom (8′) is adjoined. But $- - [p_1 \supset p_1]$ is not regular. Hence, it can be obtained only by modus ponens and axiom (8′). That is, we must have $\vdash_{\mathfrak{S}_W} W'$. Hence, $\vdash_{\mathfrak{P}_{\sigma_0}} W'$.

As in the case of completeness, a suitable use of Gödel numbers now yields

THEOREM 5.11. *The problem of determining, of a given partial propositional calculus, whether or not it is independent is recursively unsolvable.*[1]

[1] This result is due to Linial and Post [1].

PART 3

FURTHER DEVELOPMENT OF THE GENERAL THEORY

CHAPTER 9

THE KLEENE HIERARCHY

1. The Iteration Theorem. We begin by recalling (Theorem 4-2.1) that, if r is a Gödel number of a Turing machine Z, then

$$\Psi_{Z;A}^{(n)}(\mathfrak{x}^{(n)}) = U(\min_y T_n^A(r, \mathfrak{x}^{(n)}, y)).$$

We introduce the notation $[r]_n^A(\mathfrak{x}^{(n)})$ for the function of n arguments $U(\min_y T_n^A(r, \mathfrak{x}^{(n)}, y))$. We write $[r]_n(\mathfrak{x}^{(n)})$ for $[r]_n^{\emptyset}(\mathfrak{x}^{(n)})$. Thus, the cited result may be written in the form

$$\Psi_{Z;A}^{(n)}(\mathfrak{x}^{(n)}) = [r]_n^A(\mathfrak{x}^{(n)}).$$

This notation is an analogue, for A-partial recursive functions, of the notation $\{n\}_A$ of Definition 5-4.2 for A-recursively enumerable sets. In fact, the notations are distinctly related; $\{n\}_A$ is precisely the domain of the A-partial recursive function $[n]_1^A$.

We shall now prove a result which will require us (for the last time) to go back to the definition of a Turing machine in terms of quadruples and to the details of the arithmetization of the theory of Turing machines.

THEOREM 1.1. *There is a primitive recursive function $\gamma(r, y)$ such that, for $n \geqq 1$,*

$$[r]_{1+n}^A(y, \mathfrak{x}^{(n)}) = [\gamma(r, y)]_n^A(\mathfrak{x}^{(n)}).$$

Intuitively, this result may be interpreted, for $A = \emptyset$, $n = 1$, as declaring the existence of an algorithm[1] by means of which, given any Turing machine Z and number m, a Turing machine Z_m can be found such that

$$\Psi_{Z^{(2)}}(m, x) = \Psi_{Z_m}(x).$$

Now it is clear that there exist Turing machines Z_m satisfying this last relation since, for each fixed m, $\Psi_{Z^{(2)}}(m, x)$ is certainly a partial recursive function of x. Hence, the content of our theorem (in this special case) is that Z_m can be found effectively in terms of Z and m. However, such a Z_m can readily be described as a Turing machine which, beginning at $\alpha = q_1 1^{x+1}$, proceeds to print $\overline{m} = 1^{m+1}$ to the left, eventually arriving at $\beta = q_N 1^{m+1} B 1^{x+1}$, and then proceeds to act like Z when confronted with

[1] Actually, an algorithm given by a *primitive* recursive function.

$q_1 1^{m+1} B 1^{x+1}$. As the general case does not differ essentially from this special case, all that is required for a formal proof is a detailed construction of Z_m and a careful consideration of the Gödel numbers. The reader who wishes to omit the tedious details, and simply accept the result, may well do so.

PROOF OF THEOREM 1.1. For each value of y, let W_y be the Turing machine consisting of the following quadruples:

$$\begin{array}{l} q_1\ 1\ L\ q_1 \\ q_1\ B\ L\ q_2 \\ \left.\begin{array}{l} q_{i+1}\ B\ 1\ q_{i+1} \\ q_{i+1}\ 1\ L\ q_{i+2} \end{array}\right\}\ 1 \leqq i \leqq y \\ q_{y+2}\ B\ 1\ q_{y+3}. \end{array}$$

Then, with respect to W_y,

$$\begin{aligned} q_1(\overline{\mathfrak{x}^{(n)}}) &\rightarrow q_1B(\overline{\mathfrak{x}^{(n)}}) \\ &\rightarrow q_2BB(\overline{\mathfrak{x}^{(n)}}) \\ &\rightarrow \cdots \\ &\rightarrow q_{y+3}(\overline{y, \mathfrak{x}^{(n)}}). \end{aligned}$$

Let r be a Gödel number of a Turing machine Z, and let

$$Z_y = W_y \cup Z^{(y+2)}.\dagger$$

Then, since the quadruples of $Z^{(y+2)}$ have precisely the same effect on $q_{y+3}(\overline{y, \mathfrak{x}^{(n)}})$ that those of Z have on $q_1(\overline{y, \mathfrak{x}^{(n)}})$, we have

$$\Psi^{(n)}_{Z_y;A}(\mathfrak{x}^{(n)}) = \Psi^{(1+n)}_{Z}(y, \mathfrak{x}^{(n)}) = [r]^{A}_{1+n}(y, \mathfrak{x}^{(n)}). \tag{1}$$

We now proceed to evaluate one of the Gödel numbers of Z_y as a function of r and y. The Gödel numbers of the quadruples that make up W_y are as follows:[1]

$$\begin{aligned} a &= \text{gn}\,(q_1\ 1\ L\ q_1) = 2^9 \cdot 3^{11} \cdot 5^5 \cdot 7^9, \\ b &= \text{gn}\,(q_1\ B\ L\ q_2) = 2^9 \cdot 3^7 \cdot 5^5 \cdot 7^{13}, \\ c(i) &= \text{gn}\,(q_{i+1}\ B\ 1\ q_{i+1}) = 2^{4i+9} \cdot 3^7 \cdot 5^{11} \cdot 7^{4i+9},\ 1 \leqq i \leqq y, \\ d(i) &= \text{gn}\,(q_{i+1}\ 1\ L\ q_{i+2}) = 2^{4i+9} \cdot 3^{11} \cdot 5^5 \cdot 7^{4i+13},\ 1 \leqq i \leqq y, \\ e(y) &= \text{gn}\,(q_{y+2}\ B\ 1\ q_{i+3}) = 2^{4y+13} \cdot 3^7 \cdot 5^{11} \cdot 7^{4y+17}. \end{aligned}$$

Thus, if we let

$$\varphi(y) = 2^a \cdot 3^b \cdot 5^{e(y)} \cdot \prod_{i=1}^{y} [\text{Pr}\,(i+3)^{c(i)}\ \text{Pr}\,(i+y+3)^{d(i)}],$$

then $\varphi(y)$ is a primitive recursive function, and, for each y, $\varphi(y)$ is a Gödel number of W_y.

† Recall Definition 2-1.3.

[1] Recall the discussion at the beginning of Chap. 4.

We recall that the predicate IC (x), which is true if and only if x is the number associated with an internal configuration q_i, is primitive recursive, since

$$\text{IC }(x) \leftrightarrow \bigvee_{y=0}^{x} (x = 4y + 9).$$

Hence, the function $\iota(x)$, which is 1 when x is the number associated with a q_i and 0 otherwise, is primitive recursive. If h is the Gödel number of a quadruple, then the Gödel number of the quadruple obtained from this one by replacing each q_i by q_{i+y+2} is

$$f(h, y) = 2^{1 \text{ Gl } h+4y+8} \cdot 3^{2 \text{ Gl } h} \cdot 5^{3 \text{ Gl } h+(4y+8)\iota(3 \text{ Gl } h)} \cdot 7^{4 \text{ Gl } h+4y+8}.$$

Here, $f(h, y)$ is primitive recursive. Hence, if we let

$$\theta(r, y) = \prod_{i=1}^{\mathfrak{L}(r)} \text{Pr }(i)^{f(i \text{ Gl } r, y)},$$

then $\theta(r, y)$ is a primitive recursive function and, for each y, $\theta(r, y)$ is a Gödel number of $Z^{(y+2)}$.

Let $\tau(x) = 1$ if x is a Gödel number of a Turing machine; 0, otherwise. Then, by (11) of Chap. 4, Sec. 1, $\tau(x)$ is primitive recursive. Finally, let

$$\gamma(r, y) = (\varphi(y) * \theta(r, y))\tau(r).$$

Then $\gamma(r, y)$ is a primitive recursive function and, for each y, $\gamma(r, y)$ is a Gödel number of Z_{y}. Hence, by (1),

$$[\gamma(r, y)]_n{}^A(\mathfrak{x}^{(n)}) = [r]^A_{1+n}(y, \mathfrak{x}^{(n)}). \tag{2}$$

It remains only to consider the case where r is not a Gödel number of a Turing machine. In that case, $\gamma(r, y)$, as defined above, is 0 and, thus, is itself not the Gödel number of a Turing machine; so (2) remains correct.[1]

THEOREM 1.2 (Kleene's Iteration Theorem[2]). *For each m there is a primitive recursive function $S^m(r, \mathfrak{y}^{(m)})$ such that, for $n \geqq 1$,*

$$[r]^A_{m+n}(\mathfrak{y}^{(m)}, \mathfrak{x}^{(n)}) = [S^m(r, \mathfrak{y}^{(m)})]_n{}^A(\mathfrak{x}^{(n)}).$$

Note that Theorem 1.1 is simply Theorem 1.2 with $m = 1$.

[1] I.e., the functions whose equality is asserted are nowhere defined.

[2] Cf. Kleene [3*a*, 6]. The slight difference between the present result and Kleene's is due to the fact that a given Turing machine computes an n-ary A-partial recursive function for each value of $n \geqq 1$, whereas a Herbrand-Gödel-Kleene system of equations defines a function of a *fixed* number of arguments.

PROOF. For $m = 0$, the result holds with $S^0(r) = r$. Suppose it is known for $m = k$. Then there is a primitive recursive function $S^k(r, \mathfrak{y}^{(k)})$ with the required property. Then, using Theorem 1.1,

$$\begin{aligned}[r]^A_{1+k+n}(y, \mathfrak{y}^{(k)}, \mathfrak{x}^{(n)}) &= [\gamma(r, y)]^A_{k+n}(\mathfrak{y}^{(k)}, \mathfrak{x}^{(n)}) \\ &= [S^k(\gamma(r, y), \mathfrak{y}^{(k)})]_n{}^A(\mathfrak{x}^{(n)}).\end{aligned}$$

Hence, if we set $S^{k+1}(r, y, \mathfrak{y}^{(k)}) = S^k(\gamma(r, y), \mathfrak{y}^{(k)})$, we have the desired result.

Henceforth we reserve the notation $S^m(r, \mathfrak{y}^{(m)})$ for the primitive recursive functions defined in the above proof.

COROLLARY 1.3. *Let $f(\mathfrak{y}^{(m)}, \mathfrak{x}^{(n)})$ be an A-partial recursive function. Then there is a primitive recursive function $\varphi(\mathfrak{y}^{(m)})$ such that, for $m, n \geqq 1$,*

$$f(\mathfrak{y}^{(m)}, \mathfrak{x}^{(n)}) = [\varphi(\mathfrak{y}^{(m)})]_n{}^A(\mathfrak{x}^{(n)}).$$

PROOF. Let r be such that $[r]^A_{m+n}(\mathfrak{y}^{(m)}, \mathfrak{x}^{(n)}) = f(\mathfrak{y}^{(m)}, \mathfrak{x}^{(n)})$. The result follows on taking $\varphi(\mathfrak{y}^{(m)}) = S^m(r, \mathfrak{y}^{(m)})$.

COROLLARY 1.4. $\bigvee_y T^A_{m+n}(z, \mathfrak{y}^{(m)}, \mathfrak{x}^{(n)}, y) \leftrightarrow \bigvee_y T_n{}^A(S^m(z, \mathfrak{y}^{(m)}), \mathfrak{x}^{(n)}, y)$.

PROOF. By Theorem 1.2,

$$[z]^A_{m+n}(\mathfrak{y}^{(m)}, \mathfrak{x}^{(n)}) = [S^m(z, \mathfrak{y}^{(m)})]_n{}^A(\mathfrak{x}^{(n)}).$$

That is,

$$U(\min_y T^A_{m+n}(z, \mathfrak{y}^{(m)}, \mathfrak{x}^{(n)}, y)) = U(\min_y T_n{}^A(S^m(z, \mathfrak{y}^{(m)}), \mathfrak{x}^{(n)}, y)).$$

This equality implies that, for each choice of $z, \mathfrak{y}^{(m)}, \mathfrak{x}^{(n)}$, the left-hand side is defined if and only if the right-hand side is defined.[1] The desired result follows at once.

COROLLARY 1.5. *Let $R(\mathfrak{y}^{(m)}, \mathfrak{x}^{(n)})$ be A-semicomputable. Then there is a primitive recursive function $\varphi(\mathfrak{y}^{(m)})$ such that*[2]

$$R(\mathfrak{y}^{(m)}, \mathfrak{x}^{(n)}) \leftrightarrow \bigvee_y T_n{}^A(\varphi(\mathfrak{y}^{(m)}), \mathfrak{x}^{(n)}, y).$$

PROOF. Immediate from Corollary 1.4 and Theorem 5-1.4.

COROLLARY 1.6. *Let $R(x, z)$ be A-semicomputable. Then there is a primitive recursive function $\varphi(z)$ such that*

$$\{\varphi(z)\}_A = \{x \mid R(x, z)\}.$$

PROOF. Take $m = n = 1$ in Corollary 1.5.

[1] Of course, the equality also implies that the two sides, where defined, are equal.

[2] The author derived essentially this result directly in Davis [1]. No reference to Kleene [3a] was given there because the author was at the time unaware of the connection between this result and Kleene's published work.

2. Some First Applications of the Iteration Theorem. By Theorem 5-4.9, the union and intersection of a pair of recursively enumerable sets are themselves recursively enumerable. One might be tempted to ask:

Does there exist an algorithm for obtaining the union (or intersection) of a given pair of recursively enumerable sets?

Now it is of course impossible to apply an algorithm to an infinite set of integers. However, recursively enumerable sets can be determined by a finite object. For example, as we have seen, each recursively enumerable set can be generated by a normal system. Or any recursively enumerable set S may be written as $\{n\}$, for a suitable integer n. We prove

THEOREM 2.1. *There exist primitive recursive functions* $h(p, q)$, $k(p, q)$ *such that*

$$\begin{aligned} \{p\} \cup \{q\} &= \{h(p, q)\}, \\ \{p\} \cap \{q\} &= \{k(p, q)\}. \end{aligned}$$

PROOF.
$$\begin{aligned} \{x \mid x \in \{p\} \cup \{q\}\} &= \{x \mid x \in \{p\} \vee x \in \{q\}\} \\ &= \left\{x \mid \bigvee_y T(p, x, y) \vee \bigvee_y T(q, x, y)\right\} \\ &= \left\{x \mid \bigvee_y T(h(p, q), x, y)\right\} \\ &= \{x \mid x \in \{h(p, q)\}\}, \end{aligned}$$

for suitable primitive recursive $h(p, q)$, by taking $A = \emptyset$, $m = 2$, $n = 1$, in Corollary 1.5.

Similarly,

$$\begin{aligned} \{x \mid x \in \{p\} \cap \{q\}\} &= \{x \mid x \in \{p\} \wedge x \in \{q\}\} \\ &= \left\{x \mid \bigvee_y T(p, x, y) \wedge \bigvee_y T(q, x, y)\right\} \\ &= \left\{x \mid \bigvee_y T(k(p, q), x, y)\right\}, \end{aligned}$$

where $k(p, q)$ is primitive recursive.

In terms of Turing machines (recall the notation of Chap. 6, Sec. 2), this result implies the existence of effective processes by means of which we can obtain, for each given pair Z_1, Z_2 of simple Turing machines, new Turing machines Z, Z' such that

$$\begin{aligned} P_Z &= P_{Z_1} \cup P_{Z_2}, \\ P_{Z'} &= P_{Z_1} \cap P_{Z_2}. \end{aligned}$$

To prove this directly would require quite some effort, involving detailed construction of Turing machines.

As another typical application of the iteration theorem, we have

THEOREM 2.2. *There exists a primitive recursive function $l(p, q)$ such that*

$$[p]_1([q]_1(x)) = [l(p, q)]_1(x).$$

That is, there is an effective procedure by which, given Turing machines Z_1, Z_2, we can obtain a new Turing machine Z such that

$$\Psi_Z(x) = \Psi_{Z_1}(\Psi_{Z_2}(x)).$$

Such an effective procedure was actually developed in Chap. 2. A direct proof that this procedure is given by a recursive function, however, would be extremely complicated, involving a computation of Gödel numbers of various of the Turing machines constructed in the course of the proof. The iteration theorem yields a simple proof.

PROOF

$$\begin{aligned}[p]_1([q]_1(x)) &= U(L(\min_t [T(p, U(K(t)),L(t)) \wedge T(q, x, K(t))])) \\ &= [l(p, q)]_1(x),\end{aligned}$$

where $l(p, q)$ is primitive recursive, by Corollary 1.3, with $A = \phi$, $m = 2$, $n = 1$.

3. Predicates, Sets, and Functions. We have been dealing with *predicates that are A-computable or A-semicomputable*, with *sets that are A-recursive or A-recursively enumerable*, and with *functions that are A-computable or partially A-computable.* However, in all cases, A can only be a set. This restriction was introduced in Chap. 1, Sec. 4, where it was indicated that the limitation involved was not so severe as might be thought.

In fact, we may proceed as follows:

Let P be an n-ary predicate. Then, we write

$$P^* = \{x \mid P(1 \text{ Gl } x, \ldots, n \text{ Gl } x)\}.$$

P^* is called the *associated set* of P. We shall now say, e.g., that a function f is *partially P-computable*, meaning that f is partially P^*-computable. Analogously, we can proceed for each of the italicized phrases at the beginning of the present section.

Next, let g be an n-ary *total* function. Let Q be the $(n + 1)$-ary predicate

$$y = g(\mathfrak{x}^{(n)}).$$

Then we set $g^* = Q^*$; and g^* is called the associated set of g. Again, we shall say, for instance, that a predicate P is *g-semicomputable*, meaning that P is g^*-semicomputable. And again we proceed analogously for each of the italicized phrases at the beginning of this section.

Finally, for the sake of completeness, we take $A^* = A$, where A is a set.

We shall use lower-case Greek letters to represent objects that may be either predicates, sets, or total functions, as the context permits. In particular, we write

$$\alpha \prec \beta$$

to indicate that α is β-computable. In this case we also say that α is *recursively reducible* to β or, simply, that α is *reducible* to β.

In addition, we write $\alpha \nprec \beta$ to indicate that it is not the case that $\alpha \prec \beta$.

THEOREM 3.1. *$\alpha \prec \beta$ if and only if $\alpha^* \prec \beta^*$.*

PROOF. By the definitions introduced in the above discussion $\alpha \prec \beta$ if and only if $\alpha \prec \beta^*$. The rest of the proof follows readily on considering cases. First, suppose α is a set. Then $\alpha^* = \alpha$, and the result is immediate.

Next, let α be an n-ary predicate. Then

$$\alpha^* = \{x \mid \alpha(1 \text{ Gl } x, \ldots, n \text{ Gl } x)\};$$

$$\alpha(x_1, \ldots, x_n) \leftrightarrow \left(\prod_{k=1}^{n} \text{Pr}(k)^{x_k}\right) \in \alpha^*.$$

Hence, $\alpha \prec \beta^*$ if and only if $\alpha^* \prec \beta^*$.

Finally, let α be an n-ary total function. Then

$$\alpha^* = \{x \mid (n+1) \text{ Gl } x = \alpha(1 \text{ Gl } x, \ldots, n \text{ Gl } x)\};$$

$$\alpha(x_1, \ldots, x_n) = \min_z \left[\left(\prod_{k=1}^{n} \text{Pr}(k)^{x_k} \cdot \text{Pr}(n+1)^z\right) \in \alpha^*\right].$$

Hence, $\alpha \prec \beta^*$ if and only if $\alpha^* \prec \beta^*$.

This completes the proof.

THEOREM 3.2. $\alpha \prec \alpha$.

PROOF. By Theorem 3.1, we need only verify that $\alpha^* \prec \alpha^*$. But this we already know.

THEOREM 3.3. *If $\alpha \prec \beta$ and $\beta \prec \gamma$, then $\alpha \prec \gamma$.*

PROOF. If $\alpha \prec \beta$ and $\beta \prec \gamma$, then, by Theorem 3.1, $\alpha^* \prec \beta^*$ and $\beta^* \prec \gamma^*$. That is, α^* is β^*-recursive and β^* is γ^*-recursive. Then C_{α^*} is obtainable from C_{β^*} and functions (2) to (6) of Definition 3-1.1; and C_{β^*} is obtainable from C_{γ^*} and functions (2) to (6) by a finite number of applications of composition and of minimalization of regular functions. Hence, C_{α^*} is obtainable from C_{γ^*} and functions (2) to (6) by a finite number of applications of composition and minimalization. Thus, $\alpha^* \prec \gamma^*$.

4. Strong Reducibility. We begin with

DEFINITION 4.1. *Let A and B be sets. Then we write $A \prec\prec B$, and we* *say that A is* **strongly reducible**[1] *to B, if there exists a recursive function $f(x)$ such that*

$$A = \{x \mid f(x) \in B\}.$$

DEFINITION 4.2. *If each of α, β is a predicate, a set, or a total function, then we write $\alpha \prec\prec \beta$, and we say that α is strongly reducible to β, to mean that $\alpha^* \prec\prec \beta^*$.*[2]

COROLLARY 4.1. *If $\alpha \prec\prec \beta$, then $\alpha \prec \beta$.*

PROOF. If $\alpha \prec\prec \beta$, then $\alpha^* \prec\prec \beta^*$. Hence, for some recursive function $f(x)$,

$$\alpha^* = \{x \mid f(x) \in \beta^*\}.$$

Thus, $C_{\alpha^*}(x) = C_{\beta^*}(f(x))$. Hence, $\alpha^* \prec \beta^*$. Therefore, $\alpha \prec \beta$.

We shall see later that, even for recursively enumerable sets, the converse of Corollary 4.1 is false.

THEOREM 4.2. $\alpha \prec\prec \alpha$.

PROOF. $\alpha^* = \{x \mid x \in \alpha^*\}$.

THEOREM 4.3. *If $\alpha \prec\prec \beta$ and $\beta \prec\prec \gamma$, then $\alpha \prec\prec \gamma$.*

PROOF. If $\alpha^* = \{x \mid f(x) \in \beta^*\}$ and $\beta^* = \{x \mid g(x) \in \gamma^*\}$, then $\alpha^* = \{x \mid g(f(x)) \in \gamma^*\}$.

THEOREM 4.4. *C is A-recursively enumerable if and only if $C \prec\prec A'$.*

PROOF. If $C \prec\prec A'$, then, for suitable recursive $f(x)$,

$$\begin{aligned} C &= \{x \mid f(x) \in A'\} \\ &= \left\{x \mid \bigvee_y T^A(f(x), f(x), y)\right\}; \end{aligned}$$

so C is A-recursively enumerable.

Next, suppose that C is A-recursively enumerable. Then there is an A-recursive predicate $R(x, y)$ such that

$$C = \left\{x \mid \bigvee_y R(x, y)\right\}.$$

Now, by Corollary 1.5, with $m = n = 1$, there is a recursive function $g(x)$

[1] This notion is due to Post [3], who called it *many-one reducibility*. The term "strong reducibility" is that of Shapiro [1].

[2] If $P(\mathfrak{x}^{(n)})$, $Q(\mathfrak{x}^{(m)})$ are predicates, then, as is easily seen,

$$P(\mathfrak{x}^{(n)}) \prec\prec Q(\mathfrak{x}^{(m)})$$

is equivalent to the existence of recursive functions $g_1(\mathfrak{x}^{(n)}), g_2(\mathfrak{x}^{(n)}), \ldots, g_m(\mathfrak{x}^{(n)})$ such that

$$P(\mathfrak{x}^{(n)}) \leftrightarrow Q(g_1(\mathfrak{x}^{(n)}), g_2(\mathfrak{x}^{(n)}), \ldots, g_m(\mathfrak{x}^{(n)})).$$

What amounts to the special case $m = 1$ is proved below as Theorem 4.5.

such that

$$\bigvee_y [R(x, y) \wedge (z = z)] \leftrightarrow \bigvee_y T^A(g(x), z, y).$$

Now,

$$\begin{aligned} C &= \left\{x \mid \bigvee_y R(x, y)\right\} \\ &= \left\{x \mid \bigvee_y [R(x, y) \wedge (g(x) = g(x))]\right\} \\ &= \left\{x \mid \bigvee_y T^A(g(x), g(x), y)\right\} \\ &= \{x \mid g(x) \in A'\}. \end{aligned}$$

THEOREM 4.5. *$P(\mathfrak{x}^{(n)}) \prec\prec A$ if and only if there is a recursive function $f(\mathfrak{x}^{(n)})$ such that*

$$P(\mathfrak{x}^{(n)}) \leftrightarrow f(\mathfrak{x}^{(n)}) \in A.$$

PROOF. $P(\mathfrak{x}^{(n)}) \prec\prec A$ if and only if $P^* \prec\prec A$, which, in turn, is true if and only if, for suitable recursive $g(x)$,

$$P^* = \{x \mid g(x) \in A\}.$$

Thus, if $P(\mathfrak{x}^{(n)}) \prec\prec A$, then

$$\begin{aligned} P(\mathfrak{x}^{(n)}) &\leftrightarrow \left(\prod_{k=1}^{n} \text{Pr}\,(k)^{x_k}\right) \in P^* \\ &\leftrightarrow g\left(\prod_{k=1}^{n} \text{Pr}\,(k)^{x_k}\right) \in A. \end{aligned}$$

Conversely, if

$$P(\mathfrak{x}^{(n)}) \leftrightarrow f(\mathfrak{x}^{(n)}) \in A,$$

then

$$\begin{aligned} P^* &= \{x \mid P(1 \text{ Gl } x, \ldots, n \text{ Gl } x)\} \\ &= \{x \mid f(1 \text{ Gl } x, \ldots, n \text{ Gl } x) \in A\}. \end{aligned}$$

This completes the proof.

THEOREM 4.6. *The predicate $P(\mathfrak{x}^{(n)})$ is A-semicomputable if and only if $P(\mathfrak{x}^{(n)}) \prec\prec A'$.*

PROOF. As is easily seen, $P(\mathfrak{x}^{(n)})$ is A-semicomputable if and only if P^* is A-recursively enumerable. The result then follows from Definition 4.2 and Theorem 4.4.

THEOREM 4.7. *$A \prec\prec A'$, but $A' \not\prec A$.*

PROOF. A is A-recursively enumerable. Hence, by Theorem 4.4, $A \prec\prec A'$.

That $A' \not\prec A$ is part of the statement of Theorem 5-4.6.

THEOREM 4.8. *If $A < B$, then*[1] $A' << B'$.

PROOF. $T^A(x, x, y) < A$. Hence, since $A < B$, $T^A(x, x, y) < B$. Hence, $\bigvee_y T^A(x, x, y)$ is B-semicomputable; that is,

$$A' = \left\{x \mid \bigvee_y T^A(x, x, y)\right\}$$

is B-recursively enumerable. Hence, by Theorem 4.4, $A' << B'$.

5. Some Classes of Predicates. We are now in a position to obtain relatively explicit information about the classification of nonrecursive predicates. We begin by considering some classes of predicates.

DEFINITION 5.1. *By $\boldsymbol{P}_1^A$ we understand the class of A-semicomputable predicates. Given $\boldsymbol{P}_k^A$, we take $\boldsymbol{P}_{k+1}^A$ to be the class of all predicates R for which there exists a predicate $Q \in \boldsymbol{P}_k^A$ such that R is Q-semicomputable.*

We write $\boldsymbol{P}_n$ for $\boldsymbol{P}_n^\emptyset$.

Thus, for example, $\boldsymbol{P}_2^A$ consists of all predicates that are Q-semicomputable for a suitable A-semicomputable predicate Q.

DEFINITION 5.2. *A predicate is said to be A,n-***generable** *if it belongs to $\boldsymbol{P}_n^A$. We write n-***generable** *for $\emptyset$,n-generable, and A-***generable** *for A,1-generable.*

(Thus, "A-generable" is a new term for "A-semicomputable.")

We shall see that the classes $\boldsymbol{P}_n^A$ may be characterized in many different ways.

DEFINITION 5.3. *If A is any set, we write*

$$A^0 = A,$$
$$A^{k+1} = (A^k)'.$$

THEOREM 5.1. $A^n << A^{n+1}$, *but* $A^{n+1} \not< A^n$.

PROOF. Immediate from Theorem 4.7.

We shall see that the sets $A, A^1, A^2, A^3, \ldots$ are closely related to the classes $\boldsymbol{P}_n^A$ of predicates.

THEOREM 5.2. *For each n, the predicate $x \in A^n$ is A,n-generable.*

PROOF. For $n = 1$, the result is obvious, since $\bigvee_y T^A(x, x, y)$ is A,1-*generable.*

Suppose the result known for $n = k$. Now,

$$x \in A^{k+1} \leftrightarrow \bigvee_y T^{A^k}(x, x, y).$$

Here, $T^{A^k}(x, x, y)$ is A^k-recursive and, hence, is Q-recursive, where Q is the predicate $x \in A^k$. Hence, the predicate $x \in A^{k+1}$ is Q-semicom-

[1] Cf. Davis [1].

putable. By induction hypothesis, Q is A,k-generable. Hence, $x \in A^{k+1}$ is $A,(k + 1)$-generable.

THEOREM 5.3. *$Q \in \boldsymbol{P}_n^A$ if and only if Q is A^{n-1}-semicomputable.*

PROOF. If Q is A^{n-1}-semicomputable, then Q is R-semicomputable, where R is the predicate $x \in A^{n-1}$. By Theorem 5.2, $R \in \boldsymbol{P}_{n-1}^A$. By Definition 5.1, $Q \in \boldsymbol{P}_n^A$.

To prove the converse, we proceed by induction on n. For $n = 1$, the result follows directly from Definitions 5.1 and 5.3.

Suppose the result known for $n = k$. Suppose $Q \in \boldsymbol{P}_{k+1}^A$. Then, by Definition 5.1, for some R, $R \in \boldsymbol{P}_k^A$, Q is R-semicomputable. By induction hypothesis, R is A^{k-1}-semicomputable; hence, by Theorem 4.6, $R \prec\prec A^k$. Moreover, since Q is R-semicomputable,

$$Q \leftrightarrow \bigvee_z S,$$

where $S \prec R$. By Corollary 4.1 and Theorem 3.3, $S \prec A^k$. Hence, Q is A^k-semicomputable. This completes the proof.

COROLLARY 5.4. *$Q \in \boldsymbol{P}_n^A$ if and only if $Q \prec\prec A^n$.*

PROOF. Immediate from Theorems 5.3 and 4.6.

Theorem 5.3 suggests the following definitions:

DEFINITION 5.4. *For $n \geqq 1$, we take $\boldsymbol{Q}_n^A$ to be the class of all predicates P such that $\sim P$ is A,n-generable. We also take $\boldsymbol{R}_n^A = \boldsymbol{P}_n^A \cap \boldsymbol{Q}_n^A$. We write $\boldsymbol{Q}_n = \boldsymbol{Q}_n^\emptyset$, $\boldsymbol{R}_n = \boldsymbol{R}_n^\emptyset$.*†

DEFINITION 5.5. *The predicates of $\boldsymbol{Q}_n^A$ are called A,n-**antigenerable**; those of $\boldsymbol{R}_n^A$ are called A,n-**recursive**. We write n-**antigenerable** for $\emptyset,n$-antigenerable, n-**recursive** for $\emptyset,n$-recursive, and A-**antigenerable** for $A,1$-antigenerable.*

THEOREM 5.5. *Q is A,n-recursive if and only if Q is A^{n-1}-recursive.*

PROOF. Immediate from Theorem 5.3, Definition 5.5, and Theorem 5-1.5.

Since the predicates belonging to each $\boldsymbol{P}_n^A$ are precisely those that are A^{n-1}-semicomputable, the results of Chap. 5, Sec. 3, yield

THEOREM 5.6. *Let P, $Q \in \boldsymbol{P}_n^A$. Then*

$$(1)\ \bigvee_y P \in \boldsymbol{P}_n^A.$$

$$(2)\ \bigwedge_{y=0}^{z} P \in \boldsymbol{P}_n^A.$$

$$(3)\ P \wedge Q \in \boldsymbol{P}_n^A.$$

$$(4)\ P \vee Q \in \boldsymbol{P}_n^A.$$

† The notation is taken from Mostowski [1].

Taking negations, we have

THEOREM 5.7. *Let* $P, Q \in \boldsymbol{Q}_n{}^A$. *Then*

$$(1)\ \bigwedge_{y} P \in \boldsymbol{Q}_n{}^A.$$

$$(2)\ \bigvee_{y=0}^{z} P \in \boldsymbol{Q}_n{}^A.$$

$$(3)\ P \wedge Q \in \boldsymbol{Q}_n{}^A.$$

$$(4)\ P \vee Q \in \boldsymbol{Q}_n{}^A.$$

For $\boldsymbol{R}_n{}^A$, we have easily

THEOREM 5.8. *Let* $P, Q \in \boldsymbol{R}_n{}^A$. *Then*

$$(1)\ \bigvee_{y=0}^{z} P \in \boldsymbol{R}_n{}^A.$$

$$(2)\ \bigwedge_{y=0}^{z} P \in \boldsymbol{R}_n{}^A.$$

$$(3)\ P \wedge Q \in \boldsymbol{R}_n{}^A.$$

$$(4)\ P \vee Q \in \boldsymbol{R}_n{}^A.$$

$$(5)\ \sim P \in \boldsymbol{R}_n{}^A.$$

THEOREM 5.9. $\boldsymbol{P}_n{}^A \subset \boldsymbol{R}_{n+1}^A$; $\boldsymbol{Q}_n{}^A \subset \boldsymbol{R}_{n+1}^A$.

PROOF. If $P \in \boldsymbol{P}_n{}^A$, then, by Corollary 5.4, $P \prec\prec A^n$. By Corollary 4.1, $P \prec A^n$. Hence, by Theorem 5.5, $P \in \boldsymbol{R}_{n+1}^A$.

If $Q \in \boldsymbol{Q}_n{}^A$, then $\sim Q \in \boldsymbol{P}_n{}^A$; so $\sim Q \in \boldsymbol{R}_{n+1}^A$. By Theorem 5.8, $Q \in \boldsymbol{R}_{n+1}^A$.

THEOREM 5.10 (Kleene's Hierarchy Theorem[1]). *There are predicates* P, Q *such that* $P \in \boldsymbol{P}_n{}^A$, $Q \in \boldsymbol{Q}_n{}^A$, $P \notin \boldsymbol{R}_n{}^A$, $Q \notin \boldsymbol{R}_n{}^A$.

PROOF. Immediate from Theorem 5-1.6.

Thus, for each n, $\boldsymbol{P}_n{}^A$ contains predicates not contained in $\boldsymbol{Q}_n{}^A$, and vice versa. Moreover, by Theorem 5.9, such predicates are guaranteed not to occur in $\boldsymbol{P}_k{}^A$ or $\boldsymbol{Q}_k{}^A$ for any $k < n$.

Thus, in the sequence $\boldsymbol{P}_1{}^A, \boldsymbol{P}_2{}^A, \boldsymbol{P}_3{}^A, \ldots$ each class contains predicates not contained in any of the preceding classes.

6. A Representation Theorem for $\boldsymbol{P}_2{}^A$. We begin with

THEOREM 6.1. *If* $R(y, z, \mathfrak{x}^{(k)}) \prec A$, *then* $\bigvee_{y} \bigwedge_{z} R(y, z, \mathfrak{x}^{(k)}) \in \boldsymbol{P}_2{}^A$.

PROOF. $\bigvee_{z} \sim R(y, z, \mathfrak{x}^{(k)})$ is A-semicomputable. Therefore,

$$\bigvee_{z} \sim R(y, z, \mathfrak{x}^{(k)}) \in \boldsymbol{P}_1{}^A.$$

[1] Kleene [4, 6].

But

$$\bigwedge_z R(y, z, \mathfrak{x}^{(k)}) \prec \bigvee_z \sim R(y, z, \mathfrak{x}^{(k)}).$$

Hence, $\bigvee_y \bigwedge_z R(y, z, \mathfrak{x}^{(k)})$ is $\bigvee_z \sim R(y, z, \mathfrak{x}^{(k)})$-semicomputable. By Definition 5.1, $\bigvee_y \bigwedge_z R(y, z, \mathfrak{x}^{(k)}) \in \boldsymbol{P}_2{}^A$.

The converse of Theorem 6.1 is also true; in fact, it is a rather deep result, and we shall devote the remainder of the present section to its proof.

We let $\boldsymbol{E}^A$ *be the class of all predicates of the form* $\bigvee_y \bigwedge_z R(y, z, \mathfrak{x}^{(n)})$, *where* $R \prec A$. Then, our last theorem states that $\boldsymbol{E}^A \subset \boldsymbol{P}_2{}^A$; and we wish to show that $\boldsymbol{P}_2{}^A \subset \boldsymbol{E}^A$. We begin with a series of easy lemmas.

LEMMA 1. *If* $R(u, \mathfrak{x}^{(k)}) \in \boldsymbol{E}^A$, *then* $\bigvee_u R(u, \mathfrak{x}^{(k)}) \in \boldsymbol{E}^A$.

PROOF. Let

$$R(u, \mathfrak{x}^{(k)}) \leftrightarrow \bigvee_y \bigwedge_z Q(u, y, z, \mathfrak{x}^{(k)}),$$

where $Q \prec A$. Then

$$\bigvee_u \bigvee_y \bigwedge_z Q(u, y, z, \mathfrak{x}^{(k)}) \leftrightarrow \bigvee_y \bigwedge_z Q(K(y), L(y), z, \mathfrak{x}^{(k)}).$$

LEMMA 2. *If* $R(u, \mathfrak{x}^{(k)}) \in \boldsymbol{E}^A$, *then* $\bigwedge_{u=0}^{v} R(u, \mathfrak{x}^{(k)}) \in \boldsymbol{E}^A$.

PROOF. Let

$$R(u, \mathfrak{x}^{(k)}) \leftrightarrow \bigvee_y \bigwedge_z Q(u, y, z, \mathfrak{x}^{(k)}).$$

Then

$$\begin{aligned}\bigwedge_{u=0}^{v} R(u, \mathfrak{x}^{(k)}) &\leftrightarrow \bigwedge_{u=0}^{v} \bigvee_y \bigwedge_z Q(u, y, z, \mathfrak{x}^{(k)}) \\ &\leftrightarrow \bigvee_t \bigwedge_{u=0}^{v} \bigwedge_z Q(u, (u+1) \text{ Gl } t, z, \mathfrak{x}^{(k)}) \\ &\leftrightarrow \bigvee_t \bigwedge_z \bigwedge_{u=0}^{v} Q(u, (u+1) \text{ Gl } t, z, \mathfrak{x}^{(k)}).\end{aligned}$$

LEMMA 3. *Let* $R(\mathfrak{x}^{(n)}) \in \boldsymbol{E}^A$ *and also* $S(\mathfrak{x}^{(n)}) \in \boldsymbol{E}^A$. *Then, we have* $R(\mathfrak{x}^{(n)}) \vee S(\mathfrak{x}^{(n)}) \in \boldsymbol{E}^A$ *and* $R(\mathfrak{x}^{(n)}) \wedge S(\mathfrak{x}^{(n)}) \in \boldsymbol{E}^A$.

PROOF. Let

$$R(\mathfrak{x}^{(n)}) \leftrightarrow \bigvee_y \bigwedge_z Q_1(y, z, \mathfrak{x}^{(n)}),$$

$$S(\mathfrak{x}^{(n)}) \leftrightarrow \bigvee_y \bigwedge_z Q_2(y, z, \mathfrak{x}^{(n)}).$$

Then

$$\begin{aligned} R(\mathfrak{x}^{(n)}) \vee S(\mathfrak{x}^{(n)}) &\leftrightarrow \bigvee_y \Big[\bigwedge_z Q_1(y, z, \mathfrak{x}^{(n)}) \vee \bigwedge_z Q_2(y, z, \mathfrak{x}^{(n)})\Big] \\ &\leftrightarrow \bigvee_y \bigwedge_{z_1} \bigwedge_{z_2} [Q_1(y, z_1, \mathfrak{x}^{(n)}) \vee Q_2(y, z_2, \mathfrak{x}^{(n)})] \\ &\leftrightarrow \bigvee_y \bigwedge_z [Q_1(y, K(z), \mathfrak{x}^{(n)}) \vee Q_2(y, L(z), \mathfrak{x}^{(n)})], \end{aligned}$$

and

$$\begin{aligned} R(\mathfrak{x}^{(n)}) \wedge S(\mathfrak{x}^{(n)}) &\leftrightarrow \bigvee_{y_1} \bigvee_{y_2} \Big[\bigwedge_z Q_1(y_1, z, \mathfrak{x}^{(n)}) \wedge \bigwedge_z Q_2(y_2, z, \mathfrak{x}^{(n)})\Big] \\ &\leftrightarrow \bigvee_y \bigwedge_z [Q_1(K(y), z, \mathfrak{x}^{(n)}) \wedge Q_2(L(y), z, \mathfrak{x}^{(n)})]. \end{aligned}$$

LEMMA 4. *The predicate* $u = C_{A'}(x)$ *belongs to* $\boldsymbol{E}^A$.

PROOF. $u = C_{A'}(x)$

$$\begin{aligned} &\leftrightarrow [(u = 0) \wedge (x \in A')] \vee [(u = 1) \wedge (x \notin A')] \\ &\leftrightarrow \Big[(u = 0) \wedge \bigvee_y T^A(x, x, y)\Big] \vee \Big[(u = 1) \wedge \bigwedge_y \sim T^A(x, x, y)\Big] \\ &\leftrightarrow \bigvee_y \bigwedge_z \{[(u = 0) \wedge T^A(x, x, y)] \vee [(u = 1) \wedge \sim T^A(x, x, z)]\}. \end{aligned}$$

LEMMA 5. *If* $f(\mathfrak{x}^{(n)}) \prec A'$, *then the predicate* $u = f(\mathfrak{x}^{(n)})$ *belongs to* $\boldsymbol{E}^A$.

PROOF. We use the characterization of Definition 3-1.2.

By Lemma 4, the result holds for function 1 of the list of Definition 3-1.1 (with, of course, A replaced by A'). For functions 2 to 6, the result is obvious. Hence, it remains only to show that the property in question is preserved under composition and under minimalization of regular functions.

Thus, let the result be known for the functions $f(\mathfrak{y}^{(m)})$, $g_1(\mathfrak{x}^{(n)}), \ldots,$ $g_m(\mathfrak{x}^{(n)})$. Let $h(\mathfrak{x}^{(n)}) = f(g_1(\mathfrak{x}^{(n)}), \ldots, g_m(\mathfrak{x}^{(n)}))$. Then

$$u = h(\mathfrak{x}^{(n)}) \leftrightarrow \bigvee_{\mathfrak{y}^{(m)}} [u = f(\mathfrak{y}^{(m)}) \wedge y_1 = g_1(\mathfrak{x}^{(n)}) \wedge \cdots \wedge y_m = g_m(\mathfrak{x}^{(n)})],$$

and the result for $h(\mathfrak{x}^{(n)})$ follows from Lemmas 1 and 3.

Finally, let the result be known for the regular function $f(y, \mathfrak{x}^{(n)})$, and let

$$g(\mathfrak{x}^{(n)}) = \min_y [f(y, \mathfrak{x}^{(n)}) = 0].$$

Then

$$u = g(\mathfrak{x}^{(n)}) \leftrightarrow \bigvee_t [(f(u, \mathfrak{x}^{(n)}) = t) \wedge (t = 0)] \wedge \bigwedge_{y=0}^{u} \bigvee_z \{(y = u) \vee [(f(y, \mathfrak{x}^{(n)}) = z) \wedge (z \neq 0)]\},$$

and the result for $g(\mathfrak{x}^{(n)})$ follows from Lemmas 1 to 3.

LEMMA 6. *If $P(\mathfrak{x}^{(n)}) \prec A'$, then $P(\mathfrak{x}^{(n)}) \in \boldsymbol{E}^A$.*

PROOF. By Lemma 5, the predicate $u = C_P(\mathfrak{x}^{(n)})$ belongs to $\boldsymbol{E}^A$. Hence, so does the predicate $0 = C_P(\mathfrak{x}^{(n)})$. But

$$P(\mathfrak{x}^{(n)}) \leftrightarrow (0 = C_P(\mathfrak{x}^{(n)})).$$

LEMMA 7. *If $P(\mathfrak{x}^{(n)})$ is A'-semicomputable, then $P(\mathfrak{x}^{(n)}) \in \boldsymbol{E}^A$.*

PROOF. Immediate from Lemmas 1 and 6.

THEOREM 6.2 (Representation Theorem for $\boldsymbol{P}_2{}^A$). *If $P(\mathfrak{x}^{(n)}) \in \boldsymbol{P}_2{}^A$, then there is a predicate $R(y, z, \mathfrak{x}^{(n)})$ such that $R \prec A$, and*

$$P(\mathfrak{x}^{(n)}) \leftrightarrow \bigvee_y \bigwedge_z R(y, z, \mathfrak{x}^{(n)}).$$

PROOF. This is but a restatement of Lemma 7.

7. Post's Representation Theorem. The result of Sec. 6 suggests that a similar result might be derivable for $\boldsymbol{P}_n{}^A$. This is indeed the case. In fact, as we shall see, Theorem 6.2 really furnishes the inductive step in the proof of such a result. Actually, Theorem 6.2 is used in the following form:

THEOREM 7.1. *$P(\mathfrak{x}^{(n)}) \in \boldsymbol{P}_2{}^A$ if and only if there exists a predicate $Q(y, \mathfrak{x}^{(n)}) \in \boldsymbol{Q}_1{}^A$ such that*

$$P(\mathfrak{x}^{(n)}) \leftrightarrow \bigvee_y Q(y, \mathfrak{x}^{(n)}).$$

PROOF. By Theorems 6.1 and 6.2, it suffices to show that a predicate $P(\mathfrak{x}^{(n)})$ can be written $\bigvee_y \bigwedge_z R(y, z, \mathfrak{x}^{(n)})$, with $R \prec A$, if and only if $P(\mathfrak{x}^{(n)})$ can be written $\bigvee_y Q(y, \mathfrak{x}^{(n)})$, with $Q \in \boldsymbol{Q}_1{}^A$. But this is obvious, since

$$\bigvee_y \bigwedge_z R(y, z, \mathfrak{x}^{(n)}) \leftrightarrow \bigvee_y \sim \bigvee_z \sim R(y, z, \mathfrak{x}^{(n)}).$$

THEOREM 7.2. *For $0 < k \leq m$, $\boldsymbol{P}_m{}^A = \boldsymbol{P}_k^{A^{m-k}}$; $\boldsymbol{Q}_m{}^A = \boldsymbol{Q}_k^{A^{m-k}}$. Also, $\boldsymbol{R}_m{}^A = \boldsymbol{R}_k^{A^{m-k}}$.*

PROOF. By Corollary 5.4, $P \in \boldsymbol{P}_k^{A^{m-k}}$ if and only if $P \prec\prec (A^{m-k})^k = A^m$. But, by Corollary 5.4 again, this last holds if and only if $P \in \boldsymbol{P}_m{}^A$.

Finally, $Q \in \boldsymbol{Q}_m{}^A$ if and only if $\sim Q \in \boldsymbol{P}_m{}^A$, if and only if $\sim Q \in \boldsymbol{P}_k^{A^{m-k}}$, if and only if $Q \in \boldsymbol{Q}_k^{A^{m-k}}$.

DEFINITION 7.1. *We define the symbols* $\bigvee^n$, $\bigwedge^n$ *as representing strings of quantifiers as follows:*

$$\bigvee^1 \textit{ is } \bigvee_{x_1}; \ \bigwedge^1 \textit{ is } \bigwedge_{x_1};$$

and, *assuming* $\bigvee^m$, $\bigwedge^m$ *defined,*

$$\bigvee^{m+1} \textit{ is } \bigvee_{x_{m+1}} \bigwedge^m; \ \bigwedge^{m+1} \textit{ is } \bigwedge_{x_{m+1}} \bigvee^m.$$

Thus, for example, $\bigvee^3$ is $\bigvee_{x_3} \bigwedge_{x_2} \bigvee_{x_1}$.

THEOREM 7.3 (Post's Representation Theorem[1]). *$P(\mathfrak{y}^{(k)}) \in \boldsymbol{P}_n{}^A$ if and only if there is a predicate $R(\mathfrak{x}^{(n)}, \mathfrak{y}^{(k)}) \prec A$ such that*

$$P(\mathfrak{y}^{(k)}) \leftrightarrow \bigvee^n R(\mathfrak{x}^{(n)}, \mathfrak{y}^{(k)}).$$

Similarly, $Q(\mathfrak{y}^{(k)}) \in \boldsymbol{Q}_n{}^A$ if and only if there is a predicate $S(\mathfrak{x}^{(n)}, \mathfrak{y}^{(k)}) \prec A$ such that

$$Q(\mathfrak{y}^{(k)}) \leftrightarrow \bigwedge^n S(\mathfrak{x}^{(n)}, \mathfrak{y}^{(k)}).$$

PROOF. Our proof is by induction on n. For $n = 1$, the first part is an immediate consequence of Definitions 5.1 and 7.1. The second part (for $n = 1$) then follows by taking negations.

Suppose the result known for $n = m$. Then, by Theorem 7.2, $P(\mathfrak{y}^{(k)}) \in \boldsymbol{P}_{m+1}^A$ if and only if $P(\mathfrak{y}^{(k)}) \in \boldsymbol{P}_2^{A^{m-1}}$. By Theorem 7.1, $P(\mathfrak{y}^{(k)}) \in \boldsymbol{P}_{m+1}^A$ if and only if there is a predicate $M(x_{m+1}, \mathfrak{y}^{(k)}) \in \boldsymbol{Q}_1^{A^{m-1}}$ such that

$$P(\mathfrak{y}^{(k)}) \leftrightarrow \bigvee_{x_{m+1}} M(x_{m+1}, \mathfrak{y}^{(k)}).$$

But, by Theorem 7.2, $M(x_{m+1}, \mathfrak{y}^{(k)}) \in \boldsymbol{Q}_1^{A^{m-1}}$ will hold if and only if $M(x_{m+1}, \mathfrak{y}^{(k)}) \in \boldsymbol{Q}_m{}^A$. By induction hypothesis, $M(x_{m+1}, \mathfrak{y}^{(k)}) \in \boldsymbol{Q}_m{}^A$ if and only if

$$M(x_{m+1}, \mathfrak{y}^{(k)}) \leftrightarrow \bigwedge^m S(\mathfrak{x}^{(m)}, x_{m+1}, \mathfrak{y}^{(k)}),$$

[1] Cf. Post [7].

where $S \prec A$. Thus, $P(\mathfrak{y}^{(k)}) \in \boldsymbol{P}_{m+1}^A$ if and only if there is a predicate $S(\mathfrak{x}^{(m)}, x_{m+1}, \mathfrak{y}^{(k)}) \prec A$ such that

$$\begin{aligned} P(\mathfrak{y}^{(k)}) &\leftrightarrow \bigvee_{x_{m+1}} \bigwedge^{m} S(\mathfrak{x}^{(m)}, x_{m+1}, \mathfrak{y}^{(k)}) \\ &\leftrightarrow \bigvee^{m+1} S(\mathfrak{x}^{(m+1)}, \mathfrak{y}^{(k)}). \end{aligned}$$

Finally, $Q(\mathfrak{y}^{(k)}) \in \boldsymbol{Q}_{m+1}^A$ if and only if $\sim Q(\mathfrak{y}^{(k)}) \in \boldsymbol{P}_{m+1}^A$, hence, if and only if, for suitable $R \prec A$,

$$\sim Q(\mathfrak{y}^{(k)}) \leftrightarrow \bigvee^{m+1} R(\mathfrak{x}^{(m+1)}, \mathfrak{y}^{(k)}),$$

i.e., if and only if

$$\begin{aligned} Q(\mathfrak{y}^{(k)}) &\leftrightarrow \sim \bigvee^{m+1} R(\mathfrak{x}^{(m+1)}, \mathfrak{y}^{(k)}) \\ &\leftrightarrow \bigwedge^{m+1} \sim R(\mathfrak{x}^{(m+1)}, \mathfrak{y}^{(k)}). \end{aligned}$$

This completes the induction.

Thus, the classes $\boldsymbol{P}_n{}^A$, $\boldsymbol{Q}_n{}^A$ may be represented schematically in the following table, where, in each case, $S \prec A$:

n	1	2	3	. . .
$\boldsymbol{P}_n{}^A$	$\bigvee_{x_1} S$	$\bigvee_{x_1} \bigwedge_{x_2} S$	$\bigvee_{x_1} \bigwedge_{x_2} \bigvee_{x_3} S$	. . .
$\boldsymbol{Q}_n{}^A$	$\bigwedge_{x_1} S$	$\bigwedge_{x_1} \bigvee_{x_2} S$	$\bigwedge_{x_1} \bigvee_{x_2} \bigwedge_{x_3} S$	. . .

In tracing back the proof of Post's representation theorem, we see that the heart of the argument lies in Lemmas 4 and 5 of Sec. 6. The idea of developing the proof using, in effect, the very definition of A-recursiveness is due to Shapiro [1]. The author (Davis [1]) and Kleene [6] gave proofs which depended, instead, on an analysis of the predicate $T^{A'}(z, x, y)$. Post's original proof has never been published. It was based on the notion that a pseudo-A' can be constructed from a finite subset of A and that, as this finite subset "approaches" A as a limit, the pseudo-A' approaches A'.

THEOREM 7.4. *$P \in \boldsymbol{P}_n$ for some n if and only if P is arithmetical.*

PROOF. Immediate from Theorem 7.3 and Corollary 7-3.5.

Thus, the classes $\boldsymbol{P}_n$, $\boldsymbol{Q}_n$, $\boldsymbol{R}_n$ give a classification of the arithmetical predicates. This classification is known as **the Kleene hierarchy.**

CHAPTER 10

COMPUTABLE FUNCTIONALS

1. Functionals. In this chapter, we shall see how the notion of algorithm can be applied to operations on functions. Our development will be motivated by the consideration that, in a given finite computation, only finitely many values of a given function can be employed.

We shall consider *functionals* defined on n-tuples of functions and numbers and with numbers as values. For example, we might consider the functional $\mathbf{F}$ for which

$$\mathbf{F}(f, x) = f(x).$$

Now, the kinds of entities we shall want available as arguments for functionals include numbers, singulary functions, binary functions, etc. In order to have a uniform notation available, we note that constants, i.e., numbers, may be regarded as functions of zero arguments. Thus, each argument for a functional will be an n-ary function for some non-negative integer n. When we wish to emphasize that the function f is n-ary, we denote it by $f^{(n)}$.†

DEFINITION 1.1. *A* **functional of order** $(n_1, n_2, \ldots, n_k)$ *is a correspondence by which an integer is associated with certain of the n-tuples* $(f_1^{(n_1)}, f_2^{(n_2)}, \ldots, f_k^{(n_k)})$.

Functionals will be indicated by boldface upper-case characters. When it is desired to indicate the order of a functional explicitly, it may be used as a subscript on the letter used to denote the functional. For example,

$$\mathbf{F}_{(2,0,0)}(f^{(2)}, x, y) = f^{(2)}(x, y).$$

We use the symbol $\mathfrak{f}_k^n$ to abbreviate $f_1^{(n_1)}, \ldots, f_k^{(n_k)}$. Similarly for $\mathfrak{g}_k^n$. Also, we write $\mathfrak{n}_k$ for $(n_1, \ldots, n_k)$.

DEFINITION 1.2. *We write* $f^{(n)} \subseteq g^{(n)}$, *and we call* $g^{(n)}$ *an* **extension of** $f^{(n)}$ *if* $f^{(n)}(\mathfrak{x}^{(n)}) = k$ *implies* $g^{(n)}(\mathfrak{x}^{(n)}) = k$.

For instance, if $f(2) = 4$, $f(4) = 16$, and $f(x)$ is otherwise undefined, and if $g(x) = x^2$, then $f \subseteq g$.

† More accurately, we use lower-case letters of the Roman alphabet with superscript (n) as variables whose range is the set of all n-ary functions.

Note that, for 0-ary functions, i.e., numbers, $x^{(0)} \subseteq y^{(0)}$ if and only if $x^{(0)} = y^{(0)}$.

We write $\mathfrak{f}_k{}^n \subseteq \mathfrak{g}_k{}^n$ to mean that $f_i^{(n_i)} \subseteq g_i^{(n_i)}$, for $1 \leqq i \leqq k$.

DEFINITION 1.3. *A function is* **finite** *if its domain is a finite set.*

In particular, since the empty set is finite, numbers are finite functions.

DEFINITION 1.4. *Let $n > 0$, and let $f^{(n)}$ be the finite function such that $f^{(n)}(r_{1j}, r_{2j}, \ldots, r_{nj}) = s_j$, $j = 1, 2, \ldots, m$, and such that $f^{(n)}$ is otherwise undefined. Then we write*

$$\langle f^{(n)} \rangle = \prod_{j=1}^{m} \left[\Pr \left(\prod_{i=1}^{n} \Pr (i)^{r_{ij}+1} \right) \right]^{s_j+1}.$$

If $f^{(n)}$ has an empty domain, we write $\langle f^{(n)} \rangle = 0$. Finally, for numbers x, we take $\langle x \rangle = x$.

Thus, each finite function f has a definite number associated with it.

THEOREM 1.1. *For $n > 0$, if $f^{(n)}$ is finite, then*

$$f^{(n)}(x_1, \ldots, x_n) = \min_y \left[\left(\prod_{i=1}^{n} \Pr (i)^{x_i+1} \right) \text{Gl} \langle f^{(n)} \rangle = y + 1 \right].$$

PROOF. If $f^{(n)}$ is completely undefined (that is, has an empty domain), then both sides of the equation are undefined. For n-tuples $(x_1, \ldots, x_n)$ which do not belong to the domain of $f^{(n)}$,

$$\left(\prod_{i=1}^{n} \Pr (i)^{x_i+1} \right) \text{Gl} \langle f^{(n)} \rangle = 0;$$

so the right-hand side is undefined.

Finally, let $f^{(n)}$ be as in Definition 1.4, and let

$$t_j = \prod_{i=1}^{n} \Pr (i)^{r_{ij}+1}.$$

Then

$$\begin{aligned} t_j \text{ Gl } \langle f^{(n)} \rangle &= t_j \text{ Gl } \prod_{j=1}^{m} [\Pr (t_j)]^{s_j+1} \\ &= s_j + 1 \\ &= f^{(n)}(r_{1j}, \ldots, r_{nj}) + 1. \end{aligned}$$

Hence,

$$\min_y [t_j \text{ Gl } \langle f^{(n)} \rangle = y + 1] = f^{(n)}(r_{1j}, \ldots, r_{nj}).$$

COROLLARY 1.2. *If $\langle f^{(n)} \rangle = \langle g^{(n)} \rangle$, then $f^{(n)} = g^{(n)}$.*

DEFINITION 1.5. *For each integer p and each $n > 0$, we let $p^{[n]}$ denote the finite function such that*

$$p^{[n]}(x_1, \ldots, x_n) = \min_y \left[\left(\prod_{i=1}^{n} \text{Pr}\,(i)^{x_i+1} \right) \text{Gl}\, p = y + 1 \right].$$

We also define $p^{[0]} = p$.

THEOREM 1.3. $\langle f^{(n)} \rangle^{[n]} = f^{(n)}$.

PROOF. Using Definition 1.5 and Theorem 1.1,

$$\langle f^{(n)} \rangle^{[n]}(x_1, \ldots, x_n) = \min_y \left[\left(\prod_{i=1}^{n} \text{Pr}\,(i)^{x_i+1} \right) \text{Gl}\, \langle f^{(n)} \rangle = y + 1 \right]$$

$$= f^{(n)}(x_1, \ldots, x_n).$$

THEOREM 1.4. *For each n, the predicate $r^{[n]} \subseteq s^{[n]}$ is primitive recursive.*

PROOF. $r^{[n]} \subseteq s^{[n]}$ if and only if

$$\bigwedge_{k=1}^{\mathfrak{L}(r)} \{k \,\text{Gl}\, r = 0 \vee \sim [\mathfrak{L}(k) = n \wedge \text{GN}\,(k)] \vee (k \,\text{Gl}\, r = k \,\text{Gl}\, s)\}.$$

We write $\mathfrak{x}_k^{[n]}$ for $x_1^{[n_1]}, x_2^{[n_2]}, \ldots, x_k^{[n_k]}$, and similarly for other letters.

2. Completely Computable Functionals. Our main definition is based on the view that a functional is to be regarded as "effective" if its values can be computed for given arguments by computing on finite functions of which the arguments are extensions.

DEFINITION 2.1.† *$F_{\mathfrak{n}_k}$ is* **completely computable** *if there is a k-ary partially computable function φ such that $F(\mathfrak{f}_k{}^n) = t$ if and only if there are numbers $\mathfrak{x}^{(k)}$ such that $\mathfrak{x}_k^{[n]} \subseteq \mathfrak{f}_k{}^n$ and $\varphi(\mathfrak{x}^{(k)}) = t$.*

Note that, for the case where all of the $\mathfrak{f}_k{}^n$ are 0-ary, so that F reduces to an ordinary function, being *completely computable* is equivalent to being *partially computable.*

DEFINITION 2.2. *$F_{\mathfrak{n}_k}$ is called* **compact** *if*

(1) *Whenever $F(\mathfrak{f}_k{}^n) = t$ and $\mathfrak{f}_k{}^n \subseteq \mathfrak{g}_k{}^n$, we have $F(\mathfrak{g}_k{}^n) = t$, and*

(2) *Whenever $F(\mathfrak{f}_k{}^n) = t$, there is a k-tuple $\mathfrak{g}_k{}^n$ of* **finite** *functions such that $\mathfrak{g}_k{}^n \subseteq \mathfrak{f}_k{}^n$ and $F(\mathfrak{g}_k{}^n) = t$.*

THEOREM 2.1. *A completely computable functional is compact.*

PROOF. Let $F_{\mathfrak{n}_k}$ be a completely computable functional, let φ be as in Definition 2.1, and let $F(\mathfrak{f}_k{}^n) = t$. Then, for some $\mathfrak{x}^{(k)}$, we have $\varphi(\mathfrak{x}^{(k)}) = t$

† This definition is a special case of the definition given, for functionals of arbitrary finite type, in Davis [4]. Most of the theorems in this chapter will be found in Kleene [6], proved on the basis of Kleene's definition of recursive functional. The development of the theory for arbitrary finite types will appear in a forthcoming paper by the author and Hilary Putnam.

and $\mathfrak{x}_k^{[n]} \subseteq \mathfrak{f}_k{}^n$. But then $F(\mathfrak{x}_k^{[n]}) = t$; so condition (2) of Definition 2.2 is satisfied. Suppose $\mathfrak{f}_k{}^n \subseteq \mathfrak{g}_k{}^n$; then $\mathfrak{x}_k^{[n]} \subseteq \mathfrak{g}_k{}^n$, and $F(\mathfrak{g}_k{}^n) = t$.

DEFINITION 2.3. We define:

$$M_k(z, \mathfrak{x}^{(k)}, y) \leftrightarrow \bigvee_{\mathfrak{y}^{(k)}=0}^{y} [T_k(z, \mathfrak{y}^{(k)}, y) \wedge (\mathfrak{y}_k^{[n]} \subseteq \mathfrak{x}_k^{[n]})]$$

THEOREM 2.2. *Let $F_{\mathfrak{n}_k}$ be compact, and let θ be defined by*

$$\theta(\mathfrak{x}^{(k)}) = F(\mathfrak{x}_k^{[n]}).$$

Then F is completely computable if and only if θ is partially computable.

PROOF. If θ is partially computable, then we may take $\varphi = \theta$ in Definition 2.1.

Conversely, let F be completely computable, and let φ be as in Definition 2.1. By the normal form theorem (Corollary 4-2.2),

$$\varphi(\mathfrak{x}^{(k)}) = U(\min_y T_k(z_0, \mathfrak{x}^{(k)}, y)).$$

Then, using Definitions 2.1 and 2.3,

$$\theta(\mathfrak{x}^{(k)}) = F(\mathfrak{x}_k^{[n]}) = U(\min_y M_k(z_0, \mathfrak{x}^{(k)}, y))$$

Finally, by Theorem 1.4, θ is partially computable.

EXAMPLE. Let $F(f^{(m)}, \mathfrak{x}^{(m)}) = f^{(m)}(\mathfrak{x}^{(m)})$. Then F is completely computable. For F is obviously compact. Using Theorem 2.2,

$$\begin{aligned} \theta(y, \mathfrak{x}^{(m)}) &= F(y^{[m]}, \mathfrak{x}^{(m)}) \\ &= y^{[m]}(\mathfrak{x}^{(m)}), \end{aligned}$$

which is partial recursive, by Definition 1.5.

THEOREM 2.3. Let *$h^{(m)}$ be partial recursive. Then if $G_{\mathfrak{n}_k}^{(i)}$ is completely computable for $i = 1, 2, \ldots, m$, so is $F_{\mathfrak{n}_k}$ where*

$$F_{\mathfrak{n}_k}(\mathfrak{f}_k{}^n) = h^{(m)}(G_{\mathfrak{n}_k}^{(1)}(\mathfrak{f}_k{}^n), \ldots, G_{\mathfrak{n}_k}^{(m)}(\mathfrak{f}_k{}^n)).$$

PROOF. F is obviously compact. Let

$$\theta(\mathfrak{x}^{(k)}) = F(\mathfrak{x}_k^{[n]})$$

and

$$\psi_i(\mathfrak{x}^{(k)}) = G^{(i)}(\mathfrak{x}_k^{[n]}).$$

Then

$$\theta(\mathfrak{x}^{(k)}) = h^{(m)}(\psi_1(\mathfrak{x}^{(k)}), \ldots, \psi_m(\mathfrak{x}^{(k)})).$$

Hence, the partial recursiveness of the ψ_i implies that of θ. The result follows from Theorem 2.2.

It is possible to object to our definition of complete computability by pointing out that we have restricted ourselves to computations that require only finitely many values of functions in order to be carried out.

Would we not get a larger class of completely computable functionals if we employed Gödel numbers of partial recursive functions of which our given functions are extensions? No, as the following theorem shows.

THEOREM 2.4. *Let* $\boldsymbol{F}_{\text{n}_k}$ *be compact and let* ρ *be defined by*[1]

$$\rho(x_1, \ldots, x_k) = \boldsymbol{F}([x_1]_{n_1}, \ldots, [x_k]_{n_k}).$$

Then, $\boldsymbol{F}$ *is completely computable if and only if* ρ *is partially computable.*

PROOF. First suppose that ρ is partially computable. We note that, by Definition 1.5, for each k, $m^{[k]}(x_1, \ldots, x_k)$ is partially computable, regarded as a function of $m, x_1, \ldots, x_k$. Hence, by Corollary 9-1.3, there are primitive recursive functions $q_k(m)$ such that $[q_k(m)]_k = m^{[k]}$. We let θ be defined as in the statement of Theorem 2.2. Then

$$\begin{aligned} \theta(x_1, \ldots, x_k) &= \boldsymbol{F}(x_1^{[n_1]}, \ldots, x_k^{[n_k]}) \\ &= \boldsymbol{F}([q_{n_1}(x_1)]_{n_1}, \ldots, [q_{n_k}(x_k)]_{n_k}) \\ &= \rho(q_{n_1}(x_1), \ldots, q_{n_k}(x_k)). \end{aligned}$$

Hence, θ is partially computable; so, by Theorem 2.2, $\boldsymbol{F}$ is completely computable.

In order to prove the converse, we require the following lemma:

LEMMA. *For each* n, *there is a primitive recursive function* $r_n(t, x)$ *such that*

(1) *For each* t, $(r_n(t, x))^{[n]} \subseteq [x]_n$, *and*

(2) *Whenever* $[x]_n(s_1, \ldots, s_n) = l$, *then, for all sufficiently large* t, $(r_n(t, x))^{[n]}(s_1, \ldots, s_n) = l$.

PROOF OF LEMMA. For $n = 0$, we need only take $r_n(t, x) = x$.

For $n > 0$, we set $r_n(0, x) = 1$, and

$$r_n(t+1, x) = r_n(t, x) \text{ if } \sim T_n(x, 1 \text{ Gl } t, \ldots, n \text{ Gl } t, (n+1) \text{ Gl } t);$$

$$r_n(t+1, x) = r_n(t, x) \left[\text{Pr}\left(\prod_{i=1}^{n} \text{Pr}\,(i)^{(i \text{ Gl } t)+1} \right) \right]^{U((n+1) \text{ Gl } t)+1}, \text{ otherwise.}$$

Then it is easy to see that $r_n(t, x)$ is primitive recursive and has the required properties.

PROOF OF THEOREM 2.4 (concluded). Let θ be as in the statement of Theorem 2.2. Then, for each choice of $x_1, \ldots, x_k$, if a corresponding value of t is chosen sufficiently large,

$$\begin{aligned} \rho(x_1, \ldots, x_k) &= \boldsymbol{F}([x_1]_{n_1}, \ldots, [x_k]_{n_k}) \\ &= \boldsymbol{F}((r_{n_1}(t, x_1))^{[n_1]}, \ldots, (r_{n_k}(t, x_k))^{[n_k]}) \\ &= \theta(r_{n_1}(t, x_1), \ldots, r_{n_k}(t, x_k)). \end{aligned}$$

[1] For notation, cf. the very beginning of Chap. 9. For $n = 0$, we must take $[r]_n = r$.

Since we are supposing that $\boldsymbol{F}$ is completely computable, it follows, by Theorem 2.2, that θ is partially computable. Let

$$\theta(x_1, \ldots, x_k) = U(\min_y T_k(e, x_1, \ldots, x_k, y)).$$

Then

$$\rho(x_1, \ldots, x_k) = U(L(\min_u T_k(e, r_{n_1}(K(u), x_1), \ldots, r_{n_k}(K(u), x_k), L(u)))).$$

3. Normal Form Theorems. We require a method for "cutting down" the values of a function so that only a finite function "remains."

DEFINITION 3.1. *By* $f^{(n)} \mid y$ *(for* $n > 0$*) we understand the finite function such that* $(f^{(n)} \mid y)\ (x_1, \ldots, x_n) = t$ *if and only if* $x_i \leqq y$ *for* $i = 1, 2, \ldots, n; t \leqq y$*; and* $f^{(n)}(x_1, \ldots, x_n) = t$*. We take* $f^{(0)} \mid y = f^{(0)}$.

We write $\langle \mathfrak{f}_k{}^n \mid y\rangle$ to abbreviate $\langle f_1^{(n_1)} \mid y\rangle, \ldots, \langle f_k^{(n_k)} \mid y\rangle$.

THEOREM 3.1. *Let* $\boldsymbol{F}_{\mathfrak{n}_k}$ *be a completely computable functional. Then there is a number e such that* $\boldsymbol{F}(\mathfrak{f}_k{}^n) = U(\min_y M_k(e, \langle \mathfrak{f}_k{}^n \mid y\rangle, y))$.

PROOF. Let θ be as in Theorem 2.2, and let $\theta = [e]_k$. The result follows at once from Theorem 2.2.

As things stand, the converse of Theorem 3.1 is not true. This is because, for some values of e, the function, say $[e]_1$, will assign different values to numbers r, s, even though $r^{[k]} \subseteq s^{[k]}$. However, a converse of Theorem 3.1 can be proved by suitably modifying the T-predicate. We imagine a Turing machine with Gödel number e, modified so that, whenever it computes values for two numbers representing finite functions one of which is an extension of the other, the same value will be produced in both cases.

We begin with the predicate $\text{Comp}_k\ (b, c)$, defined as follows:

$$\text{Comp}_k\ (r, s) \leftrightarrow \bigvee_{t=0}^{rs} [r^{[k]} \subseteq t^{[k]} \wedge s^{[k]} \subseteq t^{[k]}].$$

Next, we set

$$P_{\mathfrak{n}_k}(z, \mathfrak{x}^{(k)}, y) \leftrightarrow M_k(z, \mathfrak{x}^{(k)}, y) \wedge \sim \bigvee_{\mathfrak{y}^{(k)}} \bigvee_{v=0}^{y} \Big[M_k(z, \mathfrak{y}^{(k)}, v)$$

$$\wedge \prod_{i=1}^{k} \text{Pr}\ (i)^{y_i} \leqq \prod_{i=1}^{k} \text{Pr}\ (i)^{x_i} \wedge \bigwedge_{i=1}^{k} \text{Comp}_{n_i}\ (y_i, x_i) \wedge (U(v) \neq U(y))\Big].$$

Note that, in the case where z is one of the numbers e occurring in Theorem 3.1,

$$P_{\mathfrak{n}_k}(e, \mathfrak{x}^{(k)}, y) \leftrightarrow M_k(e, \mathfrak{x}^{(k)}, y).$$

However, in other cases, $P_{\mathfrak{n}_k}$ may be false for certain arguments making T_k true, because some smaller ones may be incompatible with them.

Thus, $P_{\mathfrak{n}_k}$ is obtained from T_k by simply "throwing away" all arguments which would embarrass us in trying to regard T_k as determining a functional in the manner of Theorem 3.1. However, we have "thrown away" too much. Hence we set

$$\mathfrak{J}_{\mathfrak{n}_k}(z, \mathfrak{x}^{(k)}, y) \leftrightarrow \bigvee_{\mathfrak{u}^{(k)}=0}^{y} [P_{\mathfrak{n}_k}(z, \mathfrak{u}^{(k)}, y) \wedge (\mathfrak{u}_k^{[n]} \subseteq \mathfrak{x}_k^{[n]})].$$

Note once again that if z is one of the numbers e occurring in Theorem 3.1,

$$\mathfrak{J}_{\mathfrak{n}_k}(e, \mathfrak{x}^{(k)}, y) \leftrightarrow M_k(e, \mathfrak{x}^{(k)}, y).$$

$\mathfrak{J}_{\mathfrak{n}_k}$ is primitive recursive, by Theorem 1.4.

THEOREM 3.2 (Normal Form Theorem). *$F_{\mathfrak{n}_k}$ is completely computable if and only if there is a number e such that*

$$F(\mathfrak{f}_k{}^n) = U(\min_y \mathfrak{J}_{\mathfrak{n}_k}(e, \langle \mathfrak{f}_k{}^n \mid y\rangle, y)).$$

PROOF. That any completely computable functional can be represented in this form is an immediate consequence of the above discussion and Theorem 3.1.

Conversely, let

$$F(\mathfrak{f}_k{}^n) = U(\min_y \mathfrak{J}_{\mathfrak{n}_k}(e, \langle \mathfrak{f}_k{}^n \mid y\rangle, y)),$$

for some fixed e. It is clear from the manner of definition of $\mathfrak{J}_{\mathfrak{n}_k}$ that F is compact. We complete the proof by applying Theorem 2.2. Thus,

$$\begin{aligned}\theta(\mathfrak{x}^{(k)}) &= F(\mathfrak{x}_k^{[n]}) \\ &= U(\min_y \mathfrak{J}_{\mathfrak{n}_k}(e, \langle \mathfrak{x}_k^{[n]} \mid y\rangle, y)).\end{aligned}$$

Hence, in order to show that θ is partially computable, it suffices to prove that, for each fixed n, $\langle x^{[n]} \mid y\rangle$ is a recursive function of x and y. For $n = 0$, this is obvious, since $\langle x^{[0]} \mid y\rangle = x$; for $n > 0$, it follows from the fact that

$$\langle x^{[n]} \mid y\rangle = \underset{z=0}{\overset{x}{\mathfrak{m}}} \bigwedge_{j=1}^{x} \Big[\sim \mathrm{GN}\,(j) \vee \mathfrak{L}(j) \neq n \vee j \,\mathrm{Gl}\, x > y + 1 \vee \bigvee_{i=1}^{n} (i \,\mathrm{Gl}\, j > y + 1) \vee j \,\mathrm{Gl}\, x = j \,\mathrm{Gl}\, z \Big].$$

4. Partially Computable and Computable Functionals. Because total functions play a special role in many problems, it is natural to consider functionals defined only on total functions. In order to include the case of 0-ary functions, we shall regard all such (i.e., all integers) as total.

DEFINITION 4.1. *$F_{\mathfrak{n}_k}$ is* **contracted** *if it is defined only on k-tuples of total functions. $F_{\mathfrak{n}_k}$ is* **total** *if it is contracted and if it is defined on* **all** *k-tuples $\mathfrak{f}_k^n$ of total functions.*

An algorithm for using functions to compute answers will lead to a completely computable functional. Such functionals will of necessity be defined on nontotal functions, in particular on finite functions (because of compactness). However, one is sometimes interested in problems having to do only with total functions. What is one to understand by an algorithm applied to total functions? The following definitions are motivated by the consideration that, when one is interested in an algorithm for a given set of objects, one is not at all concerned with the behavior of the algorithm with respect to objects that do not belong to the set[1] in question.

DEFINITION 4.2. *A contracted functional $F_{\mathfrak{n}_k}$ is said to be* **partially computable** *if there is a completely computable functional $G_{\mathfrak{n}_k}$ such that $F(\mathfrak{f}_k^n) = t$ if and only if $\mathfrak{f}_k^n$ is a k-tuple of total functions such that $G(\mathfrak{f}_k^n) = t$.*

DEFINITION 4.3. *F is a* **computable** *functional if it is partially computable and total.*

Note that, for functions, i.e., functionals of 0-ary functions, the notions of partial computability and computability introduced by these definitions agree with those with which we have been working. Note also that, for functions, the notions of partial computability and complete computability coincide.

An immediate consequence of Definition 4.2 and our normal form theorem (Theorem 3.2) is

THEOREM 4.1. *A functional $F_{\mathfrak{n}_k}$ is partially computable if and only if there is a number e such that, for all k-tuples $\mathfrak{f}_k^n$ of* **total** *functions,*

$$F_{\mathfrak{n}_k}(\mathfrak{f}_k^n) = U(\min_y \mathfrak{I}_{\mathfrak{n}_k}(e, \langle \mathfrak{f}_k^n \mid y \rangle, y)).$$

5. Functionals and Relative Recursiveness. To say that a function f is recursive in a function g corresponds, intuitively, to saying that there is an algorithm for computing the values of f from suitable values of g. But this should be equivalent to saying that there is a partially computable functional whose value for g is f.

THEOREM 5.1. *$f^{(n)}$ is partially $g^{(m)}$-computable, where $g^{(m)}$ is total, if and only if there is a partially computable[2] functional F such that[3]*

$$f^{(n)}(\mathfrak{x}^{(n)}) = F(g^{(m)}, \mathfrak{x}^{(n)}). \tag{1}$$

[1] Cf. Shapiro [1].

[2] Note that the words "partially computable" could be replaced by "completely computable."

[3] Here, F is of order $(m, \underbrace{0, 0, \ldots, 0}_{n})$.

PROOF. If (1) holds for a partially computable $\boldsymbol{F}$, then, by Theorem 4.1,

$$\begin{aligned} f^{(n)}(\mathfrak{x}^{(n)}) &= U(\min_y \mathfrak{J}(e, \langle g^{(m)} \mid y \rangle, \langle \mathfrak{x}^{(n)} \mid y \rangle, y)) \\ &= U(\min_y \mathfrak{J}(e, \langle g^{(m)} \mid y \rangle, \mathfrak{x}^{(n)}, y)), \end{aligned}$$

where we have omitted the subscript on $\mathfrak{J}$. Hence, we need only prove that $\tau(y) = \langle g^{(m)} \mid y \rangle$ is $g^{(m)}$-recursive. But

$$\langle g^{(m)} \mid y \rangle = \min_z \bigwedge_{\mathfrak{x}^{(m)}=0}^{y} \Big[(g\,(\mathfrak{x}^{(m)}) > y) \vee \Big(\prod_{i=1}^{m} \mathrm{Pr}\,(i)^{x_i+1} \Big)\,\mathrm{Gl}\,z \\ = g^{(m)}(\mathfrak{x}^{(m)}) + 1 \Big].$$

The fact that τ is $g^{(m)}$-recursive now follows from Theorem 9-3.2 and the results of Chap. 3.

Next, suppose that $f^{(n)}$ is partially $g^{(m)}$-computable. Now, by the normal form theorem (Corollary 4-2.2) and by Chap. 9, Sec. 3,

$$f^{(n)}(x^{(n)}) = U(\min_y T_n^{g^*}(e, x^{(n)}, y)).$$

We define the functional $\boldsymbol{F}$ by the requirement that $\boldsymbol{F}(v^{(m)}, \mathfrak{x}^{(n)}) = t$ if and only if there is a number y such that $U(y) = t$, such that all m-tuples of numbers $\leqq y$ belong to the domain of $v^{(m)}$, and such that, whenever $v^{(m)} \subseteq h^{(m)}$, where $h^{(m)}$ is total, it is true that $T_n^{h^*}(e, \mathfrak{x}^{(n)}, y)$. It is clear that (1) holds for this $\boldsymbol{F}$. Hence it suffices to prove that $\boldsymbol{F}$ is completely computable.

It is easy to see that $\boldsymbol{F}$ is compact; for clause 1 of Definition 2.2 is obviously satisfied. To see that clause 2 is also satisfied we must refer back to the meaning of the T_n^A predicate in terms of Turing machines (Definition 4-1.5). That is, any A-computation of a Turing machine can involve only a *finite* number of questions about the set A.

Using Theorem 2.2, it now remains only to prove the partial recursiveness of θ, where

$$\theta(s, \mathfrak{x}^{(n)}) = \boldsymbol{F}(s^{[m]}, \mathfrak{x}^{(n)}).$$

Let

$$A = \{u \mid s^{[m]}(1 \,\mathrm{Gl}\, u, \ldots, m \,\mathrm{Gl}\, u) = (m+1) \,\mathrm{Gl}\, u\}.$$

Then $\theta(s, \mathfrak{x}^{(n)})$ is equal to

$$U\Big(\min_y \Big\{T_n^A(e, \mathfrak{x}^{(n)}, y) \wedge \bigwedge_{\mathfrak{u}^{(m)}=0}^{y} \Big[\Big(\prod_{i=1}^{m} \mathrm{Pr}\,(i)^{u_i+1}\Big)\,\mathrm{Gl}\,s \neq 0\Big]\Big\}\Big).$$

A glance at formulas 25_A and 26_A (Chap. 4, Sec. 1) and at the proof of Theorem 4-1.4 shows that it suffices to verify that $C_A(u)$ is a recursive function of s and u. But this clearly follows from the fact that it is the characteristic function of the predicate $u \in A$, which is a recursive predicate of s and u.

Theorem 5.1 has the consequence that the notion of relative recursiveness is definable in terms of partially computable functionals and, hence, in terms of partial recursiveness. It is of interest to note that the notions of relative recursiveness and of completely computable functional are, therefore, *intrinsic*[1] in the sense that they ultimately depend only on the class of partial recursive functions (that is, only on the notion of partial recursiveness taken in extension) and not on the special formal apparatus used to define this (Turing machines, Herbrand-Gödel-Kleene systems of equations, normal systems, etc.). In particular, *formalisms known to give equivalent notions of partial recursiveness will automatically give equivalent notions of relative partial recursiveness and of completely computable functional.*

Theorem 5.1 also suggests that the notion of relative recursiveness be extended as follows:

DEFINITION 5.1. *$f^{(n)}$ is* **partially computable** (*or, equivalently,* **partial recursive**) **in the** (not necessarily total) **functions** $\mathfrak{g}_k^m$ *if there is a completely computable functional* $\boldsymbol{F}$ *such that*

$$f^{(n)}(\mathfrak{x}^{(n)}) = \boldsymbol{F}(\mathfrak{g}_k^m, \mathfrak{x}^{(n)}).$$

If $f^{(n)}$ is also total, it is said to be **computable** (*or* **recursive**) **in the functions** $\mathfrak{g}_k^n$.

It is natural to attempt to prove a normal form theorem for this notion, analogous to Corollary 4-2.2. Actually, we are now able to do even more; our normal form theorem will show explicitly how the "T-predicates" depend on the functions in which we are considering computability. Thus, our result will give new information even for the case of A-computability, where A is a set.

DEFINITION 5.2. *We write $T_n^{\mathfrak{m}_k}$ for $\mathfrak{Z}_{(m_1,\ldots,m_k,0,\ldots,0)}$, where there are n occurrences of 0 following m_k in the subscript of $\mathfrak{Z}$.*

The following theorem is an immediate consequence of these definitions and of Theorem 3.2:

THEOREM 5.2 (Kleene's Extended Normal Form Theorem). *$f^{(n)}$ is partially computable in $\mathfrak{g}_k^m$ if and only if there is a number e such that*

$$f^{(n)}(\mathfrak{x}^{(n)}) = U(\min_y T_n^{\mathfrak{m}_k}(e, \langle \mathfrak{g}_k^m \mid y \rangle, \mathfrak{x}^{(n)}, y)).$$

6. Decision Problems. We shall now apply the theory of computable functionals to decision problems. In general, decision problems that depend on a functional's turning out to be computable (or completely computable or partially computable) are "unlikely" to be solvable.

[1] The author is indebted to Norman Shapiro for the remark that, if partial recursiveness is to be an adequate formalization of our intuitive notion of effectiveness, then *all* concepts involving effectiveness should be intrinsic in this sense. The term "intrinsic" is from Davis [4].

This is because a completely computable functional not only must be partial recursive when considered as a function defined on the numbers associated with finite functions, but must also be *compact.*

To begin with, we consider the following decision problem:

To determine, of a given real number x, $0 \leqq x \leqq 1$, *whether or not it is rational.*

Employing the representation of such real numbers as decimal fractions to the base 2, we may interpret this problem as being solvable if and only if the functional F defined below is partially computable.

The domain of F *is the set of all total functions* v, *all of whose values are either* 0 *or* 1. $F(v) = 0$ *if* v *is ultimately periodic; otherwise* $F(v) = 1$.

But, since any finite function has both ultimately periodic extensions and extensions that are not ultimately periodic, there is no compact functional whose values agree with F wherever F is defined. Hence, F is not partially computable, and the decision problem stated above is recursively unsolvable.[1]

Next, we shall consider decision problems relating to classes of recursively enumerable sets. For this purpose, we shall use the characterization of recursively enumerable sets as the domains of definition of singulary partial recursive functions. We write Dom (v) for the domain of the function v.

For the rest of this section only, we shall write ϕ for the empty set of recursively enumerable sets and Γ for the class of all recursively enumerable sets. Also, if Ψ and Φ are two classes of recursively enumerable sets, then $\Psi - \Phi$ is the class of all those recursively enumerable sets that belong to Ψ but not to Φ.

DEFINITION 6.1. *Let* Φ *be a class of recursively enumerable sets. Then,* Φ *is called* **completely recursive** *if there exists a completely computable functional* F *such that* $F(v) = 0$ *if* Dom $(v) \in \Phi$ *and* $F(v) = 1$ *whenever* Dom $(v) \in \Gamma - \Phi$.

Note that we make no demands on $F(v)$ for v's that are not partial recursive.

COROLLARY 6.1. ϕ *and* Γ *are completely recursive.*

PROOF. For the first case, we take $F(v) = 1$ for all v; for the second case, we take $F(v) = 0$ for all v.

DEFINITION 6.2. *The class* Φ *of recursively enumerable sets is* **completely recursively enumerable** *if there is a completely computable functional* F *such that* $F(v) = 0$ *if* Dom $(v) \in \Phi$ *whereas* $F(v)$ *is undefined whenever* Dom $(v) \in \Gamma - \Phi$.

THEOREM 6.2. Φ *is completely recursive if and only if* Φ *and* $\Gamma - \Phi$ *are both completely recursively enumerable.*

[1] This result is due to Shapiro [1], who has proved much stronger results in this direction.

PROOF. First, suppose that Φ is completely recursive, and let F be as in Definition 6.1. Let $g(x)$ be the partial recursive function such that $g(0) = 0$, whereas $g(x)$ is undefined for $x \neq 0$. Then the functionals $g(F(v))$, $g(1 \dot{-} F(v))$ are completely computable, by Theorem 2.3. But these functionals have the properties required by Definition 6.2, for Φ and $\Gamma - \Phi$, respectively.

Conversely, let Φ and $\Gamma - \Phi$ be completely recursively enumerable. Let the functionals F and G be completely computable, and let them satisfy the requirements

$$\begin{aligned} &F(v) = 0,\ G(v) \text{ undefined, when Dom } (v) \in \Phi; \\ &F(v) \text{ undefined, } G(v) = 0, \text{ when Dom } (v) \in \Gamma - \Phi. \end{aligned}$$

Let ρ_1, ρ_2 be defined by

$$\rho_1(x) = F([x]_1);\ \rho_2(x) = G([x]_1) + 1,$$

so that, by Theorem 2.4, ρ_1, ρ_2 are partial recursive. Since it is impossible for Dom (v) to belong simultaneously to Φ and $\Gamma - \Phi$, the domains of ρ_1 and ρ_2 have no elements in common. Let ρ_3 be defined by

$$\begin{aligned} \rho_3(x) &= \rho_1(x) \qquad \text{for } x \in \text{Dom } (\rho_1), \\ &= \rho_2(x) \qquad \text{for } x \in \text{Dom } (\rho_2); \end{aligned}$$

undefined, otherwise. Then ρ_3 is partial recursive {for, if $\rho_1 = [m]_1$, $\rho_2 = [n]_1$, then $\rho_3(x) = U(\min_y [T(m, x, y) \vee T(n, x, y)])$}. Let H be defined by the requirement that $H(v) = t$ if and only if $[e]_1 \subseteq v$ and $\rho_3(e) = t$. Then H is obviously compact and, hence, by Theorem 2.4, is completely computable. Hence, Φ is completely recursive.

THEOREM 6.3† (Rice's "Key-array" Conjecture). *Φ is completely recursively enumerable if and only if there is a recursively enumerable set of integers S such that a set $R \in \Phi$ if and only if there is a $q \in S$ such that* Dom $(q^{[1]}) \subset R$.

PROOF. First suppose there is such a recursively enumerable set S. Let $F(v) = 0$ if there is a $q \in S$ such that Dom $(q^{[1]}) \subset$ Dom (v); otherwise let F be undefined. We must show that F is completely computable. Now F is clearly compact. Hence, by Theorem 2.2, it suffices to prove the partial recursiveness of θ, where $\theta(r) = F(r^{[1]})$, or, what amounts to the same thing (since θ is constant), to prove that the domain of θ is

† Conjectured by Rice [1]. Proved in Myhill and Shepherdson [1]. It should be mentioned that our definitions of completely recursive and completely recursively enumerable classes of recursively enumerable sets differ from Rice's original formulation. The equivalence of the formulations follows from our Theorem 2.4 and the main result of Myhill and Shepherdson [1]. Although our proofs are quite easy, we have, in effect, transferred the difficulty to the equivalence proof. However, the definitions employed here are quite natural formalizations of the concepts involved.

recursively enumerable. Let $S = \left\{x \mid \bigvee_y T(e, x, y)\right\}$. Then

$$\text{Dom}\,(\theta) = \left\{r \mid \bigvee_y \bigvee_q [T(e, q, y) \wedge \text{Dom}\,(q^{[1]}) \subset \text{Dom}\,(r^{[1]})]\right\},$$

where

$$\text{Dom}\,(q^{[1]}) \subset \text{Dom}\,(r^{[1]}) \leftrightarrow \bigwedge_{j=1}^{q+r} [(2^j \text{ Gl } q = 0) \vee (2^j \text{ Gl } r \neq 0)].$$

Conversely, let Φ be completely recursively enumerable. Let $\boldsymbol{F}(v) = 0$ if Dom $(v) \in \Phi$; let $\boldsymbol{F}(v)$ be undefined if Dom $(v) \in \Gamma - \Phi$, where $\boldsymbol{F}$ is completely computable. Let S be the set of all numbers q such that $\boldsymbol{F}(q^{[1]})$ is defined. Then $R \in \Phi$ if and only if $R =$ Dom (v) and $\boldsymbol{F}(v) = 0$, which is true if and only if there is a number q such that $q^{[1]} \subseteq v$ and $\boldsymbol{F}(q^{[1]}) = 0$, and this implies that $q \in S$ and Dom $(q^{[1]}) \subset$ Dom (v). If, on the other hand, $q \in S$ and Dom $(q^{[1]}) \subset R$, then v can be chosen so that Dom $(v) = R$ and $q^{[1]} \subseteq v$. That S is recursively enumerable is an immediate consequence of Theorem 2.2.

The following corollary is an immediate consequence of Theorem 6.3.

COROLLARY 6.4. *If $\Phi \neq \emptyset$ and Φ is completely recursively enumerable, then $I \in \Phi$, where I is the set of nonnegative integers.*

THEOREM 6.5 (Rice's Theorem[1]). *There are no completely recursive sets other than Γ and $\emptyset$.*

PROOF. Let Φ be completely recursive. Then Φ and $\Gamma - \Phi$ are completely recursively enumerable. But, by Corollary 6.4, one of Φ, $\Gamma - \Phi$ must be empty, since otherwise both would have to contain the set of all integers. That is, either $\Phi = \emptyset$ or $\Phi = \Gamma$.

7. The Recursion Theorems. By a *solution* of the equation

$$\boldsymbol{F}(f^{(m)}, \mathfrak{x}^{(m)}) = f^{(m)}(\mathfrak{x}^{(m)}) \tag{1}$$

we understand a specific function $f^{(m)}$ with respect to which (1) holds for all $\mathfrak{x}^{(m)}$.† Kleene [3a, 6] has proved theorems bearing on the existence of partially computable solutions of (1).

DEFINITION 7.1. *Let r_n be defined for $n \geqq n_0$. We write $\lim r_n = r$, and we call r the* **limit** *of the sequence r_n, if there is an N such that $r_n = r$ for $n \geqq N$.*

DEFINITION 7.2. *The sequence f_n of functions is called* **monotone** *if $f_n \subseteq f_{n+1}$.*

[1] Cf. Rice [1].

† Of course, this equality is understood in the sense that both sides are defined or undefined together and are equal where defined.

THEOREM 7.1. *If f_n is monotone, then there is a function $f(\mathfrak{x}^{(m)})$ such that, for each $\mathfrak{x}^{(m)}$ in the domain of any f_n, we have* $\lim f_n(\mathfrak{x}^{(m)}) = f(\mathfrak{x}^{(m)})$.

PROOF. By monotonicity, for any numbers $\mathfrak{x}^{(m)}$ in the domains of f_r and f_s, we have $f_r(\mathfrak{x}^{(m)}) = f_s(\mathfrak{x}^{(m)})$. Hence, we may define f by the stipulation that $f(\mathfrak{x}^{(m)}) = t$ holds if and only if $f_n(\mathfrak{x}^{(m)}) = t$ for some n. Then, for $\mathfrak{x}^{(m)} \in \text{Dom}(f_N)$, we have $\mathfrak{x}^{(m)} \in \text{Dom}(f_n)$ for all $n \geqq N$. Hence, $\lim f_n(\mathfrak{x}^{(m)}) = f(\mathfrak{x}^{(m)})$.

THEOREM 7.2. *Let $\boldsymbol{F}$ be a compact functional of order* $(m, \underbrace{0, 0, \ldots, 0}_{m})$.

Let f_0 be completely undefined, and let f_{n+1} be defined by

$$f_{n+1}(\mathfrak{x}^{(m)}) = \boldsymbol{F}(f_n, \mathfrak{x}^{(m)}).$$

Then the sequence $f_n(\mathfrak{x}^{(m)})$ has a limit $f(\mathfrak{x}^{(m)})$, which is a solution of (1). *Moreover, all solutions of* (1) *are extensions of f.*

PROOF. We first note that the sequence f_n is monotone. We prove by mathematical induction that, for each n, $f_n \subseteq f_{n+1}$. For $n = 0$, this is obvious. Suppose it known for $n = k$, and let $\mathfrak{x}^{(m)} \in \text{Dom}(f_{k+1})$. Then, using the induction hypothesis and the compactness of $\boldsymbol{F}$,

$$\begin{aligned} f_{k+1}(\mathfrak{x}^{(m)}) &= \boldsymbol{F}(f_k, \mathfrak{x}^{(m)}) \\ &= \boldsymbol{F}(f_{k+1}, \mathfrak{x}^{(m)}) \\ &= f_{k+2}(\mathfrak{x}^{(m)}). \end{aligned}$$

Thus, $f_{k+1} \subseteq f_{k+2}$, and the induction is complete. Hence, by Theorem 7.1, there is a function f such that $\lim f_n(\mathfrak{x}^{(m)}) = f(\mathfrak{x}^{(m)})$.

We next show that f is a solution of (1). First suppose that $f(\mathfrak{x}^{(m)})$ is defined. Then, for some n, $f(\mathfrak{x}^{(m)}) = f_{n+1}(\mathfrak{x}^{(m)})$. Thus, using the compactness of $\boldsymbol{F}$,

$$\begin{aligned} f(\mathfrak{x}^{(m)}) &= f_{n+1}(\mathfrak{x}^{(m)}) \\ &= \boldsymbol{F}(f_n, \mathfrak{x}^{(m)}) \\ &= \boldsymbol{F}(f, \mathfrak{x}^{(m)}). \end{aligned}$$

Next, suppose that $\boldsymbol{F}(f, \mathfrak{x}^{(m)})$ is defined. Then, since $\boldsymbol{F}$ is compact, there is a *finite* function $g \subseteq f$ such that

$$\boldsymbol{F}(f, \mathfrak{x}^{(m)}) = \boldsymbol{F}(g, \mathfrak{x}^{(m)}).$$

But, for each $\mathfrak{x}^{(m)} \in \text{Dom}(g)$, there is some n such that $\mathfrak{x}^{(m)} \in \text{Dom}(f_n)$ and $f_n(\mathfrak{x}^{(m)}) = g(\mathfrak{x}^{(m)})$. Letting N be the largest of these n's (necessarily finite in number), we have $g \subseteq f_N$. Hence, by the compactness of $\boldsymbol{F}$,

$$\begin{aligned} \boldsymbol{F}(g, \mathfrak{x}^{(m)}) &= \boldsymbol{F}(f_N, \mathfrak{x}^{(m)}) \\ &= f_{N+1}(\mathfrak{x}^{(m)}) \\ &= f(\mathfrak{x}^{(m)}). \end{aligned}$$

This completes the proof that f is a solution of (1).

Finally, let h be any other solution of (1). We shall show by induction that, for each n, $f_n \subseteq h$, which will suffice to prove that $f \subseteq h$. For $n = 0$, this is obvious. Suppose it known for $n = k$, and let $\mathfrak{x}^{(m)} \in \text{Dom}(f_{k+1})$. Then, using the induction hypothesis and the compactness of $\boldsymbol{F}$, we have

$$\begin{aligned} f_{k+1}(\mathfrak{x}^{(m)}) &= \boldsymbol{F}(f_k, \mathfrak{x}^{(m)}) \\ &= \boldsymbol{F}(h, \mathfrak{x}^{(m)}) \\ &= h(\mathfrak{x}^{(m)}). \end{aligned}$$

THEOREM 7.3 (Kleene's First Recursion Theorem). *If, in addition to the hypothesis of Theorem 7.2, $\boldsymbol{F}$ is completely computable, then the function f there determined is partial recursive.*

PROOF. Each of the functions f_n is partially computable, as we shall demonstrate by constructing Gödel numbers of Turing machines which compute the f_n. Namely, let ρ be as in Theorem 2.4, so that ρ is partially computable, and

$$\rho(r, \mathfrak{x}^{(m)}) = \boldsymbol{F}([r]_m, \mathfrak{x}^{(m)}).$$

By Corollary 9-1.3, there is a primitive recursive function $\sigma(r)$ such that for all $\mathfrak{x}^{(m)}$,

$$\rho(r, \mathfrak{x}^{(m)}) \doteq [\sigma(r)]_m(\mathfrak{x}^{(m)}).$$

Let

$$\begin{aligned} \tau(0) &= 0; \\ \tau(n+1) &= \sigma(\tau(n)). \end{aligned}$$

Then τ is a primitive recursive function. We shall prove by induction that $[\tau(n)]_m = f_n$. This is clearly true for $n = 0$. Suppose it known for $n = k$. Then

$$\begin{aligned} f_{k+1}(\mathfrak{x}^{(m)}) &= \boldsymbol{F}(f_k, \mathfrak{x}^{(m)}) \\ &= \boldsymbol{F}([\tau(k)]_m, \mathfrak{x}^{(m)}) \\ &= \rho(\tau(k), \mathfrak{x}^{(m)}) \\ &= [\sigma(\tau(k))]_m(\mathfrak{x}^{(m)}) \\ &= [\tau(k+1)]_m(\mathfrak{x}^{(m)}), \end{aligned}$$

which completes the induction.

Now it is easy to see that f is partial recursive, since

$$f(\mathfrak{x}^{(m)}) = U(L(\min_t T_m(\tau(K(t)), \mathfrak{x}^{(m)}, L(t)))).$$

Next we shall see that a partial recursive solution to (1) can be found even when $\boldsymbol{F}$ depends not only on f but also on the Gödel number of a Turing machine which computes f. Our first version will not explicitly involve functionals.

THEOREM 7.4 (Kleene's Second Recursion Theorem). *If $g^{(m+1)}$ is a partial recursive function, then there is a number e such that*

$$[e]_m(\mathfrak{x}^{(m)}) = g(e, \mathfrak{x}^{(m)}).$$

PROOF. The function $g(S^1(y, y), \mathfrak{x}^{(m)})$ is partial recursive. Let z_0 be a Gödel number of a Turing machine which computes this function. Then, using Kleene's iteration theorem (Theorem 9-1.2), we have

$$\begin{aligned} g(S^1(y, y), \mathfrak{x}^{(m)}) &= [z_0]_{m+1}(y, \mathfrak{x}^{(m)}) \\ &= [S^1(z_0, y)]_m(\mathfrak{x}^{(m)}). \end{aligned}$$

Hence, setting $y = z_0$ and $e = S^1(z_0, z_0)$, we have

$$g(e, \mathfrak{x}^{(m)}) = [e]_m(\mathfrak{x}^{(m)}).$$

The following consequence of Theorem 7.4 is closely related to Theorem 7.3; in some ways it is stronger, in other ways weaker.

THEOREM 7.5. *If* $\boldsymbol{F}$ *is a completely computable functional of order* $(m, \underbrace{0, 0, \ldots, 0}_{m+1})$, *then there is a partial recursive function* f *such that*

$$f(\mathfrak{x}^{(m)}) = \boldsymbol{F}(f, e, \mathfrak{x}^{(m)}), \tag{2}$$

and e *is a Gödel number of* f.

PROOF. Let ρ be as in Theorem 2.4, so that

$$\rho(r, s, \mathfrak{x}^{(m)}) = \boldsymbol{F}([r]_m, s, \mathfrak{x}^{(m)}),$$

and let

$$g(z, \mathfrak{x}^{(m)}) = \rho(z, z, \mathfrak{x}^{(m)}).$$

By Theorem 7.4, there is a number e such that

$$\begin{aligned} [e]_m(\mathfrak{x}^{(m)}) &= g(e, \mathfrak{x}^{(m)}) \\ &= \rho(e, e, \mathfrak{x}^{(m)}) \\ &= \boldsymbol{F}([e]_m, e, \mathfrak{x}^{(m)}). \end{aligned}$$

The result then holds with $f = [e]_m$.

Although Theorem 7.5 is stronger than Theorem 7.3 in that $\boldsymbol{F}$ is permitted to depend not only on f but also on the Gödel number of a Turing machine for computing f, it is also weaker in that there is no reason to suppose that other solutions of (2) will necessarily be extensions of the function we obtain.

THEOREM 7.6 (Implicit-function Theorem). *For* $i = 1, 2, \ldots, k$, *let* $\boldsymbol{G}^{(i)}$ *be completely computable functionals of order* $(m, \underbrace{0, 0, \ldots, 0}_{m+1})$ *and let* $P_i(\mathfrak{x}^{(m)})$ *be mutually exclusive semicomputable predicates. Then there is a partial recursive function* f *and a number* e *such that* $f = [e]_m$ *and for each* $i = 1, 2, \ldots, k$, $P_i(\mathfrak{x})^{(m)})$ *implies* $\boldsymbol{G}^{(i)}(f, e, \mathfrak{x}^{(m)}) = f(\mathfrak{x}^{(m)})$.

PROOF. Let

$$\boldsymbol{F}(f, z, \mathfrak{x}^{(m)}) = \boldsymbol{G}^{(i)}(f, z, \mathfrak{x}^{(m)}) \text{ when } P_i(\mathfrak{x}^{(m)}) \text{ holds},$$

so that, by Theorem 2.2, as the reader will readily verify, $\boldsymbol{F}$ is completely computable. Then, by Theorem 7.5, there is a function f and a number e, where $[e]_m = f$, such that $\boldsymbol{F}(f, e, \mathfrak{x}^{(m)}) = f(\mathfrak{x}^{(m)})$. That is,

$$P_i(\mathfrak{x}^{(m)}) \text{ implies } \boldsymbol{G}^{(i)}(f, e, \mathfrak{x}^{(m)}) = f(\mathfrak{x}^{(m)}),$$

or, otherwise expressed, for $i = 1, 2, \ldots, k$, we have

$$\boldsymbol{G}^{(i)}(f, e, \mathfrak{x}^{(m)}) = f(\mathfrak{x}^{(m)}), \text{ when } P_i(\mathfrak{x}^{(m)}) \text{ holds.}$$

This last result is very useful. It has the consequence that we can impose varied sets of effective conditions on a function and its Gödel number and find a partial recursive function that satisfies all of them.

CHAPTER 11

THE CLASSIFICATION OF UNSOLVABLE DECISION PROBLEMS

1. Reducibility and the Kleene Hierarchy. The development in Chap. 9 suggests two different approaches to the question of classifying the decision problems for arithmetical predicates.

One approach is given by the *Kleene hierarchy* consisting of the classes $\boldsymbol{P}_n$, $\boldsymbol{Q}_n$, $\boldsymbol{R}_n$. Each arithmetical predicate is in all but a finite number of these classes; to state explicitly which classes for a given predicate is to give a measure of just how unsolvable the decision problem for that predicate is.

DEFINITION 1.1. *If $R \in \boldsymbol{P}_n$ and $R \notin \boldsymbol{Q}_n$, we say that R is* **properly** n**-generable.**

If $R \in \boldsymbol{Q}_n$ and $R \notin \boldsymbol{P}_n$, we say that R is **properly** n**-antigenerable.**

If $R \in \boldsymbol{R}_n$, but $R \notin \boldsymbol{P}_{n-1}$ and $R \notin \boldsymbol{Q}_{n-1}$, we say that R is **properly** n**-recursive.**

Note that every predicate satisfies just one of these three conditions and for precisely one value of n.

Another approach to the classification of decision problems is given by the relation "$<$" of *recursive reducibility.* We may ask, of a given problem P,

If we could solve P, what else could we solve?

And, we may ask,

The solutions to which problems would also furnish solutions to P?

Of course, these two approaches are connected. Thus, the predicates of $\boldsymbol{P}_n$ are precisely those that are strongly reducible to $\emptyset^n$ (cf. Corollary 9-5.4).

THEOREM 1.1. *If $Q_0 \in \boldsymbol{P}_n$ and if, for every predicate $Q \in \boldsymbol{P}_n$, we have $Q < < Q_0$, then Q_0 is properly n-generable.*

PROOF. For suppose otherwise. Then $Q_0 \in \boldsymbol{Q}_n$. Hence, $\sim Q_0 \in \boldsymbol{P}_n$, that is, $\sim Q_0 < < \emptyset^n$ (Corollary 9-5.4). Now, by hypothesis, for each $Q \in \boldsymbol{P}_n$, $\sim Q < < \sim Q_0$. By Theorem 9-4.3, $\sim Q < < \emptyset^n$, that is, $\sim Q \in \boldsymbol{P}_n$. But this contradicts Kleene's hierarchy theorem (Theorem 9-5.10).

COROLLARY 1.2. *If $Q_0 \in \boldsymbol{Q}_n$ and if, for every predicate $Q \in \boldsymbol{Q}_n$, we have $Q < < Q_0$, then Q_0 is properly n-antigenerable.*

PROOF. Follows from Theorem 1.1 on taking negations.

DEFINITION 1.2. *We write $\alpha \simeq \beta$ to mean that $\alpha < \beta$ and $\beta < \alpha$.*

Now predicates that belong to $\boldsymbol{P}_n$ or $\boldsymbol{Q}_n$ are $< \emptyset^n$. Theorem 1.1 and its corollary show that, under certain conditions, $Q \simeq \emptyset^n$ can imply that Q is properly n-generable or n-antigenerable. Of course, the predicates that are $< \emptyset^n$ will be found in $\boldsymbol{R}_{n+1}$.

As a simple application of these methods, we shall consider the predicate of Theorem 5-6.1, $\bigwedge_x \bigvee_y T(z, x, y)$. Theorem 5-6.1 asserts of this predicate that it is not 1-generable. By Post's representation theorem (Theorem 9-7.3), this predicate is 2-antigenerable. Consider an arbitrary predicate $\bigwedge_x \bigvee_y R(z, x, y)$ that belongs to $\boldsymbol{Q}_2$. By Corollary 9-1.5, there is a recursive function $\varphi(z)$ such that

$$\bigvee_y R(z, x, y) \leftrightarrow \bigvee_y T(\varphi(z), x, y).$$

Hence,

$$\bigwedge_x \bigvee_y R(z, x, y) \leftrightarrow \bigwedge_x \bigvee_y T(\varphi(z), x, y).$$

Thus,

$$\bigwedge_x \bigvee_y R(z, x, y) < < \bigwedge_x \bigvee_y T(z, x, y).$$

Now using Corollary 1.2, we have

THEOREM 1.3. *The predicate $\bigwedge_x \bigvee_y T(z, x, y)$ is properly 2-antigenerable.*[1]

2. Incomparability. In this section we shall see that "pathological" decision problems can be constructed for which the two approaches to the classification problem suggested above diverge.

THEOREM 2.1. *There exist total functions $v(x)$, $w(x)$ such that:*

(1) $v(x) < \emptyset'$.
(2) $w(x) < \emptyset'$.
(3) $v(x) \not< w(x)$.
(4) $w(x) \not< v(x)$.

PROOF. For the purpose of this proof, we introduce the notation

$$\boldsymbol{F}_n(f, x) = U(\min_y \mathfrak{J}_{(1,0)}(n, \langle f \mid y \rangle, x, y)).$$

[1] Cf. Davis [1].

We shall construct the functions v, w in such a way that for no n do either of the following equations hold for all x:

$$v(x) = \boldsymbol{F}_n(w, x),$$
$$w(x) = \boldsymbol{F}_n(v, x).$$

Then the fact that $v \not\prec w$ and $w \not\prec v$ will follow at once from Theorems 10-3.2 and 10-5.1.

Our construction of v and w will consist of an approximating procedure; that is, we shall approximate to v and w by functions whose domains consist of the sets of all numbers $\leqq q$ for suitable values of q. These functions will be recognized as finite functions in the sense of Chap. 10 and will be represented by integers.

Our construction then appears in the guise of constructing a pair of infinite sequences of numbers, a_n, b_n. We set $a_0 = b_0 = 0$, and we suppose that a_n, b_n have been defined for $n < 2k + 1$. We shall show how to define a_{2k+1}, b_{2k+1}, a_{2k+2}, b_{2k+2}.

In defining a_{2k+1}, b_{2k+1}, we distinguish two cases:

CASE I. *It is possible to determine functions α, β and a number x such that $a_{2k}^{[1]} \subseteq \alpha$, $b_{2k}^{[1]} \subseteq \beta$, $x \in$ Dom (α), $(\beta, x) \in$ Dom $(\boldsymbol{F}_k)$, and*[1]

$$\alpha(x) \neq \boldsymbol{F}_k(\beta, x).$$

We note that

$$x \in \text{Dom}\,(r^{[1]}) \leftrightarrow 2^{x+1}\ \text{Gl}\ r \neq 0,$$

and we set

$$\text{Consec}\,(r) \leftrightarrow \bigwedge_{x=0}^{r} \left\{x \notin \text{Dom}\,(r^{[1]}) \vee \bigwedge_{y=0}^{x} [y \in \text{Dom}\,(r^{[1]})]\right\}.$$

Consec (r) holds, for given r, if and only if the domain of $r^{[1]}$ consists of all $x \leqq q$ for some q. Next, we set

$$\begin{aligned} z_0 = \min_t \Big[& (K(t) \neq a_{2k}) \wedge (L(t) \neq b_{2k}) \\ & \wedge\ a_{2k}^{[1]} \subseteq K(t)^{[1]} \wedge b_{2k}^{[1]} \subseteq L(t)^{[1]} \wedge \text{Consec}\,(K(t)) \wedge \text{Consec}\,(L(t)) \\ & \wedge \bigvee_{x=0}^{t} \Big\{x \in \text{Dom}\,(K(t)^{[1]}) \wedge \bigvee_{y} [\mathfrak{I}_{(1,0)}(k, \langle L(t)^{[1]} \mid y\rangle, x, y) \\ & \qquad \wedge\ (K(t)^{[1]}(x) \neq U(y))]\Big\}\Big]. \end{aligned}$$

Finally, we set

$$a_{2k+1} = K(z_0),\ b_{2k+1} = L(z_0).$$

Note that our construction ensures the existence of a number $x \in$ Dom $(a_{2k+1}^{[1]})$ such that $(b_{2k+1}^{[1]}, x) \in$ Dom $(\boldsymbol{F}_k)$ and $a_{2k+1}^{[1]}(x) \neq \boldsymbol{F}_k(b_{2k+1}^{[1]}, x)$.

[1] Here, Dom $(\boldsymbol{F}_k)$ is the set of pairs (f, x) such that $\boldsymbol{F}_k(f, x)$ is defined.

CASE II. *The conditions of Case I do not hold.* In this case we let

$$z_0 = \min_t [(K(t) \neq a_{2k}) \wedge (L(t) \neq b_{2k}) \wedge a_{2k}^{[1]} \subseteq K(t)^{[1]} \wedge b_{2k}^{[1]} \subseteq L(t)^{[1]} \wedge \text{Consec}\,(K(t)) \wedge \text{Consec}\,(L(t))],$$

and again set

$$a_{2k+1} = K(z_0),\ b_{2k+1} = L(z_0).$$

The construction of a_{2k+2}, b_{2k+2} is identical with that of a_{2k+1}, b_{2k+1} except that the roles of the a's and b's are reversed.

Now, we define the functions $v(x)$, $w(x)$ concerning which the desired conclusions will be demonstrated.

$$\begin{aligned} v(x) &= K(\min_t a_{L(t)}^{[1]}(x) = K(t)), \\ w(x) &= K(\min_t b_{L(t)}^{[1]}(x) = K(t)). \end{aligned}$$

Note that, for each n, $a_n^{[1]} \subseteq v$, $b_n^{[1]} \subseteq w$.

By our construction, a_n and b_n, considered as functions of n, are both recursive in the semicomputable predicate

$$\bigvee_y [\mathfrak{I}_{(1,0)}(k, \langle L(t)^{[1]} \mid y\rangle, x, y) \wedge (K(t)^{[1]}(x) \neq U(y))],$$

and hence, by Theorem 9-4.6, a_n and b_n are ϕ'-recursive. Hence, by the defining equations for $v(x)$, $w(x)$, we have

$$v(x) \prec \phi',\ w(x) \prec \phi'.$$

Suppose now that $v(x) \prec w(x)$. Then, by Theorems 10-3.2 and 10-5.1, there is a partial recursive functional $\boldsymbol{F}_n$ such that

$$v(x) = \boldsymbol{F}_n(w, x).$$

We show this to be impossible by considering two cases.

CASE I. *In the construction of a_{2n+1}, b_{2n+1}, Case I above arose.*

Then there is a number x_0 such that $x_0 \in \text{Dom}\,(a_{2n+1}^{[1]})$, $(b_{2n+1}^{[1]}, x_0) \in \text{Dom}\,(\boldsymbol{F}_n)$, and

$$a_{2n+1}^{[1]}(x_0) \neq \boldsymbol{F}_n(b_{2n+1}^{[1]}, x_0).$$

But

$$\begin{aligned} a_{2n+1}^{[1]}(x_0) &= v(x_0) \\ &= \boldsymbol{F}_n(w, x_0) \\ &= \boldsymbol{F}_n(b_{2n+1}^{[1]}, x_0), \end{aligned}$$

since $\boldsymbol{F}_n$ is compact. This is a contradiction.

CASE II. *In the construction of a_{2n+1}, b_{2n+1}, Case II above arose.*

Choose $x_0 \in \text{Dom}\,(v)$, $x_0 \notin \text{Dom}\,(a_{2n}^{[1]})$, and let $\alpha(x) = a_{2n}^{[1]}(x)$ for $x \neq x_0$, $\alpha(x_0) = v(x_0) + 1$. We have $a_{2n}^{[1]} \subseteq \alpha$, $b_{2n}^{[1]} \subseteq w$, $x_0 \in \text{Dom}\,(\alpha)$,

$(w, x_0) \in \text{Dom}(F_n)$, and

$$F_n(w, x_0) = v(x_0) \neq \alpha(x_0),$$

which is a contradiction.

Thus we have shown that $v(x) \not< w(x)$. The proof that $w(x) \not< v(x)$ is similar.

THEOREM 2.2. *There exist sets A, B such that*

(1) $A < \phi'$.
(2) $B < \phi'$.
(3) $A \not< B$.
(4) $B \not< A$.†

PROOF. Let v, w be constructed satisfying Theorem 2.1. Let $A = v^*$, $B = w^*$. The result follows at once from Theorem 9-3.1.

THEOREM 2.3. *There exists a nonrecursive set A such that $A < \phi'$ but $\phi' \not< A$.*†

PROOF. Let A, B be constructed satisfying Theorem 2.2. Then A is clearly nonrecursive (otherwise, we should have $A < B$). Moreover, if $\phi' < A$, then, since $B < \phi'$, we should have $B < A$.

All the decision problems for recursively enumerable sets that we have dealt with in Chaps. 5 to 7 have been concerned with sets S for which $S \simeq \phi'$. And yet Theorem 2.3 shows that there are decision problems which, though unsolvable, are less unsolvable than that of ϕ'. Of course, the set A of Theorem 2.3 is not necessarily recursively enumerable. However, since $A < \phi'$, we must have

$$A = \left\{z \mid \bigvee_x \bigwedge_y R(x, y, z)\right\}$$

and

$$A = \left\{z \mid \bigwedge_x \bigvee_y S(x, y, z)\right\},$$

where R and S are recursive.

The question of whether there exists a recursively enumerable set A satisfying Theorem 2.3 was posed by Post in his [3]. This problem has since become known as *Post's problem.* A solution to it was found by Friedberg [1] and also, independently, by Mucnik [1]. They each showed how the proof of Theorem 2.1 could be modified so that $u(x)$ and $v(x)$ would be the characteristic functions of recursively enumerable sets.

3. Creative Sets and Simple Sets. In attempting to settle Post's problem, Post was led, in his [3], to consider various types of reducibility intermediate in strength between strong reducibility and reducibility.

† Cf. Kleene and Post [1].

In this section, we shall outline part of this development. The reader should be familiar with Chap. 5, Sec. 5.

THEOREM 3.1. *If C is A-creative, then there is a recursively enumerable set R such that $R \subset \bar{C}$ and R is infinite.*

PROOF. Since C is A-creative, there is a recursive function $f(n)$ such that, whenever $\{n\}_A \subset \bar{C}$, then $f(n) \in \bar{C}$ and $f(n) \notin \{n\}_A$. By Corollary 9-1.5, there is a recursive function $\varphi(r)$ such that

$$\bigvee_y [T^A(r, x, y) \vee x = f(r)] \leftrightarrow \bigvee_y T^A(\varphi(r), x, y),$$

that is,

$$[x \in \{r\}_A \vee x = f(r)] \leftrightarrow x \in \{\varphi(r)\}_A.$$

Now, let

$$g(0) = 0,$$
$$g(r + 1) = \varphi(g(r)),$$

and let

$$R = \left\{x \mid \bigvee_r \bigvee_y T(g(r), x, y)\right\}.$$

Then it is easily seen that R has the required properties. (Cf. the discussion following Corollary 5-5.2.)

COROLLARY 3.2. *An A-creative set is not A-simple.*

THEOREM 3.3. *If $A' << C$ and C is A-recursively enumerable, then C is A-creative.*[1]

PROOF. Since $\bigvee_y T^A(z, x, y) << A'$, there is a recursive function $f(z, x)$ such that

$$\bigvee_y T^A(z, x, y) \leftrightarrow f(z, x) \in C$$

Also, by Corollary 9-1.5, there is a recursive function $g(n)$ such that

$$\bigvee_y T^A(n, f(x, x), y) \leftrightarrow \bigvee_y T^A(g(n), x, y).$$

Now suppose that, for some fixed n, we have $\{n\}_A \subset \bar{C}$. Then

$$\bigvee_y T^A(n, x, y) \text{ implies } x \notin C;$$

so

$$\bigvee_y T^A(n, f(z, x), y) \text{ implies } f(z, x) \notin C.$$

[1] Cf. Davis [1]. The converse of this theorem was proved by Myhill in his [2].

But

$$f(z, x) \notin C \text{ implies } \sim \bigvee_y T^A(z, x, y).$$

Hence, setting $z = x$,

$$\bigvee_y T^A(n, f(x, x), y) \text{ implies } \sim \bigvee_y T^A(x, x, y).$$

Thus,

$$\bigvee_y T^A(g(n), x, y) \text{ implies } \sim \bigvee_y T^A(x, x, y).$$

Setting $x = g(n)$, we have

$$\bigvee_y T^A(g(n), g(n), y) \text{ implies } \sim \bigvee_y T^A(g(n), g(n), y).$$

This can be the case only if

$$\sim \bigvee_y T^A(g(n), g(n), y),$$

that is, if

$$\sim \bigvee_y T^A(n, f(g(n), g(n)), y),$$

that is,

$$f(g(n), g(n)) \notin \{n\}_A.$$

On the other hand,

$$\sim \bigvee_y T^A(g(n), g(n), y) \leftrightarrow f(g(n), g(n)) \notin C;$$

so

$$f(g(n), g(n)) \notin C.$$

Hence, C is A-creative.

THEOREM 3.4. *If S is A-simple, then it is not the case that $A' \prec\prec S$.*

PROOF. Otherwise, by Theorem 3.3, S would be A-creative.

Theorem 3.4 suggests that perhaps, if S is A-simple, then $A' \not\prec S$. However, we have

THEOREM 3.5. *For every A-creative set C, there is an A-simple set S_1 such that $C \prec S_1$.*†

PROOF. We begin with the A-simple set S^A of Theorem 5-5.3. The *proof* of Lemma 4, Chap. 5, Sec. 5, showed that, of the first $2n + 2$ integers, at least $n + 1$ belong to $\overline{S^A}$. Hence, for each n, at least one of the integers

$$n, n + 1, \ldots, 2n + 2$$

† Cf. Post [3].

belongs to $\overline{S^A}$, or, setting $n = 3 \cdot 2^m - 3$, for each m, at least one of the consecutive integers

$$3 \cdot 2^m - 3,\ 3 \cdot 2^m - 2,\ \ldots,\ 3 \cdot 2^{m+1} - 4$$

belongs to $\overline{S^A}$.

Now, let

$$S_1 = S^A \cup \left\{x \mid \bigvee_{y} \bigvee_{z=0}^{3 \cdot 2^x \dot{-} 1} [y = 3(2^x \dot{-} 1) + z \wedge x \in C]\right\}.$$

Clearly, S_1 is A-recursively enumerable. Also, since $S^A \subset S_1$, we have $\overline{S_1} \subset \overline{S^A}$, so $\overline{S_1}$ contains no infinite A-recursively enumerable set.

Next we show that $\overline{S_1}$ is infinite. For, since C is A-creative, $\bar{C}$ is infinite. But, for each $x \in \bar{C}$, each of the numbers $3(2^x - 1) + z$, $z \leqq 3 \cdot 2^x - 1$, will belong to S_1 if and only if it belongs to S^A. But, for each such x, as we have remarked, at least one of these numbers must belong to $\overline{S^A}$ and, hence, to $\overline{S_1}$. Thus, $\overline{S_1}$ is infinite.

Therefore, S_1 is A-simple. It remains to be shown that $C < S_1$. But this follows at once from the following observation:

$$C = \left\{x \mid \bigwedge_{z=0}^{3 \cdot 2^x \dot{-} 1} [3(2^x \dot{-} 1) + z] \in S_1\right\}.$$

4. Constructive Ordinals.[1] Let us recall the situation that was discussed following Corollary 5-5.2 and formalized in Theorem 3.1. That is, let C be a creative set and let $f(n)$ be a recursive function such that, whenever $\{n\} \subset \bar{C}$, then $f(n) \in \bar{C}$ and $f(n) \notin \{n\}$. Then, letting $\{n_0\} = \emptyset$, we are able to define integers $n_1, n_2, \ldots$ such that

$$\{n_{i+1}\} = \{x \mid x \in \{n_i\} \vee x = f(n_i)\},$$

thus obtaining larger and larger finite subsets of $\bar{C}$. Finally, as we have seen in the proof of Theorem 3.1, the elements of the $\{n_i\}$ can be combined into an infinite recursively enumerable subset R of $\bar{C}$. But let $R = \{n_\omega\}$. Then we may determine integers $n_{\omega+1}, n_{\omega+2}, \ldots$ such that $\{n_{\omega+i+1}\} = \{x \mid x \in \{n_{\omega+i}\} \vee x = f(n_{\omega+i})\}$. It is quite clear that by using larger transfinite ordinals as subscripts this process can be extended even further. How far? In order to answer this question, it is necessary to introduce Kleene's notion of *constructive ordinal*.[2]

Before discussing constructive ordinals, it may be worthwhile to comment briefly on the significance of the process that we have been dis-

[1] In this section we have followed closely the presentation given by Hartley Rogers in a seminar at the Massachusetts Institute of Technology.

[2] Cf. Church and Kleene [1], Church [2a], Kleene [3a, 4a, 8], Markwald [1], Spector [1].

cussing in the case where we are seeking to "represent" C within some logic $\mathfrak{L}$. In this case (cf. the discussion following Theorem 8-3.7), what we obtain is successive extensions of $\mathfrak{L}$ in which more and more of the undecidable propositions concerning C become decided. And now what our question amounts to is: How far into the transfinite can this process of extending $\mathfrak{L}$ be carried?

We are going to define a set $\mathfrak{O}$ and a relation $<_o$ so as to have the following properties:

(1) $1 \in \mathfrak{O}$.

(2) *If* $y \in \mathfrak{O}$, *then* $2^y \in \mathfrak{O}$ *and* $y <_o 2^y$.

(3) *If* $[y]_1(x)$ *is a recursive function and if, for each* n, $[y]_1(n) \in \mathfrak{O}$ *and* $[y]_1(n) <_o [y]_1(n+1)$, *then* $3 \cdot 5^y \in \mathfrak{O}$ *and, for each* n, $[y]_1(n) <_o 3 \cdot 5^y$.†

(4) *If* $x <_o y$ *and* $y <_o z$, *then* $x <_o z$.

(5) *If* $x <_o y$, *then* $x \in \mathfrak{O}$ *and* $y \in \mathfrak{O}$.

We wish to define $\mathfrak{O}$ so as to be as "small" as possible consistent with conditions (1) to (5). For the purpose of defining $\mathfrak{O}$ and $<_o$, we consider arbitrary sets R of ordered pairs (x, y) of integers, where R has the following properties:

(*a*) $(1, 2) \in R$.

(*b*) *If* $(1, y) \in R$, *then* $(y, 2^y) \in R$.

(*c*) *If* $[y]_1(x)$ *is a recursive function* (i.e., if $[y]_1$ is total) *and if, for each* n, $([y]_1(n), [y]_1(n+1)) \in R$, *then, for each* n, $([y]_1(n), 3 \cdot 5^y) \in R$.

(*d*) *If* $(x, y) \in R$ *and* $(y, z) \in R$, *then* $(x, z) \in R$.

Now we define $x <_o y$ to mean that $(x, y) \in R$ for *all* sets R that satisfy (*a*) to (*d*), and we define

$$\mathfrak{O} = \{x \mid x = 1 \lor 1 <_o x\}.$$

We see at once that conditions (1) to (5) are satisfied.

With each $x \in \mathfrak{O}$ we associate an ordinal number $|x|$ as follows:

If $x = 1$, $|x| = 0$.

If $x = 2^y$, then $|x| = |y| + 1$.

If $x = 3 \cdot 5^y$, then $|x| = \lim_{n \to \infty} |[y]_1(n)|$.

DEFINITION 4.1. *An ordinal number* α *is called* **constructive** *if* $\alpha = |x|$, *where* $x \in \mathfrak{O}$.

THEOREM 4.1. *There are partial recursive functions* $M(x)$, $P(x)$, $Q(x,n)$ *such that:*

(1) *If* $x \in \mathfrak{O}$, *then* $M(x) = 0$, 1, *or* 2 *according as* $|x|$ *is* 0, *a successor ordinal, or a limit ordinal,*

† Kleene's original version differs in an inessential manner in this clause.

(2) *If* $|x|$ *is a successor ordinal, then* $|x| = |P(x)| + 1$, *and*

(3) *If* $|x|$ *is a limit ordinal, then for each n we have* $Q(x, n) \in \mathcal{O}$. *Moreover,* $Q(x, n) <_o Q(x, n + 1)$, *and* $|x| = \lim_{n\to\infty} |Q(x, n)|$.

PROOF. We need only put

$$M(x) = \underset{y=0}{\overset{2}{\mathfrak{m}}} (y \text{ Gl } x \neq 0),$$

$$P(x) = 1 \text{ Gl } x,$$

and

$$Q(x, n) = [3 \text{ Gl } x]_1(n).$$

THEOREM 4.2. *There is a partial recursive function* $g(x, n)$ *such that, if* $x \in \mathcal{O}$, *then for each* $\alpha < |x|$, *there is at least one integer n for which* $|g(x, n)| = \alpha$.

PROOF. We seek a function $g(x, n)$ that satisfies the following conditions:

$$g(x, 0) = P(x) \qquad \text{if } M(x) = 1,$$

$$g(x, n + 1) = g(P(x), n) \qquad \text{if } M(x) = 1,$$

$$g(x, n) = g(Q(x, K(n)), L(n)) \qquad \text{if } M(x) = 2.$$

These conditions may be written as follows:

$$g(x, y) = P(x) \text{ if } M(x) = 1 \wedge y = 0,$$

$$g(x, y) = g(P(x), y \dot{-} 1) \text{ if } M(x) = 1 \wedge y > 0,$$

$$g(x, y) = g(Q(x, K(y)), L(y)) \text{ if } M(x) = 2.$$

But the existence of a partial recursive function $g(x, n)$ satisfying these conditions follows at once from Theorem 10-7.6 (the implicit-function theorem).

To see that such a function satisfies the remaining conditions of the theorem, we proceed by transfinite induction. For $|x| = 0$, the result holds vacuously.

If $|x|$ is a successor ordinal, that is, $M(x) = 1$, then the set of $\alpha < |x|$ consists of $|P(x)|$ and the α's that are $< |P(x)|$. By induction hypothesis, for each $\alpha < |P(x)|$, there is an n such that $\alpha = |g(P(x), n)|$, so that $\alpha = |g(x, n + 1)|$. If $\alpha = |P(x)|$, then $\alpha = |g(x, 0)|$.

Finally, if $|x|$ is a limit ordinal, that is, $M(x) = 2$, then

$$|x| = \lim_{n\to\infty} |Q(x, n)|.$$

Let $\alpha < |x|$. Then, for some p, $\alpha < |Q(x, p)|$. By induction hypothesis, for some q, $\alpha = |g(Q(x, p), q)|$. Letting $n = J(p, q)$, we have

$$|g(x, n)| = |g(Q(x, p), q)| = \alpha.$$

THEOREM 4.3. *There are ordinal numbers belonging to the second number class which are not constructive.*

PROOF. The set of constructive ordinals is obviously denumerable, whereas, as is well known, the second number class is not.

DEFINITION 4.2. *We let ω_1 be the least ordinal number that is not constructive.*

THEOREM 4.4. *The constructive ordinals are precisely the ordinals $< \omega_1$.*

PROOF. If $\alpha < \omega_1$, then, by Definition 4.2, α must be constructive. Conversely, let α be constructive and suppose $\alpha > \omega_1$. Let $\alpha = |x|$. Then, by Theorem 4.2, for suitable n, $\omega_1 = |g(x, n)|$, which is impossible since ω_1 is not constructive.

The class $\mathfrak{O}$ and hence the constructive ordinals may seem to have been defined in a rather arbitrary manner. However, as the following result makes quite clear, this arbitrariness is only apparent.

THEOREM 4.5. *Let $\mathfrak{O}'$ be some set of integers, and for each $x \in \mathfrak{O}'$ let $|x|'$ be an ordinal number. Moreover, let there be given partial recursive functions $M'(x)$, $P'(x)$, $Q'(x, n)$ satisfying the analogues of conditions* (1), (2), *and* (3)† *of Theorem* 4.1. *Finally, let ξ be the least ordinal number which cannot be written as $|x|'$. Then $\xi \leqq \omega_1$.*

PROOF. We seek a function $h(x)$ that associates with each element $x \in \mathfrak{O}'$ an element $h(x) \in \mathfrak{O}$ such that $|x|' = |h(x)|$. This will clearly be accomplished if we can find a function $h(x)$ and a number w that satisfy the following conditions:

$$\begin{aligned} h(x) &= 1 \qquad \text{if } M'(x) = 0, \\ h(x) &= 2^{h(P'(x))} \qquad \text{if } M'(x) = 1, \\ h(x) &= 3 \cdot 5^{w}, \end{aligned}$$

where $[w]_1(n) = h(Q'(x, n))$, if $M'(x) = 2$.

We shall show how to find such a function (in fact, it will be partial recursive) by making use of the implicit-function theorem.

To begin with, let $\alpha(z, x, n) = [z]_1(Q'(x, n))$. Then, by the iteration theorem (Corollary 9-1.3), there is a primitive recursive function $q(z, x)$ such that $\alpha(z, x, n) = [q(z, x)]_1(n)$. Thus, the third condition on h can be rewritten as follows:

$$h(x) = 3 \cdot 5^{q(z,x)},$$

where $[z]_1 = h$ if $M'(x) = 2$.

To bring the conditions into the form required by the implicit-function theorem (Theorem 10-7.6), we rewrite them as follows:

† An analogue of $<_o$ must also be available.

$$h(x) = 1 \text{ if } M'(x) = 0,$$
$$h(x) = 2^{h(P'(x))} \text{ if } M'(x) = 1,$$
$$h(x) = 3 \cdot 5^{\,q(z,x)} \text{ if } M'(x) = 2.$$

Then Theorem 10-7.6 guarantees the existence of a partial recursive function h and a number z which satisfies these conditions and for which $h = [z]_1$.

Now we shall return to the problem that we considered at the beginning of this section, that of extending into the transfinite the process of determining ever larger recursively enumerable subsets of the complement of a creative set. Thus, let C be a creative set and let $f(n)$ be a recursive function such that, whenever $\{n\} \subset \bar{C}$, then $f(n) \in \bar{C}$ and $f(n) \notin \{n\}$. We seek a function h having the following properties:

$$h(x) = 0 \quad \text{if } M(x) = 0,$$
$$\{h(x)\} = \{t \mid t \in \{h(P(x))\} \vee t = f(h(P(x)))\} \quad \text{if } M(x) = 1,$$
$$\{h(x)\} = \left\{t \mid \bigvee_n [t \in \{h(Q(x, n))\}]\right\} \quad \text{if } M(x) = 2.$$

Each of these conditions can be put into the form required for using the implicit-function theorem. For the first condition this is obvious. For the second, it suffices that, if $M(x) = 1$, then $h(x) = S^1(s, h(P(x)))$ where

$$t \in \{r\} \vee t = f(r) \leftrightarrow \bigvee_y T_2(s, r, t, y).$$

That this condition has the required form can be seen by the technique used above. The third condition states that, if $M(x) = 2$, then

$$h(x) = S^2(p, z, x) \text{ where}$$

$$\bigvee_n \bigvee_y T([z](Q(x, n)), t, y) \leftrightarrow \bigvee_y T_3(p, z, x, t, y),$$

and once again this can be handled as above. We have thus proved

THEOREM 4.6. *Let C be a creative set. Then there is a partial recursive function $h(x)$ such that:*

(1) *$x \in \mathfrak{O}$ implies $\{h(x)\} \subset \bar{C}$, and*
(2) *$x <_o y$ implies $\{h(x)\} \subset \{h(y)\}$ and $\{h(x)\} \neq \{h(y)\}$.*

Thus the process of obtaining larger and larger recursively enumerable subsets of the complement of a creative set may be extended into the transfinite through the constructive ordinals. Can it be continued further? No! For otherwise the Gödel numbers of the recursively enumerable sets would themselves form a set $\mathfrak{O}'$ whose properties would contradict Theorem 4.5.

5. Extensions of the Kleene Hierarchy. The Kleene hierarchy provides a classification of predicates beginning with the very "constructive" recursive and semicomputable predicates and proceeding upward through ever less constructive predicates. Of course, it is easy to construct predicates that do not belong to any of the levels of the Kleene hierarchy. For example, we may construct the predicate

$$R(x, y) \leftrightarrow x \in \emptyset^y.$$

For, if $R(x, y) \in \boldsymbol{P}_n$, we should have $R(x, y) < < \emptyset^n$, that is,

$$x \in \emptyset^y \leftrightarrow f(x, y) \in \emptyset^n,$$

where $f(x, y)$ is recursive. Thus,

$$x \in \emptyset^{n+1} \leftrightarrow f(x, n + 1) \in \emptyset^n,$$

and, if $R(y) \in \boldsymbol{P}_{n+1}$, so that $R(y) < < \emptyset^{n+1}$, we should have

$$\begin{aligned} R(y) &\leftrightarrow g(y) \in \emptyset^{n+1} \\ &\leftrightarrow f(g(y), n + 1) \in \emptyset^n, \end{aligned}$$

so that $R(y) \in \boldsymbol{P}_n$. This would imply $\boldsymbol{P}_{n+1} \subset \boldsymbol{P}_n$, which contradicts Kleene's hierarchy theorem (Theorem 9-5.10).

This suggests that the Kleene hierarchy be extended, presumably by transfinite induction. There have been various extensions of the Kleene hierarchy which have been discussed. In this final section, we shall attempt to do no more than briefly outline this field, which is the subject of much current research.

Any attempt to employ transfinite induction is faced with an immediate difficulty. The transfinite ordinals themselves are extremely nonconstructive, and it is difficult to see how one could hope to preserve the gradually increasing level of nonconstructivity of the Kleene hierarchy if one were to admit as indices, say, the entire second number class. The theory of constructive ordinals developed in Sec. 4 may be used to advantage here. In fact, equivalent extensions of the Kleene hierarchy, using the theory of constructive ordinals, were given by Davis [1] and by Mostowski [3].

DEFINITION 5.1. *We define* $L_n = \emptyset$ *if* $|n| = 0$. *Suppose* L_n *defined for all n for which* $|n| < \alpha$. *Then, if* α *is a successor ordinal and if* $|m| = \alpha$, *we define* $L_m = L'_{P(m)}$. *If* α *is a limit ordinal and if* $|m| = \alpha$, *we define*

$$L_m = \{x \mid K(x) \in L_{Q(m, L(x))}\}.$$

The sequence L_n may be regarded as a transfinite extension of the sequence $\emptyset, \emptyset^1, \emptyset^2, \emptyset^3, \ldots$. We may use this sequence to define an extended Kleene hierarchy consisting of classes $\mathcal{P}_n, \mathcal{Q}_n, \mathcal{R}_n$.

DEFINITION 5.2. *If n is a number such that* $n \in \mathcal{O}$, $\mathcal{P}_n$ *is the class of all predicates that are* L_n*-semicomputable,* $\mathcal{Q}_n$ *is the class of all negations of members of* $\mathcal{P}_n$, $\mathcal{R}_n = \mathcal{P}_n \cap \mathcal{Q}_n$.

It is not difficult to see that the properties developed for $\boldsymbol{P}_n$, $\boldsymbol{Q}_n$, $\boldsymbol{R}_n$ in Chap. 9, Sec. 5, carry over to $\mathcal{P}_n$, $\mathcal{Q}_n$, $\mathcal{R}_n$. One obvious question raised by Definition 5.2 is:

If $|m| = |n|$, does it follow that $\mathcal{P}_m = \mathcal{P}_n$?

In Davis [1], it is proved that it does indeed so follow if $|m| = |n| < \omega^2$. In Spector [1], it is proved that the result holds for all constructive ordinals, that is, for all ordinals $< \omega_1$.

The class of all predicates belonging to at least one $\mathcal{P}_n$ is called the class of *hyperarithmetical predicates*. If one begins the development with a fixed set A, instead of $\emptyset$, one obtains the notion of a predicate's being *hyperarithmetical in A*.

Kleene has considered a hierarchy based on function quantifiers. That is, numerical predicates are obtained by suitable use of quantifiers over functions applied to computable functionals, for example,

$$\bigvee_f \bigwedge_g \bigvee_y [F(f, g, x, y) = 0].$$

The predicates that can be represented in this way were called *analytic* by Kleene in his [7]. They fall into a hierarchy reminiscent of the Kleene hierarchy for arithmetical predicates, as follows:

$$\bigvee_f \bigwedge_y [F(f, x, y) = 0], \bigvee_f \bigwedge_g \bigvee_y [F(f, g, x, y) = 0], \ldots,$$

$$\bigwedge_f \bigvee_y [F(f, x, y) = 0], \bigwedge_f \bigvee_g \bigwedge_y [F(f, g, x, y) = 0], \ldots.$$

The predicates that can be represented in both one-quantifier forms turn out to be precisely the hyperarithmetical predicates.[1] This suggests an analogue of Post's representation theorem, namely, that an analytic predicate is expressible in both n-quantifier forms if and only if it is hyperarithmetical in an $(n - 1)$-quantifier form. But this has turned out to be false.[2]

[1] Cf. Kleene [9].

[2] Cf. Addison and Kleene [1].

APPENDIX 1

SOME RESULTS FROM THE ELEMENTARY THEORY OF NUMBERS

In this Appendix we shall derive two results from the elementary theory of numbers (the unique-factorization theorem and the Chinese remainder theorem) which are used in the text proper. All relevant definitions are included so as to make the presentation virtually self-contained. Numbers are understood to be nonnegative integers except where the contrary is explicitly stated.

DEFINITION 1. *We say that a is* **divisible** *by b, or that b is a* **divisor** *of a, and write* $b \mid a$, *if* $b \neq 0$ *and if there exists a number c such that* $a = bc$. *If b is* **not a divisor** *of a we write* $b \nmid a$.

Thus, for any number $a \neq 0$ whatever, $1 \mid a$ and $a \mid a$.

DEFINITION 2. *p is a* **prime** *if* $p > 1$ *and if p has no divisors other than* 1 *and p.*

Thus, 2, 3, and 5 are primes. $6 = 2 \cdot 3$ is not. Also, according to our definition, 1 is not a prime.

COROLLARY 1. *If* $c \mid b$, *then* $c \mid ab$.

PROOF. *If* $b = ck$, *then* $ab = c(ka)$.

On the other hand, it is quite possible to have $c \mid ab$, where $c \nmid a$ and $c \nmid b$. An example is given by $c = 6$, $a = 3$, $b = 4$.

COROLLARY 2. *If* $c \neq 0$ *and* $b \mid c$, *then* $b \leqq c$.

PROOF. Let $c = ab$. Then, $a \neq 0$ (since otherwise we should have $c = 0$). If $a = 1$, then $b = c$, and we are through. Hence we may suppose that $a > 1$, that is, that $a = 1 + k$, $k > 0$. Then, $c = ab = (1 + k)b = b + kb > b$.

COROLLARY 3. *Every number* > 1 *is divisible by at least one prime.*

PROOF. Suppose there were some number > 1, divisible by no prime. Then there would be a least number m having this property. That is, we should have:

(1) $m > 1$.
(2) m is divisible by no prime.
(3) Every number n for which $1 < n < m$ is divisible by a prime.

Now, by (2), m is not a prime. Hence, m has a divisor n, where $1 < n < m$. Then, by (3), n is divisible by a prime. But this implies that m is divisible by a prime, which contradicts (2).

THEOREM 4. *If p is a prime, then there exists a prime q such that* $p < q \leqq p! + 1$ *(that is, there are infinitely many primes).*

PROOF. Let $P = p!$, and let $N = P + 1$. Clearly, $N > p$. Hence, if N is a prime, we are through. If N is not a prime, then, by Corollary 3, N is divisible by some prime q. We shall see that $q > p$.

For suppose that $q \leqq p$. Then, by the manner of construction of P, $q \mid P$. On the other hand, q was chosen such that $q \mid N$. Let us write $P = qm_1$, $N = qm_2$.

Then, $1 = N - P = q(m_1 - m_2)$. Therefore, $q \mid 1$. By Corollary 2, it follows that $q \leqq 1$; so q cannot be a prime.

This elegant proof is due to Euclid.

THEOREM 5. *Every number $x > 1$ can be written in the form $p_1^{m_1}p_2^{m_2} \cdots p_k^{m_k}$, where $p_1, p_2, \ldots, p_k$ are distinct primes.*

PROOF. If the result were false, there would be a least number $x_0 > 1$ that could not be written in the desired form. Clearly, x_0 is not a prime. Therefore, by Corollary 3, x_0 is divisible by a prime p, $x_0 = pr$. But, by Corollary 2, $r < x_0$. Hence, by the choice of x_0, we may write

$$r = p_1^{m_1}p_2^{m_2} \cdots p_k^{m_k};$$

so

$$x_0 = pp_1^{m_1}p_2^{m_2} \cdots p_k^{m_k},$$

which contradicts our choice of x_0.

We shall now seek to demonstrate the much deeper fact that a number can be written in this manner in a unique way (except, of course, for permutations of the factors). We begin with

THEOREM 6 (Euclidean Algorithm). *If $n > 0$ and if m is any number, then there exists one and only one pair of numbers q, r such that $m = nq + r$ and $0 \leqq r < n$.*

PROOF. Let S be the set of all (nonnegative) numbers that can be written in the form $m - nq$. This set is nonempty, since

$$m \in S.$$

Therefore, there is a least number belonging to S. Let r be this least number. Then, for some number q we have

$$m - nq = r,$$

that is,

$$m = nq + r.$$

We must show that $r < n$. Suppose, to the contrary, that $r \geqq n$. Then $r = n + k$, where $0 \leqq k < r$. But

$$k = r - n = m - n(q + 1) \in S,$$

and this contradicts our choice of r as the least element of S.

It remains to show that the numbers q, r are uniquely determined by m and n. Suppose to the contrary that, in addition to the q, r we have obtained, there are numbers q', r' such that

$$m = nq' + r', \quad 0 \leqq r' < n.$$

Suppose that $q' < q$. Then,

$$q' + 1 \leqq q, \quad q' \leqq q - 1.$$

Hence,

$$\begin{aligned} r' &= m - nq' \\ &\geqq m - n(q - 1) \\ &= r + n \\ &\geqq n, \end{aligned}$$

which is a contradiction.

Next, suppose that $q' > q$. Then $r' = m - nq' < m - nq = r$. But, since $r' \in S$, this contradicts our choice of r as the least element of S.

We conclude that $q' = q$. Hence,

$$r' = m - nq' = m - nq = r.$$

This completes the proof.

DEFINITION 3. *We say that the numbers a and b are* **relatively prime** *if they have no divisor in common except* 1.

Now, let a and b be relatively prime. Let S be the set of all nonnegative integers of the form $ax + by$, where x and y are integers *positive, negative,* or *zero.* We proceed to prove the following lemmas concerning this set S.

LEMMA 1. $0 \in S, a \in S, b \in S$. *If* $u \in S, v \in S$, *then* $u + v \in S, u \dot{-} v \in S$. *If* $u \in S$, *and n is any (positive) integer,* $nu \in S$.

PROOF. $0 = a \cdot 0 + b \cdot 0 \in S$,
$a = a \cdot 1 + b \cdot 0 \in S$,
and $b = a \cdot 0 + b \cdot 1 \in S$.

If $u \in S$ and $v \in S$, say,

$$u = ax_1 + by_1,$$
$$v = ax_2 + by_2,$$

then

$$u + v = a(x_1 + x_2) + b(y_1 + y_2) \in S.$$

If $u < v$, then

$$u \dot{-} v = 0 \in S.$$

If $u \geqq v$, then

$$u \dot{-} v = a(x_1 - x_2) + b(y_1 - y_2) \in S.$$

Finally,

$$nu = a(nx_1) + b(ny_1) \in S.$$

LEMMA 2. *There is a number d such that S is the set of all multiples of d.*

PROOF. S contains positive (i.e., nonzero) elements, for example, a and b. Let d be the least positive element of S.

Now, let $x \in S$. By Theorem 6, $x = qd + r$, $0 \leqq r < d$. Now, $x \in S$, $d \in S$. Therefore, by Lemma 1, $x \dot{-} qd = r \in S$. But, since $r < d$, and d is the least positive element of S, we must conclude that $r = 0$. Hence $x = qd$.

LEMMA 3. $1 \in S$.

PROOF. By Lemma 2, S is the set of all multiples of a number d. By Lemma 1, a and b are in S, and hence are multiples of d. That is,

$$d \mid a, \quad d \mid b.$$

Since a and b are relatively prime, this entails $d = 1$.

Lemma 3 may be restated as follows:

THEOREM 7. *Let a, b be relatively prime. Then there exist integers x, y (positive, negative, or zero) such that* $1 = ax + by$.

THEOREM 8. *If p is a prime and* $p \mid ab$, *then* $p \mid a$ *or* $p \mid b$.

PROOF. Suppose $p \nmid a$. Then, p and a are relatively prime (for, if p and a had a common factor $d \neq 1$, it could only be p). Hence, by Theorem 7, there exist integers x, y (positive, negative, or zero) such that $1 = ax + py$. Thus, $b = abx + pby$. But $p \mid ab$; so, for suitable k, $ab = pk$. Then, $b = p(kx + by)$; that is, $p \mid b$.

We have immediately

COROLLARY 9. *If p is a prime and* $p \mid a_1a_2 \cdots a_k$, *then* $p \mid a_i$ *for some* i, $1 \leqq i \leqq k$.

THEOREM 10 (Fundamental Theorem of Arithmetic; Unique-factorization Theorem). *Every number* $x > 1$ *can be represented in the form* $p_1^{m_1}p_2^{m_2} \cdots p_k^{m_k}$, *where*

$p_1, \ldots, p_k$ are distinct primes. Moreover, this representation is unique except for the order of the factors.

PROOF. The existence of such a representation is the assertion of Theorem 5. Suppose that some number $x > 1$ can be represented in the required manner in two different ways:

$$x = p_1^{m_1}p_2^{m_2} \cdots p_k^{m_k} = q_1^{n_1}q_2^{n_2} \cdots q_l^{n_l}.$$

Then, by Corollary 9, for each p_i there is some q_j such that $p_i \mid q_j$, which, since p_i and q_j are primes, implies that $p_i = q_j$. Similarly, for each q_j there is some p_i such that $p_i = q_j$. Hence, we may write (with a possible rearrangement of terms)

$$x = p_1^{m_1}p_2^{m_2} \cdots p_k^{m_k} = p_1^{n_1}p_2^{n_2} \cdots p_k^{n_k}.$$

It remains to be shown that $m_i = n_i$. Suppose that $m_1 > n_1$. Then,

$$p_1^{m_1-n_1}p_2^{m_2} \cdots p_k^{m_k} = p_2^{n_2} \cdots p_k^{m_k}.$$

Here, the left-hand side is divisible by p_1, although, by Corollary 9, the right-hand side is not. This is a contradiction. Similarly, we may dispose of the case $m_1 < n_1$. Hence, $m_1 = n_1$. Similarly, $m_2 = n_2, \ldots, m_k = n_k$.

DEFINITION 4. *Let a, b, m be numbers, and, using Theorem 6, let us write*

$$a = q_1m + r_1 \qquad 0 \leqq r_1 < m,$$
$$b = q_2m + r_2 \qquad 0 \leqq r_2 < m.$$

Then we write $a \equiv b \pmod m$, and we say that a is **congruent to** *b* **modulo** *m if $r_1 = r_2$.*

It is easy to see that $a \equiv b \pmod m$ if and only if $m \mid |b - a|$. For if, for example, $b > a$, then

$$|b - a| = b - a$$
$$= (q_2 - q_1)m + (r_2 - r_1).$$

Hence, $m \mid |b - a|$ if and only if $m \mid |r_2 - r_1|$. Since $|r_2 - r_1| < m$, this is the case only when $r_1 = r_2$.

COROLLARY 11. *If $a \equiv b \pmod m$ and $b \equiv c \pmod m$, then $a \equiv c \pmod m$.*

PROOF. We write

$$a = q_1m + r_1,$$
$$b = q_2m + r_2,$$
$$c = q_3m + r_3.$$

By hypothesis, $r_1 = r_2$, $r_2 = r_3$. Hence, $r_1 = r_3$.

DEFINITION 5. *The numbers $a_1, \ldots, a_m$ are said to be* **relatively prime in pairs** *if a_i and a_j are relatively prime for $1 \leqq i \leqq m$, $1 \leqq j \leqq m$, $i \neq j$.*

THEOREM 12 (Chinese Remainder Theorem). *Let $a_1, a_2, \ldots a_k$ be any numbers, and let $m_1, m_2, \ldots, m_k$ be relatively prime in pairs. Then there exists a number x such that*

$$x \equiv a_i \pmod{m_i} \qquad i = 1, 2, \ldots, k.$$

PROOF. The proof is by induction on k. For $k = 1$, the theorem asserts the existence of a number x such that $x \equiv a_1 \pmod{m_1}$. For this it suffices to take $x = a_1$.

Suppose that the result has been demonstrated for $k = n$. We show that it follows for $k = n + 1$. Consider the system of congruences:

$$x \equiv a_1 \pmod{m_1},$$
$$x \equiv a_2 \pmod{m_2},$$
$$\ldots,$$
$$x \equiv a_n \pmod{m_n},$$
$$x \equiv a_{n+1} \pmod{m_{n+1}}.$$

By induction hypothesis, there is a number x_0 that satisfies the first n of these congruences.

Now the numbers $m_1m_2 \cdots m_n$ and m_{n+1} are relatively prime. For, if they had a divisor $d > 1$ in common, they would have a prime divisor p in common (cf. Corollary 3). But, by Corollary 9, from $p \mid m_1m_2 \cdots m_n$ we infer that $p \mid m_j$ for some j, $1 \leqq j \leqq n$. But then m_{n+1} and m_j would have a common divisor $p > 1$, which contradicts our assumption that the m_i are relatively prime in pairs. Hence, by Theorem 7, there exist integers r and s (positive, negative, or zero) such that

$$rm_1m_2 \cdots m_n + sm_{n+1} = 1.$$

Therefore,

$$r(a_{n+1} - x_0)m_1m_2 \cdots m_n + s(a_{n+1} - x_0)m_{n+1} = a_{n+1} - x_0.$$

Let $q = r(a_{n+1} - x_0)$, $p = -s(a_{n+1} - x_0)$. Then,

$$x_0 + qm_1m_2 \cdots m_n = a_{n+1} + pm_{n+1}.$$

Let t be chosen so that

$$tm_{n+1} + q$$

is a nonnegative number (if q is nonnegative, we can take $t = 0$; if q is negative, we can take $t = -q + 1$). Then,

$$x_0 + (tm_{n+1} + q)m_1m_2 \cdots m_n = a_{n+1} + m_{n+1}(p + tm_1m_2 \cdots m_n).$$

Hence,

$$x_0 + (tm_{n+1} + q)m_1m_2 \cdots m_n \equiv a_{n+1} (\text{mod } m_{n+1}).$$

But, by induction hypothesis,

$$x_0 \equiv a_i (\text{mod } m_i) \qquad i = 1, 2, \ldots, n.$$

Hence,

$$x_0 + (tm_{n+1} + q)m_1m_2 \cdots m_n \equiv a_i (\text{mod } m_i) \qquad i = 1, 2, \ldots, n.$$

This completes the proof.

APPENDIX 2

HILBERT'S TENTH PROBLEM IS UNSOLVABLE

When a long outstanding problem is finally solved, every mathematician would like to share in the pleasure of discovery by following for himself what has been done. But too often he is stymied by the abstruseness of so much of contemporary mathematics. The recent negative solution to Hilbert's tenth problem given by Matiyasevič (cf. [**23**], [**24**]) is a happy counterexample. In this article, a complete account of this solution is given; the only knowledge a reader needs to follow the argument is a little number theory: specifically basic information about divisibility of positive integers and linear congruences. (The material in Chapter 1 and the first three sections of Chapter 2 of [**25**] more than suffices.)

Hilbert's tenth problem is to give a computing algorithm which will tell of a given polynomial Diophantine equation with integer coefficients whether or not it has a solution in integers. Matiyasevič proved that *there is no such algorithm.*

Hilbert's tenth problem is the tenth in the famous list which Hilbert gave in his 1900 address before the International Congress of Mathematicians (cf. [**18**]). The way in which the problem has been resolved is very much in the spirit of Hilbert's address in which he spoke of the conviction among mathematicians "that every definite mathematical problem must necessarily be susceptible of a precise settlement, either in the form of an actual answer to the question asked, or by *the proof of the impossibility of its solution* ..." (italics added). Concerning such impossibility proofs Hilbert commented:

"Sometimes it happens that we seek the solution under unsatisfied hypotheses or in an inappropriate sense and are therefore unable to reach our goal. Then the task arises of proving the impossibility of solving the problem under the given hypotheses and in the sense required. Such impossibility proofs were already given by the ancients, in showing, e.g., that the hypotenuse of an isosceles right triangle has an irrational ratio to its leg. In modern mathematics the question of the impossibility of certain solutions has played a key role, so that we have acquired the knowledge that such old and difficult problems as to prove the parallel axiom, to square the circle, or to solve equations of the fifth degree in radicals have no solution in the originally intended sense, but nevertheless have been solved in a precise and completely satisfactory way."

Matiyasevič's negative solution of Hilbert's tenth problem is of just this character. It is not a solution in Hilbert's "originally intended sense" but rather a "precise and completely satisfactory" proof that no such solution is possible. The methods needed to make it possible to prove the non-existence of algorithms had not been developed in 1900. These methods are part of the theory of recursive (or computable) functions, developed by logicians much later ([6] is an exposition of recursive function theory). In this article no previous knowledge of recursive function theory is assumed. The little that is needed is developed in the article itself.

What will be proved in the body of this article is that no algorithm exists for testing a polynomial with integer coefficients to determine whether or not it has *positive integer* solutions (Hilbert inquired about arbitrary integer solutions). But then it will follow at once that there can be no algorithm for integer solutions either. For one could test the equation

$$P(x_1, \cdots, x_n) = 0$$

for possession of positive solutions $\langle x_1, \cdots, x_n \rangle$ by testing

$$P(1 + p_1^2 + q_1^2 + r_1^2 + s_1^2, \cdots, 1 + p_n^2 + q_n^2 + r_n^2 + s_n^2) = 0$$

for possession of integer solutions $\langle p_1, q_1, r_1, s_1, \cdots, p_n, q_n, r_n, s_n \rangle$. This is because (by a well-known theorem of Lagrange) every non-negative integer is the sum of four squares. (Just this once the stated prerequisite is exceeded! Cf. [17], p. 302.) In the body of this article, only positive integers will be dealt with—except when the contrary is explicitly stated.

When Matiyasevič announced his beautiful and ingenious solution in January 1970, it had been known for a decade that the unsolvability of Hilbert's tenth problem would follow if one could construct a Diophantine equation whose solutions were such that one of its components grew roughly exponentially with another of its components. (In §9, this is explained more precisely.) Matiyasevič showed how the Fibonacci numbers could be used to construct such an equation. In this article the historical development of the subject will not be followed; the aim has rather been to give as smooth and straightforward an account of the main results as seems currently feasible. A brief appendix gives the history.

1. Diophantine Sets. In this article the usual problem of Diophantine equations will be inverted. Instead of being given an equation and seeking its solutions, one will begin with the set of "solutions" and seek a corresponding Diophantine equation. More precisely:

DEFINITION. A set S of ordered n-tuples of positive integers is called **Diophantine** if there is a polynomial $P(x_1, \cdots, x_n, y_1, \cdots, y_m)$, where $m \geqq 0$, with integer coefficients such that a given n-tuple $\langle x_1, \cdots, x_n \rangle$ belongs to S if and only if there exist positive integers $y_1, \cdots, y_m$ for which

$$P(x_1, \cdots, x_n, y_1, \cdots, y_m) = 0.$$

Borrowing from logic the symbols "$\exists$" for "there exists" and "$\Leftrightarrow$" for "if and only if", the relation between the set S and the polynomial P can be written succinctly as:

$$\langle x_1, \cdots, x_n \rangle \in S \Leftrightarrow (\exists y_1, \cdots, y_m) [P(x_1, \cdots, x_n, y_1, \cdots, y_m) = 0],$$

or equivalently:

$$S = \{\langle x_1, \cdots, x_n \rangle \mid (\exists y_1, \cdots, y_m) [P(x_1, \cdots, x_n, y_1, \cdots, y_m) = 0]\}.$$

Note that P may (and in non-trivial cases always will) have negative coefficients. *The word* "**polynomial**" *should always be so construed in the article* except where the contrary is explicitly stated. Also *all numbers in this article are positive integers* unless the contrary is stated.

The main question which will be discussed (and settled) in this article is:

Which sets are Diophantine? A vague paraphrase of the eventual answer is: *any set which could possibly be Diophantine is Diophantine.* What does the phrase "which could possibly be Diophantine" mean? And how is all this related to Hilbert's tenth problem? These quite reasonable questions will only be answered much later. In the meantime, the task will be developing techniques for showing that various sets are indeed Diophantine.

A few very simple examples:

(i) *the numbers which are not powers of* 2:

$$x \in S \Leftrightarrow (\exists y, z)[x = y(2z + 1)],$$

(ii) *the composite numbers*:

$$x \in S \Leftrightarrow (\exists y, z) [x = (y + 1)(z + 1)],$$

(iii) *the ordering relation on the positive integers*; *that is the sets* $\{\langle x, y \rangle \mid x < y\}$, $\{\langle x, y \rangle \mid x \leqq y\}$:

$$x < y \Leftrightarrow (\exists z) (x + z = y),$$

$$x \leqq y \Leftrightarrow (\exists z) (x + z - 1 = y),$$

(iv) *the divisibility relation*; *that is* $\{\langle x, y \rangle \mid x \mid y\}$:

$$x \mid y \Leftrightarrow (\exists z) (xz = y).$$

Examples (i) and (ii) suggest, as other sets to consider, the set of powers of 2 and of primes respectively. As we shall eventually see, these sets are Diophantine; but the proof is not at all easy.

Another example:

(v) *the set W of $\langle x, y, z \rangle$ for which $x \mid y$ and $x < z$*: Here

$$x \mid y \Leftrightarrow (\exists u) (y = xu) \text{ and } x < z \Leftrightarrow (\exists v) (z = x + v).$$

Hence,

$$\langle x, y, z\rangle \in W \Leftrightarrow (\exists u, v)\,[(y - xu)^2 + (z - x - v)^2 = 0].$$

Note that the technique just used is perfectly general. So, in defining a Diophantine set one may use a *simultaneous system* $P_1 = 0$, $P_2 = 0, \cdots, P_k = 0$ of polynomial equations since this system can be replaced by the equivalent single equation:

$$P_1^2 + P_2^2 + \cdots + P_k^2 = 0.$$

By a "function" a positive integer valued function of one or more positive integer arguments will always be understood.

DEFINITION. A function f of n arguments is called **Diophantine** if

$$\{\langle x_1, \cdots, x_n, y\rangle \mid y = f(x_1, \cdots, x_n)\}$$

is a Diophantine set, (i.e., f is Diophantine if its "graph" is Diophantine).

Another question that will be answered here is: *which functions are Diophantine*?

An important Diophantine function is associated with the *triangular numbers*, that is numbers of the form:

$$T(n) = 1 + 2 + \cdots + n = \frac{n(n+1)}{2}.$$

Since $T(n)$ is an increasing function, for each positive integer z, there is a unique $n \geqq 0$ such that

$$T(n) < z \leqq T(n+1) = T(n) + n + 1.$$

Hence each z is *uniquely representable* as:

$$z = T(n) + y; \quad y \leqq n + 1,$$

or equivalently, uniquely representable as:

$$z = T(x + y - 2) + y.$$

In this case, one writes $x = L(z)$, $y = R(z)$; also one sets

$$P(x, y) = T(x + y - 2) + y.$$

Note that $L(z)$, $R(z)$ and $P(x, y)$ are Diophantine functions since

$$z = P(x, y) \Leftrightarrow 2z = (x + y - 2)(x + y - 1) + 2y$$

$$x = L(z) \Leftrightarrow (\exists y)\,[2z = (x + y - 2)(x + y - 1) + 2y]$$

$$y = R(z) \Leftrightarrow (\exists x)\,[2z = (x + y - 2)(x + y - 1) + 2y].$$

The function $P(x, y)$ maps the set of ordered pairs of positive integers one-one onto the set of positive integers. And, for each z, the ordered pair which is mapped into z by $P(x, y)$ is $(L(z), R(z))$. ("P" is for "pair", "L" for "left", and "R" for "right".) Note also that $L(z) \leqq z$, $R(z) \leqq z$. To summarize:

THEOREM 1.1 (Pairing Function Theorem[1]). *There are Diophantine functions* $P(x, y)$, $L(z)$, $R(z)$ *such that*

(1) *for all* x, y, $L(P(x, y)) = x$, $R(P(x, y)) = y$, *and*

(2) *for all* z, $P(L(z), R(z)) = z$, $L(z) \leqq z$, $R(z) \leqq z$.

Another useful Diophantine function is related to the Chinese Remainder Theorem, stated below:

DEFINITION. The numbers $m_1, \cdots, m_N$ are called an **admissible sequence of moduli** if $i \neq j$ implies that m_i and m_j are relatively prime.

THEOREM 1.2 (Chinese Remainder Theorem). *Let* $a_1, \cdots, a_N$ *be any positive integers and let* $m_1 \cdots, m_N$ *be an admissible sequence of moduli. Then there is an* x *such that*:

$$x \equiv a_1 \mod m_1$$

$$x \equiv a_2 \mod m_2$$

$$\cdots\cdots$$

$$x \equiv a_N \mod m_N.$$

The Chinese remainder theorem is proved for example in [25], p. 33. (That x can be assumed positive is not ordinarily stated. But since the product of the moduli added to a solution gives another solution, this is obvious.)

Now let the function $S(i, u)$ be defined as follows:

$$S(i, u) = w,$$

where w is the unique positive integer for which:

$$w \equiv L(u) \mod 1 + iR(u)$$

$$w \leqq 1 + i\, R(u).$$

Here w is simply the least *positive* remainder when $L(u)$ is divided by $1 + i\, R(u)$.

THEOREM 1.3 (Sequence Number Theorem). *There is a Diophantine function* $S(i, u)$ *such that*

(1) $S(i, u) \leqq u$, *and*

(2) *for each sequence* $a_1, \cdots, a_N$, *there is a number* u *such that*

$$S(i, u) = a_i \text{ for } 1 \leqq i \leqq N.$$

Proof. The first task is to show that $S(i, u)$ as defined just above, is a Diophantine function. The claim is that $w = S(i,u)$ if and only if the following system of equations has a solution:

$$2u = (x + y - 2)(x + y - 1) + 2y$$
$$x = w + (z - 1)(1 + iy)$$
$$1 + iy = w + v - 1.$$

This is because (by the discussion leading to the Pairing Function Theorem), the first equation is equivalent to:

$$x = L(u) \text{ and } y = R(u).$$

Then (using a technique already noted) one needs only sum the squares of the three equations to see that $S(i,u)$ is Diophantine.

Now $S(i,u) \leqq L(u) \leqq u$. So finally, let $a_1, \cdots, a_N$ be given numbers. Choose y to be some number greater than each of $a_1, \cdots, a_N$ and divisible by each of $1, 2, \cdots, N$. Then the numbers $1 + y, 1 + 2y, \cdots, 1 + Ny$ are an admissible sequence of moduli. (For, if $d \mid 1 + iy$ and $d \mid 1 + jy$, $i < j$, then $d \mid [j(1 + iy) - i(1 + jy)]$, i.e., $d \mid j - i$ so that $d \leqq N$; but this is impossible unless $d = 1$ because $d \mid y$.) This being the case, the Chinese Remainder Theorem can be applied to obtain a number x such that

$$x \equiv a_1 \mod 1 + y$$
$$x \equiv a_2 \mod 1 + 2y$$
$$\cdots\cdots$$
$$x \equiv a_N \mod 1 + Ny.$$

Let $u = P(x, y)$, so that $x = L(u)$ and $y = R(u)$. Then, for $i = 1, 2, \cdots, N$

$$a_i \equiv L(u) \mod 1 + iR(u)$$

and $a_i < y = R(u) < 1 + iR(u)$. But then by definition, $a_i = S(i, u)$.

A striking characterization of Diophantine sets of positive integers (cf. [**26**]) is given by:

THEOREM 1.4. *A set S of positive integers is Diophantine if and only if there is a polynomial P such that S is precisely the set of* **positive** *integers in the range of P.*

Proof. If S is related to $P(x_1, \cdots, x_m)$ as in the theorem then

$$x \in S \Leftrightarrow (\exists x_1, \cdots, x_m)\, [x = P(x_1, \cdots, x_m)].$$

Conversely, let

$$x \in S \Leftrightarrow (\exists x_1, \cdots, x_m)\, [Q(x, x_1, \cdots, x_m) = 0].$$

Let $P(x, x_1, \cdots, x_m) = x[1 - Q^2(x, x_1, \cdots, x_m)]$. Then, if $x \in S$, choose $x_1, \cdots, x_m$ such that $Q(x, x_1, \cdots, x_m) = 0$. Then $P(x, x_1, \cdots, x_m) = x$; so x is in the range of P. On the other hand, if $z = P(x, x_1, \cdots, x_m)$, $z > 0$, then $Q(x, x_1, \cdots, x_m)$ must vanish (otherwise $1 - Q^2 \leqq 0$) so that $z = x$ and $x \in S$.

2. Twenty-four easy lemmas. The first major task is to prove that the exponen ial function $h(n, k) = n^k$ is Diophantine. This is the hardest thing we shall have to do. The proof is in §3. In this section we develop the methods we shall need, using the so-called Pell equation:

where
$$\left.\begin{aligned} x^2 - dy^2 = 1, &\quad x, y \geqq 0, \\ d = a^2 - 1, &\quad a > 1. \end{aligned}\right\} \qquad (*)$$

Although this is a famous equation with a considerable literature,[2] a self-contained treatment is given. Note the obvious solutions to (*):

$$x = 1 \qquad y = 0$$

$$x = a \qquad y = 1.$$

LEMMA 2.1. *There are no integers x , y, positive, negative, or zero, which satisfy (*) for which $1 < x + y\sqrt{d} < a + \sqrt{d}$.*

Proof. Let x, y satisfy (*). Since

$$1 = (a + \sqrt{d})(a - \sqrt{d}) = (x + y\sqrt{d})(x - y\sqrt{d}),$$

the inequality implies (taking negative reciprocals) $-1 < -x + y\sqrt{d} < -a + \sqrt{d}$. Adding the inequalities: $0 < 2y\sqrt{d} < 2\sqrt{d}$, i.e., $0 < y < 1$, a contradiction.

LEMMA 2.2. *Let x, y and x', y' be integers, positive, negative, or zero which satisfy (*). Let*

$$x'' + y''\sqrt{d} = (x + y\sqrt{d})\,(x' + y'\sqrt{d}).$$

Then, x'', y'' satisfies (*).

Proof. Taking conjugates: $x'' - y''\sqrt{d} = (x - y\sqrt{d})\,(x' - y'\sqrt{d})$. Multiplying gives:

$$(x'')^2 - d(y'')^2 = (x^2 - dy^2)\,((x')^2 - d(y')^2) = 1.$$

DEFINITION. $x_n(a)$, $y_n(a)$ are defined for $n \geqq 0$, $a > 1$, by setting

$$x_n(a) + y_n(a)\sqrt{d} = (a + \sqrt{d})^n.$$

Where the context permits, the dependence on a is not explicitly shown, writing x_n, y_n.

LEMMA 2.3. x_n, y_n *satisfy (*).*

Proof. This follows at once by induction using Lemma 2.2.

LEMMA 2.4. *Let x, y be a non-negative solution of* (*). *Then for some n, $x = x_n$, $y = y_n$.*

Proof. To begin with $x + y\sqrt{d} \geqq 1$. On the other hand the sequence $(a + \sqrt{d})^n$ increases to infinity. Hence for some $n \geqq 0$,

$$(a + \sqrt{d})^n \leqq x + y\sqrt{d} < (a + \sqrt{d})^{n+1}.$$

If there is equality, the result is proved; so suppose otherwise:

$$x_n + y_n\sqrt{d} < x + y\sqrt{d} < (x_n + y_n\sqrt{d})\,(a + \sqrt{d}).$$

Since $(x_n + y_n\sqrt{d})\,(x_n - y_n\sqrt{d}) = 1$, the number $x_n - y_n\sqrt{d}$ is positive. Hence, $1 < (x + y\sqrt{d})\,(x_n - y_n\sqrt{d}) < a + \sqrt{d}$. But this contradicts Lemmas 2.1 and 2.2.

The defining relation:

$$x_n + y_n\sqrt{d} = (a + \sqrt{d})^n$$

is a formal analogue of the familiar formula:

$$(\cos u) + (\sin u)\sqrt{-1} = e^{iu} = (\cos 1 + (\sin 1)\sqrt{-1})^u,$$

with x_n playing the role of cos, y_n playing the role of sin and d playing the role of -1. Thus, the familiar trigonometric identities have analogues in which -1 is replaced by d at appropriate places. For example the Pell equation itself

$$x_n^2 - dy_n^2 = 1$$

is just the analogue of the Pythagorean identity. Next analogues of the familiar addition formulas are obtained.

LEMMA 2.5. $x_{m\pm n} = x_m x_n \pm d y_n y_m$ *and* $y_{m\pm n} = x_n y_m \pm x_m y_n$.

Proof.

$$\begin{aligned} x_{m+n} + y_{m+n}\sqrt{d} &= (a + \sqrt{d})^{m+n} \\ &= (x_m + y_m\sqrt{d})\,(x_n + y_n\sqrt{d}) \\ &= (x_m x_n + d y_n y_m) + (x_n y_m + x_m y_n)\sqrt{d}. \end{aligned}$$

Hence,

$$\begin{aligned} x_{m+n} &= x_m x_n + d y_n y_m \\ y_{m+n} &= x_n y_m + x_m y_n. \end{aligned}$$

Similarly, $(x_{m-n} + y_{m-n}\sqrt{d})\,(x_n + y_n\sqrt{d}) = x_m + y_m\sqrt{d}$. So

$$x_{m-n} + y_{m-n}\sqrt{d} = (x_m + y_m\sqrt{d})\,(x_n - y_n\sqrt{d}),$$

and one proceeds as above.

LEMMA 2.6. *$y_{m\pm 1} = a\, y_m \pm x_m$, and $x_{m\pm 1} = ax_m \pm dy_m$.*

Proof. Take $n = 1$ in Lemma 2.5.

The familiar notation (x, y) is used to symbolize the g.c.d. of x and y.

LEMMA 2.7. $(x_n, y_n) = 1$.

Proof. If $s \mid x_n$ and $s \mid y_n$, then $s \mid x_n^2 - dy_n^2$, i.e., $s \mid 1$.

LEMMA 2.8. $y_n \mid y_{nk}$.

Proof. This is obvious when $k = 1$. Proceeding by induction, using the addition formula (Lemma 2.5),

$$y_{n(m+1)} = x_n y_{nm} + x_{nm} y_n.$$

By the induction hypothesis $y_n \mid y_{nm}$. Hence, $y_n \mid y_{n(m+1)}$.

LEMMA 2.9. *$y_n \mid y_t$ if and only if $n \mid t$.*

Proof. Lemma 2.8 gives the implication in one direction. For the converse suppose $y_n \mid y_t$ but $n \nmid t$. So one can write $t = nq + r$, $0 < r < n$. Then,

$$y_t = x_r y_{nq} + x_{nq} y_r.$$

Since (by Lemma 2.8) $y_n \mid y_{nq}$, it follows that $y_n \mid x_{nq} y_r$. But $(y_n, x_{nq}) = 1$. (If $s \mid y_n$, $s \mid x_{nq}$, then by Lemma 2.8 $s \mid y_{nq}$ which, by Lemma 2.7, implies $s = 1$.) Hence $y_n \mid y_r$. But, since $r < n$, we have $y_r < y_n$ (e.g., by Lemma 2.6). This is a contradiction.

LEMMA 2.10. $y_{nk} \equiv k\, x_n^{k-1} y_n \bmod (y_n)^3$.

Proof.

$$\begin{aligned} x_{nk} + y_{nk}\sqrt{d} &= (a + \sqrt{d})^{nk} \\ &= (x_n + y_n\sqrt{d})^k \\ &= \sum_{j=0}^{k} \binom{k}{j} x_n^{k-j} y_n^j d^{j/2}. \end{aligned}$$

So,

$$y_{nk} = \sum_{\substack{j=1 \\ j \text{ odd}}}^{k} \binom{k}{j} x_n^{k-j} y_n^j d^{(j-1)/2}.$$

But all terms of this expansion for which $j > 1$ are $\equiv 0 \bmod (y_n)^3$.

LEMMA 2.11. $y_n^2 \mid y_{ny_n}$.

Proof. Set $k = y_n$ in Lemma 2.10.

LEMMA 2.12. *If $y_n^2 \mid y_t$, then $y_n \mid t$.*

Proof. By Lemma 2.9, $n|t$. Set $t = nk$. Using Lemma 2.10, $y_n^2 | k\, x_n^{k-1} y_n$, i.e., $y_n | kx_n^{k-1}$. But by Lemma 2.7, $(y_n, x_n) = 1$. So, $y_n | k$ and hence $y_n | t$.

LEMMA 2.13. *$x_{n+1} = 2ax_n - x_{n-1}$ and $y_{n+1} = 2ay_n - y_{n-1}$.*

Proof. By Lemma 2.6,

$$x_{n+1} = ax_n + dy_n, \qquad y_{n+1} = ay_n + x_n,$$

$$x_{n-1} = ax_n - dy_n, \qquad y_{n-1} = ay_n - x_n.$$

So, $x_{n+1} + x_{n-1} = 2ax_n$, $y_{n+1} + y_{n-1} = 2ay_n$.

These second order difference equations, together with the initial values $x_0 = 1$, $x_1 = a$, $y_0 = 0$, $y_1 = 1$, determine the values of all the x_n, y_n. Various properties o, these sequences are easily established by checking them for $n = 0$, 1 and using these difference equations to show that the property for $n + 1$ can be inferred from its holding for n and $n - 1$. Some simple (but important) examples follow:

LEMMA 2.14. $y_n \equiv n \bmod a - 1$.

Proof. For $n = 0, 1$ equality holds. Proceeding inductively, using $a \equiv 1$, mod $a-1$:

$$\begin{aligned} y_{n+1} &= 2ay_n - y_{n-1} \\ &\equiv 2n - (n-1) \mod a - 1. \end{aligned}$$

LEMMA 2.15. *If $a \equiv b \bmod c$, then for all n,*

$$x_n(a) \equiv x_n(b), \quad y_n(a) \equiv y_n(b) \mod c.$$

Proof. Again for $n = 0, 1$ the congruence is an equality. Proceeding by induction:

$$\begin{aligned} y_{n+1}(a) &= 2ay_n(a) - y_{n-1}(a) \\ &\equiv 2by_n(b) - y_{n-1}(b) \mod c \\ &= y_{n+1}(b). \end{aligned}$$

LEMMA 2.16. *When n is even y_n is even and when n is odd y_n is odd.*

Proof. $y_{n+1} = 2ay_n - y_{n-1} \equiv y_{n-1} \bmod 2$. So when n is even, $y_n \equiv y_0 = 0 \bmod 2$f and when n is odd, $y_n \equiv y_1 = 1 \bmod 2$.

LEMMA 2.17. $x_n(a) - y_n(a)(a - y) \equiv y^n \bmod 2ay - y^2 - 1$.

Proof. $x_0 - y_0(a - y) = 1$ and $x_1 - y_1(a - y) = y$, so the result holds for $n = 0$ and 1. Using Lemma 2.13 and proceeding by induction:

$$\begin{aligned} x_{n+1} - y_{n+1}(a - y) &= 2a[x_n - y_n(a - y)] - [x_{n-1} - y_{n-1}(a - y)] \\ &\equiv 2ay^n - y^{n-1} \end{aligned}$$

$$= y^{n-1}(2ay - 1)$$
$$\equiv y^{n-1}\,y^2$$
$$= y^{n+1}.$$

LEMMA 2.18. *For all n*, $y_{n+1} > y_n \geqq n$.

Proof. By Lemma 2.6, $y_{n+1} > y_n$. Since $y_0 = 0 \geqq 0$, it follows by induction that $y_n \geqq n$ for all n.

LEMMA 2.19. *For all n*, $x_{n+1}(a) > x_n(a) \geqq a^n$; $\quad x_n(a) \leqq (2a)^n$.

Proof. By Lemmas 2.6 and 2.13 $a\,x_n(a) \leqq x_{n+1}(a) \leqq (2a)x_n(a)$. The result follows by induction.

Next some periodicity properties of the sequence x_k are obtained.

LEMMA 2.20. $x_{2n\pm j} \equiv -x_j \bmod x_n$.

Proof. By the addition formulas (Lemma 2.5)

$$\begin{aligned} x_{2n\pm j} &= x_n x_{n\pm j} \pm d y_n y_{n\pm j} \\ &\equiv d y_n(y_n x_j \pm x\; y_j) \mod x_n \\ &\equiv d y_n^2 x_j \mod x_n \\ &= (x_n^2 - 1)x_j \\ &\equiv -x_j \mod x_n. \end{aligned}$$

LEMMA 2.21. $x_{4n\pm j} \equiv x_j \mod x_n$.

Proof. By Lemma 2.20

$$x_{4n\pm j} \equiv -x_{2n\pm j} \equiv x_j \mod x_n.$$

LEMMA 2.22. *Let* $x_i \equiv x_j \bmod x_n$, $i \leqq j \leqq 2n$, $n > 0$. *Then* $i = j$, *unless* $a = 2$, $n = 1$, $i = 0$ *and* $j = 2$.

Proof. First suppose x_n is odd and let $q = (x_n - 1)/2$. Then the numbers $-q, -q+1, -q+2, \cdots, -1, 0, 1, \cdots, q-1, q$ are a complete set of mutually incongruent residues modulo x_n. Now by Lemma 2.19,

$$1 = x_0 < x_1 < \cdots < x_{n-1}.$$

Using Lemma 2.6, $x_{n-1} \leqq x_n/a \leqq \frac{1}{2}x_n$; so $x_{n-1} \leqq q$. Also by Lemma 2.20, the numbers

$$x_{n+1}, x_{n+2}, \cdots, x_{2n-1}, x_{2n}$$

are congruent modulo x_n respectively to:

$$-x_{n-1}, \ -x_{n-2}, \cdots, -x_1, \ -x_0 = -1.$$

Thus the numbers $x_0, x_1, x_2, \cdots, x_{2n}$ are mutually incongruent modulo x_n. This gives the result.

Next suppose x_n is even and let $q = x_n/2$. In this case, it is the numbers

$$-q+1, \ -q+2, \cdots, -1, 0, 1, \cdots, q-1, q$$

which are a complete set of mutually incongruent residues modulo x_n. (For, $-q \equiv q$ mod x_n.) As above, $x_{n-1} \leqq q$. So the result will follow as above, unless $x_{n-1} = q = x_n/2$, so that $x_{n+1} \equiv -q$ mod x_n, in which case $i = n-1$, $j = n+1$ would contradict our result. But, by Lemma 2.6,

$$x_n = ax_{n-1} + dy_{n-1},$$

so that $x_n = 2x_{n-1}$ implies $a = 2$ and $y_{n-1} = 0$, i.e., $n = 1$. So the result can fail only for $a = 2$, $n = 1$ and $i = 0$, $j = 2$.

LEMMA 2.23. *Let* $x_j \equiv x_i$ mod x_n, $n > 0$, $0 < i \leqq n$, $0 \leqq j < 4n$, *then either* $j = i$ *or* $j = 4n - i$.

Proof. First suppose $j \leqq 2n$. Then by Lemma 2.22, $j = i$ unless the exceptional case occurs. Since $i > 0$, this can only happen if $j = 0$. But then

$$i = 2 > 1 = n.$$

Otherwise, let $j > 2n$ and set $\bar{j} = 4n - j$ so $0 < \bar{j} < 2n$. By Lemma 2.21, $x_{\bar{j}} \equiv x_j \equiv x_i$ mod x_n. Again $\bar{j} = i$ unless the exceptional case of Lemma 2.22 occurs. But this last is out of the question because $i, \bar{j} > 0$.

LEMMA 2.24. *If* $0 < i \leqq n$ *and* $x_j \equiv x_i$ mod x_n, *then* $j \equiv \pm i$ mod $4n$.

Proof. Write $j = 4nq + \bar{j}$, $0 \leqq \bar{j} < 4n$. By Lemma 2.21,

$$x_i \equiv x_j \equiv x_{\bar{j}} \mod x_n.$$

By Lemma 2.23 $i = \bar{j}$ or $i = 4n - \bar{j}$. So, $j \equiv \bar{j} \equiv \pm i$ mod $4n$.

3. The exponential function. Consider the system of Diophantine equations:

(I) $$x^2 - (a^2-1)y^2 = 1$$

(II) $$u^2 - (a^2-1)v^2 = 1$$

(III) $$s^2 - (b^2-1)t^2 = 1$$

(IV) $$v = ry^2$$

(V) $$b = 1 + 4py = a + qu$$

(VI) $$s = x + cu$$

(VII) $$t = k + 4(d-1)y$$

(VIII) $$y = k + e - 1.$$

Then it is possible to prove:

THEOREM 3.1. *For given* a, x, k, $a > 1$, *the system* I-VIII *has a solution in the remaining arguments* $y, u, v, s, t, b, r, p, q, c, d, e$ *if and only if* $x = x_k(a)$.

Proof. First let there be given a solution of I-VIII. By V, $b > a > 1$. Then I, II, III imply (by Lemma 2.4) that there are i, j, $n > 0$ such that

$$x = x_i(a),\ y = y_i(a),\ u = x_n(a),\ v = y_n(a),\ s = x_j(b),\ t = y_j(b).$$

By I V, $y \leqq v$ so that $i \leqq n$. V and VI yield the congruences

$$b \equiv a \mod x_n(a);\quad x_j(b) \equiv x_i(a) \mod x_n(a)$$

and by Lemma 2.15 one gets also

$$x_j(b) \equiv x_j(a) \mod x_n(a).$$

Thus,

$$x_i(a) \equiv x_j(a) \mod x_n(a).$$

By Lemma 2.24,

(1) $$j \equiv \pm i \mod 4n.$$

Next, equation IV yields

$$(y_i(a))^2 \mid y_n(a).$$

so that by Lemma 2.12,

$$y_i(a) \mid n$$

and (1) yields:

(2) $$j \equiv \pm i \mod 4y_i(a).$$

By equation V

$$b \equiv 1 \mod 4y_i(a),$$

so by Lemma 2.14,

(3) $$y_j(b) \equiv j \mod 4y_i(a).$$

By equation VII,

(4) $$y_j(b) \equiv k \mod 4y_i(a).$$

Combining (2), (3), (4),

(5) $$k \equiv \pm i \mod 4y_i(a).$$

Equation VIII yields

$$k \leqq y_i(a)$$

and by Lemma 2.18,

$$i \leqq y_i(a).$$

Since the numbers

$$-2y+1,\ -2y+2,\cdots,-1,0,1,\cdots,2y$$

form a complete set of mutually incongruent residues modulo $4y = 4y_i(a)$, these inequalities show that (5) implies $k = i$. Hence

$$x = x_i(a) = x_k(a).$$

Conversely, let $x = x_k(a)$. Set $y = y_k(a)$ so that I holds. Let $m = 2ky_k(a)$ and let $u = x_m(a)$, $v = y_m(a)$. Then II is satisfied. By Lemmas 2.9 and 2.11 $y^2 | v$. Hence one can choose r satisfying IV. Moreover by Lemma 2.16, v is even so that u is odd. By Lemma 2.7, $(u,v) = 1$. Hence$(u, 4y) = 1$. (If p is a prime divisor of u and of $4y$, then $p | y$ because u is odd, and hence $p | v$ since $y | v$.) So by the Chinese Remainder Theorem (Theorem 1.2), one can find b_0 such that

$$b_0 \equiv 1 \mod 4y$$

$$b_0 \equiv a \mod u.$$

Since $b_0 + 4juy$ will also satisfy these congruences, b, p, q satisfying V can be found. III is satisfied by setting $s = x_k(b)$, $t = y_k(b)$. Since $b > a$, $s = x_k(b) > x_k(a) = x$. By Lemma 2.15 (using V), $s \equiv x \bmod u$. So c can be chosen to satisfy VI. By Lemma 2.18, $t \geqq k$ and by Lemma 2.14, $t \equiv k \bmod b-1$ and hence using V, $t \equiv k \bmod 4y$. So d can be chosen to satisfy VII. By Lemma 2.18 again, $y \geqq k$, so VIII can be satisfied by setting $e = y - k + 1$.

COROLLARY 3.2. *The function*

$$g(z,k) = x_k(z+1)$$

is Diophantine.

Proof. Adjoin to the system I-VIII:

(A) $$a = z + 1.$$

By the theorem, the system (A), I-VIII has a solution if and only if $x = x_k(a) = g(z,k)$. Thus a Diophantine definition of g can be obtained in the usual way by summing the squares of 9 polynomials.

Now at last it is possible to prove:

THEOREM 3.3. *The exponential function $h(n,k) = n^k$ is Diophantine.*

First, a simple inequality:

LEMMA 3.4. *If* $a > y^k$, *then* $2ay - y^2 - 1 > y^k$.

Proof. Set $g(y) = 2ay - y^2 - 1$. Then (since $a \geqq 2$) $g(1) = 2a - 2 \geqq a$. For $1 \leqq y < a$, $g'(y) = 2a - 2y > 0$. So $g(y) \geqq a$ for $1 \leqq y < a$. Then for $a > y^k \geqq y$, $2ay - y^2 - 1 \geqq a > y^k$.

Now, adjoin to equations I-VIII:

IX $$(x - y(a - n) - m)^2 = (f - 1)^2(2an - n^2 - 1)^2$$

X $$m + g = 2an - n^2 - 1$$

XI $$w = n + h = k + l$$

XII $$a^2 - (w^2 - 1)(w - 1)^2z^2 = 1.$$

Theorem 3.3 then follows at once from:

LEMMA 3.5. $m = n^k$ *if and only if equations* I-XII *have a solution in the remaining arguments.*

Proof. Suppose I-XII hold. By XI, $w > 1$. Hence $(w - 1)z > 0$ and so by XII $a > 1$. So Theorem 3.1 applies and it follows that $x = x_k(a)$, $y = y_k(a)$. By IX and Lemma 2.17,

$$m \equiv n^k \mod 2an - n^2 - 1.$$

XI yields

$$k, n < w.$$

By XII (using Lemma 2.4), for some j, $a = x_j(w)$, $(w - 1)z = y_j(w)$. By Lemma 2.14,

$$j \equiv 0 \mod w - 1$$

so that $j \geqq w - 1$. So by Lemma 2.19,

$$a \geqq w^{w-1} > n^k.$$

Now by X, $m < 2an - n^2 - 1$, and by Lemma 3.4,

$$n^k < 2an - n^2 - 1.$$

Since m and n^k are congruent and both less than the modulus, they must be equal.

Conversely, suppose that $m = n^k$. Solutions must be found for I-XII. Choose any number w such that $w > n$ and $w > k$. Set $a = x_{w-1}(w)$ so that $a > 1$. By Lemma 2.14,

$$y_{w-1}(w) \equiv 0 \mod w - 1.$$

So one can write

$$y_{w-1}(w) = z(w-1);$$

thus XII is satisfied. XI can be satisfied by setting

$$h = w - n, \quad l = w - k.$$

As before, $a > n^k$ so that again by Lemma 3.4,

$$m = n^k < 2an - n^2 - 1$$

and X can be satisfied. Setting $x = x_k(a)$, $y = y_k(a)$, Lemma 2.17 permits one to define f such that

$$x - y(a-n) - m = \pm(f-1)(2an - n^2 - 1),$$

so that IX is satisfied. Finally, I-VIII can be satisfied by Theorem 3.1.

4. The language of Diophantine predicates. Now that it has been proved that the *exponential function is* Diophantine, many other functions and sets can be handled. As an example, let

$$h(u, v, w) = u^{v^w}.$$

The claim is that h is a Diophantine function. For:

$$y = u^{v^w} \Leftrightarrow (\exists z)\,(y = u^z \,\&\, z = v^w),$$

where "&" is the logician's symbol for "and". Using Theorem 3.3, there is a polynomial P such that:

$$y = u^z \Leftrightarrow (\exists r_1, \cdots, r_n)[P(y, u, z, r_1, \cdots, r_n) = 0],$$
$$z = v^w \Leftrightarrow (\exists s_1, \cdots, s_n)\,[P(z, v, w, s_1, \cdots, s_n) = 0].$$

Then,

$$y = u^{v^w} \Leftrightarrow (\exists z, r_1, \cdots, r_n, s_1, \cdots, s_n)\,[P^2(y, u, z, r_1, \cdots, r_n) + P^2(z, v, w, s_1, \cdots, s_n) = 0].$$

Now this procedure is perfectly general: Expressions which are already known to yield Diophantine sets may be combined freely using the logical operations of "&" and "($\exists$)"; the resulting expression will again define a Diophantine set. (Such expressions are sometimes called *Diophantine predicates.*) In this "language" it is also permissible to use the logician's "$\vee$" for "or", since:

$$(\exists r_1, \cdots, r_n)\,[P_1 = 0] \vee (\exists s_1, \cdots, s_m)\,[P_2 = 0]$$
$$\Leftrightarrow (\exists r_1, \cdots, r_n, s_1, \cdots, s_m)\,[P_1 P_2 = 0].$$

Three important Diophantine functions are given by:

THEOREM 4.1. *The following functions are Diophantine:*

(1) $$f(n,k) = \binom{n}{k}$$

(2) $$g(n) = n!$$

(3) $$h(a,b,y) = \prod_{k=1}^{y} (a+bk).$$

In proving this theorem the familiar notation $[\alpha]$, where α is a real number, will be used to mean the unique integer such that

$$[\alpha] \leqq \alpha < [\alpha] + 1.$$

LEMMA 4.1. *For* $0 < k \leqq n$, $u > 2^n$

$$[(u+1)^n/u^k] = \sum_{i=k}^{n} \binom{n}{i} u^{i-k}.$$

Proof.

$$(u+1)^n/u^k = \sum_{i=0}^{n} \binom{n}{i} u^{i-k} = S + R$$

where

$$S = \sum_{i=k}^{n} \binom{n}{i} u^{i-k} \qquad R = \sum_{i=0}^{k-1} \binom{n}{i} u^{i-k}.$$

Then S is an integer and

$$\begin{aligned} R &\leqq u^{-1} \sum_{i=0}^{k-1} \binom{n}{i} \\ &< u^{-1} \sum_{i=0}^{n} \binom{n}{i} \\ &= u^{-1}(1+1)^n \\ &< 1. \end{aligned}$$

So,

$$S \leqq (u+1)^n/u^k < S + 1$$

which gives the result.

Lemma 4.2. *For* $0 < k \leqq n$, $u > 2^n$,

$$[(u+1)^n/u^k] \equiv \binom{n}{k} \quad \text{mod } u.$$

Proof. In Lemma 4.1 all terms of the sum for which $i > k$ are divisible by u.

Lemma 4.3. $f(n,k) = \binom{n}{k}$ *is Diophantine.*

Proof. Since

$$\binom{n}{k} \leqq \sum_{i=0}^{n} \binom{n}{i} = 2^n, \text{ if } u > 2^n, \text{ then}$$

Lemma 4.2 determines $\binom{n}{k}$ as the unique positive integer congruent to $[(u+1)^n/u^k]$ modulo u and $< u$. Thus,

$$z = \binom{n}{k} \Leftrightarrow (\exists u, v, w)\,(v = 2^n \;\&\; u > v$$

$$\& \; w = [(u+1)^n/u^k] \;\&\; z \equiv w \quad \text{mod } u \;\&\; z < u).$$

To see that $\binom{n}{k}$ is Diophantine, it then suffices to note that each of the above expressions separated by "&" are Diophantine predicates; $v = 2^n$ is of course Diophantine by Theorem 3. The inequality $u > v$ is of course Diophantine since $u > v \Leftrightarrow (\exists x)(u = v + x)$. Also,

$$z \equiv w \text{ mod } u \;\&\; z < u \Leftrightarrow (\exists x, y)\,(w = z + (x-1)u \;\&\; u = z + y).$$

Finally

$$w = [(u+1)^n/u^k]$$

$$\Leftrightarrow$$

$$(\exists x, y, t)\,(t = u + 1 \;\&\; x = t^n \;\&\; y = u^k \;\&\; w \leqq x/y < w + 1),$$

and $w \leqq x/y < w + 1 \Leftrightarrow wy \leqq x < (w+1)y$.

Lemma 4.4. *If* $r > (2x)^{x+1}$ *then*

$$x! = \left[r^x \Big/ \binom{r}{x}\right].$$

Proof. Let $r > (2x)^{x+1}$. Then,

$$r^x \Big/ \binom{r}{x} = \frac{r^x x!}{r(r-1)\cdots(r-x+1)}$$

$$= x! \left\{ \frac{1}{\left(1 - \frac{1}{r}\right)\cdots\left(1 - \frac{x-1}{r}\right)} \right\}$$

$$< x! \cdot \frac{1}{\left(1 - \frac{x}{r}\right)^x}.$$

Now,

$$\begin{aligned} \frac{1}{1 - \frac{x}{r}} &= 1 + \frac{x}{r} + \left(\frac{x}{r}\right)^2 + \cdots \\ &= 1 + \frac{x}{r}\left\{1 + \frac{x}{r} + \left(\frac{x}{r}\right)^2 + \cdots\right\} \\ &< 1 + \frac{x}{r}\{1 + \tfrac{1}{2} + \tfrac{1}{4} + \cdots\} \\ &= 1 + \frac{2x}{r}. \end{aligned}$$

And,

$$\begin{aligned} \left(1 + \frac{2x}{r}\right)^x &= \sum_{j=0}^{x} \binom{x}{j}\left(\frac{2x}{r}\right)^j \\ &\leqq 1 + \frac{2x}{r}\sum_{j=1}^{x}\binom{x}{j} \\ &< 1 + \frac{2x}{r}\cdot 2^x. \end{aligned}$$

So,

$$\begin{aligned} r^x/\binom{r}{x} &< x! + \frac{2x}{r}\cdot x!\, 2^x \\ &\leqq x! + \frac{2^{x+1}x^{x+1}}{r} \\ &< x! + 1. \end{aligned}$$

LEMMA 4.5. *$n!$ is a Diophantine function.*

Proof. $m = n! \Leftrightarrow$

$$(\exists r, s, t, u, v)\{s = 2n + 1 \,\&\, t = n + 1 \,\&\, r = s^t$$

$$\& \, u = r^n \,\&\, v = \binom{r}{n} \,\&\, mv \leqq u < (m + 1)v\}.$$

LEMMA 4.6. *Let $bq \equiv a \bmod M$. Then,*

$$\prod_{k=1}^{y}(a + bk) \equiv b^y y! \binom{q + y}{y} \mod M.$$

Proof.

$$\begin{aligned} b^y y!\binom{q+y}{y} &= b^y(q+y)(q+y-1)\cdots(q+1) \\ &= (bq+yb)(bq+(y-1)b)\cdots(bq+b) \\ &\equiv (a+yb)(a+(y-1)b)\cdots(a+b) \pmod{M}. \end{aligned}$$

LEMMA 4.7. *$h(a,b,y) = \prod_{k=1}^{y} (a+bk)$ is a Diophantine function.*

Proof. In Lemma 4.6 choose $M = b(a+by)^y + 1$. Then, $(M,b) = 1$ and $M > \prod_{k=1}^{y} (a+bk)$. Hence the congruence $bq \equiv a \bmod M$ is solvable for q and then $\prod_{k=1}^{y} (a+bk)$ is determined as the unique number which is congruent modulo M to $b^y y!\binom{q+y}{y}$ and is also $< M$. I.e.,

$$\begin{aligned} z = \prod_{k=1}^{y} (a+bk) \Leftrightarrow (\exists M,p,q,r,s,t,u,v,w,x) \\ \Big\{ r = a+by \,\&\, s = r^y \,\&\, M = bs+1 \\ \&\, bq = a+Mt \,\&\, u = b^y \,\&\, v = y! \,\&\, z < M \\ \&\, w = q+y \,\&\, x = \binom{w}{y} \,\&\, z+Mp = uvx \Big\}. \end{aligned}$$

Using the previous expressions for the exponential function, for $v = y!$ and for $x = \binom{w}{y}$, we obtain the result.

The assertion of Theorem 4.1 is contained in Lemmas 4.3, 4.5, and 4.7.

5. Bounded quantifiers. The language of Diophantine predicates permits use of &, $\vee$, and $\exists$. Other operations used by logicians are:

$\sim$ for "not"

$(\forall x)$ for "for all x"

$\rightarrow$ for "if···, then ···"

However, as will be clear later, the use of any of these other operations can lead to expressions which define sets that are not Diophantine. There are also the *bounded existential quantifiers*:

"$(\exists y)_{\leqq x}\cdots$" which means "$(\exists y)\,(y \leqq x \,\&\cdots)$"

and the *bounded universal quantifiers*:

"$(\forall y)_{\leqq x}\cdots$" which means "$(\forall y)\,(y > x \vee \cdots)$".

It turns out that these operations may be adjoined to the language of Diophantine predicates; that is, the sets defined by expressions of this extended language will still be Diophantine. I.e.,

THEOREM 5.1. *If P is a polynomial,*

$$R = \{\langle y, x_1, \cdots, x_n\rangle \mid (\exists z)_{\leqq y}(\exists y_1, \cdots, y_m)\,[P(y, z, x_1, \cdots, x_n, y_1, \cdots, y_m) = 0]\}$$

and

$$S = \{\langle y, x_1, \cdots, x_n\rangle \mid (\forall z)_{\leqq y}(\exists y_1, \cdots, y_m)\,[P(y, z, x_1, \cdots, x_n, y_1, \cdots, y_m) = 0]\},$$

then R and S are Diophantine.

That R is Diophantine is trivial. Namely,

$$\langle y, x_1, \cdots, x_n\rangle \in R \Leftrightarrow (\exists z, y_1, \cdots, y_m)\,(z \leqq y \,\&\, P = 0).$$

The proof of the other half of the theorem is far more complicated.

LEMMA 5.1.

$$(\forall k)_{\leqq y}(\exists y_1, \cdots, y_m)\,[P(y, k, x_1, \cdots, x_n, y_1, \cdots, y_m) = 0]$$

$$\Leftrightarrow$$

$$(\exists u)\,(\forall k)_{\leqq y}(\exists y_1, \cdots, y_m)_{\leqq u}[P(y, k, x_1, \cdots, x_n, y_1, \cdots, y_m) = 0].$$

Proof. The right side of the equivalence trivially implies the left side. For the converse, suppose the left side is true for given $y, x_1, \cdots, x_n$. Then for each $k = 1, 2, \cdots, y$ there are definite numbers $y_1^{(k)}, \cdots, y_m^{(k)}$ for which:

$$P(y, k, x_1, \cdots, x_n, y_1^{(k)}, \cdots, y_m^{(k)}) = 0.$$

Taking u to be the maximum of the my numbers

$$\{y_j^{(k)} \mid j = 1, \cdots, m;\ k = 1, 2, \cdots, y\},$$

it follows that the right side of the equivalence is likewise true.

LEMMA 5.2. *Let $Q(y, u, x_1, \cdots, x_n)$ be a polynomial with the properties:*

(1) $Q(y, u, x_1, \cdots, x_n) > u$, (2) $Q(y, u, x_1, \cdots, x_n) > y$,
(3) $k \leqq y$ *and* $y_1, \cdots, y_m \leqq u$ *imply* $|P(y, k, x_1, \cdots, x_n, y_1, \cdots, y_m)| \leqq Q(y, u, x_1, \cdots, x_n)$.
Then,

$$(\forall k)_{\leqq y}(\exists y_1, \cdots, y_m)_{\leqq u}[P(y, k, x_1, \cdots, x_n, y_1, \cdots, y_m) = 0]$$

$$\Leftrightarrow$$

$$(\exists c, t, a_1, \cdots, a_m)\ \left[1 + ct = \prod_{k=1}^{y} (1 + kt)\right.$$

$$\& t = Q(y,u,x_1,\cdots,x_n)! \,\&\, 1 + ct \Big| \prod_{j=1}^{u} (a_1 - j)$$

$$\& \cdots \& \, 1 + ct \Big| \prod_{j=1}^{u} (a_m - j)$$

$$\& P(y,c,x_1,\cdots,x_n,a_1,\cdots,a_m) \equiv 0 \mod 1 + ct].$$

The point of this lemma is that while the right side of the equivalence seems the more complicated of the two, it is free of bounded universal quantifiers.

Proof. First the implication in the $\Leftarrow$ direction:

For each $k = 1,2,\cdots,y$, let p_k be a prime factor of $1 + kt$. Let $y_i^{(k)}$ be the remainder when a_i is divided by p_k ($k = 1,2,\cdots,y$; $i = 1,2,\cdots,m$). It will follow that for each k, i:

(a) $$1 \leqq y_i^{(k)} \leqq u$$

(b) $$P(y,k,x_1,\cdots,x_n,y_1^{(k)},\cdots,y_m^{(k)}) = 0.$$

To demonstrate (a), note that $p_k \big| 1 + kt$, $1 + kt \big| 1 + ct$ and $1 + ct \big| \prod_{j=1}^{u}(a_i - j)$. I.e., $p_k \big| \prod_{j=1}^{u}(a_i - j)$. Since p_k is a prime, $p_k \big| a_i - j$ for some $j = 1,2,\cdots,u$. That is

$$j \equiv a_i \equiv y_i^{(k)} \mod p_k.$$

Since $t = Q(y,u,x_1,\cdots,x_n)!$, by (2) every prime divisor of $1 + kt$ must be $> Q(y,u,x_1,\cdots,x_n)$. So $p_k > Q(y,u,x_1,\cdots,x_n)$ and by (1), $p_k > u$. Hence $j \leqq u < p_k$. Since $y_i^{(k)}$ is the *remainder* when a_i is divided by p_k, also $y_i^{(k)} < p_k$. So,

$$y_i^{(k)} = j.$$

To demonstrate (b), first note that

$$1 + ct \equiv 1 + kt \equiv 0 \mod p_k.$$

Hence

$$k + kct \equiv c + kct \mod p_k,$$

i.e., $k \equiv c \mod p_k$. We have already obtained

$$y_i^{(k)} \equiv a_i \mod p_k.$$

Thus,

$$P(y,k,x_1,\cdots,x_n,y_1^{(k)},\cdots,y_m^{(k)}) \equiv P(y,c,x_1,\cdots,x_n.\,a_1,\cdots,a_m)$$
$$\equiv 0 \mod p_k.$$

Finally

$$\left| P(y, k, x_1, \cdots, x_n, y_1^{(k)}, \cdots, y_m^{(k)}) \right| \leqq Q(y, u, x_1, \cdots, x_n) < p_k .$$

This proves (b) and completes the proof of the $\Leftarrow$ implication.

To prove the $\Rightarrow$ implication, let

$$P(y, k, x_1, \cdots, x_n, y_1^{(k)}, \cdots, y_m^{(k)}) = 0,$$

for each $k = 1, 2, \cdots, y$, where each $y_j^{(k)} \leqq u$. We set $t = Q(y, u, x_1, \cdots, x_n)!$, and since $\prod_{k=1}^{y} (1 + kt) \equiv 1 \bmod t$, we can find c such that

$$1 + ct = \prod_{k=1}^{y} (1 + kt).$$

Now, it is claimed that for $1 \leqq k < l \leqq y$,

$$(1 + kt, 1 + lt) = 1.$$

For, let $p \mid 1 + kt, p \mid 1 + lt$. Then $p \mid l - k$, so $p < y$. But since $Q(y, u, x_1, \cdots, x_n) > y$ this implies $p \mid t$ which is impossible. Thus the numbers $1 + kt$ form an *admissible sequence of moduli* and the Chinese Remainder Theorem (Theorem 1.2) may be applied to yield, for each i, $1 \leqq i \leqq m$, a number a_i such that

$$a_i \equiv y_i^{(k)} \bmod 1 + kt, \quad k = 1, 2, \cdots, y.$$

As above, $k \equiv c \bmod 1 + kt$. So

$$\begin{aligned} P(y, c, x_1, \cdots, x_n, a_1, \cdots, a_m) &\equiv P(y, k, x_1, \cdots, x_n, y_1^{(k)}, \cdots, y_m^{(k)}) \bmod 1 + kt, \\ &= 0. \end{aligned}$$

Since the numbers $1 + kt$ are relatively prime in pairs and each divides $P(y, c, x_1, \cdots, x_n, a_1, \cdots, a_m)$ so does their product. I.e.,

$$P(y, c, x_1, \cdots, x_n, a_1, \cdots, a_m) \equiv 0 \bmod 1 + ct.$$

Finally,

$$a_i \equiv y_i^{(k)} \bmod 1 + kt,$$

e.,i.

$$1 + kt \mid a_i - y_i^{(k)}.$$

Since $1 \leqq y_i^{(k)} \leqq u$,

$$1 + kt \mid \prod_{j=1}^{u} (a_i - j).$$

And again since the $1 + kt$'s are relatively prime to one another,

$$1 + ct \mid \prod_{j=1}^{u} (a_i - j).$$

Now it is easy to complete the proof of Theorem 5.1 using Lemmas 5.1 and 5.2. First find a polynomial Q satisfying (1), (2), (3) of Lemma 5.2. This is easy to do: Write

$$P(y,k,x_1,\cdots,x_n,y_1,\cdots,y_m) = \sum_{r=1}^{N} t_r$$

where each t_r has the form

$$t_r = c\,y^a k^b x_1^{q_1} x_2^{q_2} \cdots x_n^{q_n} y_1^{s_1} y_2^{s_2} \cdots y_m^{s_m}$$

for c an integer positive or negative. Set $u_r = |c| y^{a+b} x_1^{q_1} x_2^{q_2} \cdots x_n^{q_n} u^{s_1+s_2+\cdots+s_m}$ and let

$$Q(y,u,x_1,\cdots,x_n) = u + y + \sum_{r=1}^{N} u_r.$$

Then (1), (2), and (3) of Lemma 5.2 hold trivially. Thus:

$$(\forall k)_{\leq y}(\exists y_1,\cdots,y_m)\,[P(y,k,x_1,\cdots,x_n,y_1,\cdots,y_m)=0]$$

$$\Leftrightarrow$$

$$(\exists u,c,t,a_1,\cdots,a_m)\Big[1+ct=\prod_{k=1}^{y}(1+kt)$$

$$\&\, t = Q(y,u,x_1,\cdots,x_n)!\ \&\, 1+ct \,\Big|\, \prod_{j=1}^{u}(a_1-j)$$

$$\&\cdots\&\, 1+ct \,\Big|\, \prod_{j=1}^{u}(a_m-j)$$

$$\&\, P(y,c,x_1,\cdots,x_n,a_1,\cdots,a_m) \equiv 0 \mod 1+ct\Big]$$

$$\Leftrightarrow$$

$$(\exists u,c,t,a_1,\cdots,a_n,e,f,g_1,\cdots,g_m,h_1,\cdots,h_n,l)$$

$$\Big[e=1+ct\ \&\, e=\prod_{k=1}^{y}(1+kt)\ \&\, f=Q(y,u,x_1,\cdots,x_n)$$

$$\&\, t=f!\ \&\, g_1=a_1-u-1\ \&\, g_2=a_2-u-1\ \&\cdots\&\, g_m=a_m-u-1$$

$$\&\, h_1=\prod_{k=1}^{u}(g_1+k)\ \&\, h_2=\prod_{k=1}^{u}(g_2+k)$$

$$\&\cdots\&\, h_m=\prod_{k=1}^{u}(g_m+k)\ \&\, e \,|\, h_1\ \&\, e \,|\, h_2\ \&\cdots\&\, e \,|\, h_m$$

$$\&\, l = P(y,c,x_1,\cdots,x_n,a_1,\cdots,a_n)\ \&\, e \,|\, l\Big]$$

and this is Diophantine by Theorem 4.1.

6. Recursive functions. So far one trick after another has been used to show that various sets are Diophantine. But now very powerful methods are available: it turns out that the expanded version of the language of Diophantine predicates, permitting the use of bounded quantifiers (sanctioned by Theorem 5.1) together with the Sequence Number Theorem (Theorem 1.3) enables one to show in quite a straightforward way that almost any set we please is Diophantine.

Some examples are in order:

(i) *the set P of prime numbers*:

$$x \in P \Leftrightarrow x > 1 \,\&\, (\forall y, z)_{\leq x}[yz < x \vee yz > x \vee y = 1 \vee z = 1].$$

Another Diophantine definition of the primes is:

$$x \in P \Leftrightarrow x > 1 \,\&\, ((x-1)!\,, x) = 1$$

$$\Leftrightarrow x > 1 \,\&\, (\exists y, z, u, v)\, [y = x - 1 \,\&\, z = y! \,\&\, (uz - vx)^2 = 1];$$

but the first definition is the more natural one.

From Theorem 1.4 it follows that *there is a "prime-representing" polynomial* P, i.e., a positive integer is prime if and only if it is in the range of P. For an explicit construction of such a polynomial P, cf. [**23a**].

(ii) *the function* $g(y) = \prod_{k=1}^{y} (1 + k^2)$. Here we use the Sequence Number Theorem to "encode" the sequence $g(1), g(2), \cdots, g(y)$ into a single number u, i.e., so that

$$S(i, u) = g(i), \qquad i = 1, 2, \cdots, y.$$

Thus, $z = g(y)$

$$\Leftrightarrow (\exists u) \;\{S(1, u) = 2 \,\&\, (\forall k)_{\leq y}[k = 1 \vee (S(k, u) = (1 + k^2)S(k-1, u))] \,\&\, z = S(y, u)\}$$

$$\Leftrightarrow (\exists u)\, \{S(1, u) = 2 \,\&\, (\forall k)_{\leq y}[k = 1 \vee (\exists a, b, c)\, (a = k - 1$$

$$\&\, b = S(a, u) \,\&\, c = S(k, u) \,\&\, c = (1 + k^2)b)] \,\&\, z = S(y, u)\}.$$

By now it is clear that the available methods are quite general. They are so powerful that the question becomes: how can any "reasonable" set or function escape these methods, i.e., not be Diophantine?

The strength of the methods can be tested by considering the class of all *computable* or *recursive* functions. These are the functions which can be computed by a finite program or computing machine having arbitrarily large amounts of time and memory at its disposal. Many rigorous definitions of this class (all of them equivalent) are available. One of the simplest is as follows:

The *recursive functions*[3] are all those functions obtainable from the initial functions

$$c(x) = 1,\ s(x) = x + 1;\ U_i^n(x_1, \cdots, x_n) = x_i, \qquad 1 \leqq i \leqq n;$$

$$S(i, u) \text{ (The sequence number function)}^4$$

iteratively applying the three operations: *composition*, *primitive recursion*, and *minimalization* defined below:

Composition yields the function

$$h(x_1, \cdots, x_n) = f(g_1(x_1, \cdots, x_n), \cdots, g_m(x_1, \cdots, x_n))$$

from the given functions $g_1, \cdots, g_m$ and $f(t_1, \cdots, t_m)$.

Primitive Recursion yields the function $h(x_1, \cdots, x_n, z)$ which satisfies the equations:

$$h(x_1, \cdots, x_n, 1) = f(x_1, \cdots, x_n)$$

$$h(x_1, \cdots, x_n, t + 1) = g(t, h(x_1, \cdots, x_n, t), x_1, \cdots, x_n),$$

rom the given functions f, g.

When $n = 0$, f becomes a constant so that h is obtained directly from g.

Minimalization yields the function:

$$h(x_1, \cdots, x_n) = \min_y[f(x_1, \cdots, x_n, y) = g(x_1, \cdots, x_n, y)]$$

from the given functions f, g assuming that f, g are such that for each $x_1, \cdots, x_n$ there is at least one y satisfying the equation $f(x_1, \cdots, x_n, y) = g(x_1, \cdots, x_n, y)$; (i.e., h must be everywhere defined).

The main result of this article is:

Theorem 6.1. *A function is Diophantine if and only if it is recursive.*

To begin with, consider the following short list of recursive functions:

(1) $x + y$ is recursive since

$$x + 1 = s(x),$$

$$x + (t + 1) = s(x + t) = g(t, x + t, x),$$

where $g(u, v, w) = s(U_2^3(u, v, w))$.

(2) $x \cdot y$ is recursive since

$$x \cdot 1 = U_1^1(x)$$

$$x \cdot (t + 1) = (x \cdot t) + x = g(t, x \cdot t, x),$$

where $g(u, v, w) = U_2^3(u, v, w) + U_3^3(u, v, w)$.

(3) For each fixed k, the constant function $c_k(x) = k$ is recursive, since $c_1(x)$ is one of the initial functions and $c_{k+1}(x) = c_k(x) + c(x)$.

(4) Any polynomial $P(x_1, \cdots, x_n)$ with *positive* integer coefficients is recursive, tince any such function can be expressed by a finite iteration of additions and mulsiplications of variables and $c(x)$. E.g.,

$$2x^2y + 3xz^3 + 5 = c_2(x) \cdot x \cdot x \cdot y + c_3(x) \cdot x \cdot z \cdot z \cdot z + c_5(x).$$

So (1), (2), (3), and composition gives the result.

Now it is easy to see that every Diophantine function is recursive:

Let f be Diophantine, and write:

$$y = f(x_1, \cdots, x_n) \Leftrightarrow (\exists t_1, \cdots, t_m) \, [P(x_1, \cdots, x_n, y, t_1, \cdots, t_m) = Q(x_1, \cdots, x_n, y, t_1, \cdots, t_m)],$$

where P, Q are polynomials with *positive* integer coefficients. Then, by the sequence number theorem:

$$\begin{aligned} f(x_1, \cdots, x_n) &= S(1, \min_u[P(x_1, \cdots, x_n, S(1,u), S(2,u), \cdots, S(m+1,u)) \\ &= Q(x_1, \cdots, x_n, S(1,u), S(2,u), \cdots, S(m+1,u))]). \end{aligned}$$

Since P, Q, $S(i,u)$ are recursive, so is f (using composition and minimalization).

To obtain the converse: $S(i,u)$ is known to be Diophantine; the other initial functions are trivially Diophantine. Hence it suffices to prove that the Diophantine functions are closed under composition, primitive recursion and minimalization.

Composition: If $h(x_1, \cdots, x_n) = f(g_1(x_1, \cdots, x_n), \cdots, g_m(x_1, \cdots, x_n))$, where $f, g_1, \cdots, g_m$ are Diophantine, then so is h since

$$y = h(x_1, \cdots, x_n) \Leftrightarrow (\exists t_1, \cdots, t_m) \, [t_1 = g_1(x_1, \cdots, x_n) \, \& \cdots \, \& \, t_m = g_m(x_1, \cdots . x_n) \, \& \, y = f(t_1, \cdots, t_m)].$$

Primitive Recursion: If

$$\begin{aligned} h(x_1, \cdots, x_n, 1) &= f(x_1, \cdots, x_n) \\ h(x_1, \cdots, x_n, t+1) &= g(t, h(x_1, \cdots, x_n, t), x_1, \cdots, x_n), \end{aligned}$$

and f, g are Diophantine, then (using the sequence number theorem to "code" the numbers $h(x_1, \cdots, x_n, 1), h(x_1, \cdots, x_n, 2), \cdots, h(x_1, \cdots, x_n, z)$):

$$\begin{aligned} & y = h(x_1, \cdots, x_n, z) \Leftrightarrow \\ & (\exists u) \, \{(\exists v) \, [v = S(1,u) \, \& \, v = f(x_1, \cdots, x_n)] \\ & \quad \& \, (\forall t)_{\leq z}[(t = z) \vee (\exists v) \, (v = S(t+1, u) \\ & \quad \& \, v = g(t, S(t,u), x_1, \cdots, x_n))] \, \& \, y = S(z,u)\} \end{aligned}$$

so that (using Theorem 5.1) h is Diophantine.

Minimalization: If

$$h(x_1, \cdots, x_n) = \min_y [f(x_1, \cdots, x_n, y) = g(x_1, \cdots, x_n, y)],$$

where f, g are Diophantine, then so is h since,

$$\begin{aligned} & y = h(x_1, \cdots, x_n) \Leftrightarrow \\ & (\exists z)[z = f(x_1, \cdots, x_n, y) \;\&\; z = g(x_1, \cdots, x_n, y)] \\ & \&\, (\forall t)_{\leqq y} [(t = y) \vee (\exists u, v)\, (u = f(x_1, \cdots, x_n, t) \\ & \&\, v = g(x_1, \cdots, x_n, t) \;\&\; (u < v \vee v < u)]. \end{aligned}$$

7. A universal Diophantine set. An explicit enumeration of all the Diophantine sets of positive integers will now be described. Any polynomial with positive integer coefficients can be built up from 1 and variables by successive additions and multiplications. We fix the alphabet

$$x_0, x_1, x_2, x_3, \cdots$$

of variables and then set up the following enumeration of all such polynomials (using the pairing functions):

$$\begin{aligned} P_1 &= 1 \\ P_{3i-1} &= x_{i-1} \\ P_{3i} &= P_{L(i)} + P_{R(i)} \\ P_{3i+1} &= P_{L(i)} \cdot P_{R(i)}. \end{aligned}$$

Write $P_i = P_i(x_0, x_1, \cdots, x_n)$, where n is large enough so that all variables occurring in P_i are included. (Of course P_i will not in general depend on all of these variables.) Finally, let

$$D_n = \{x_0 \,|\, (\exists x_1, \cdots, x_n)\, [P_{L(n)}(x_0, x_1, \cdots, x_n) = P_{R(n)}(x_0, x_1, \cdots, x_n)]\}.$$

Here, $P_{L(n)}$ and $P_{R(n)}$ do not actually involve all of the variables x_0, x_1, $\cdots, x_n$—but clearly cannot involve any others. (Recall that $L(n)$, $R(n) \leqq n$.) By the way the sequence P_i has been constructed, it is seen that the sequence of sets:

$$D_1, D_2, D_3, D_4, \cdots$$

includes all Diophantine sets. Moreover:

THEOREM 7.1 (Universality Theorem[5]).

$$\{\langle n, x\rangle \,|\, x \in D_n\} \textit{ is Diophantine.}$$

Proof. Once again using the sequence number theorem, it is claimed that:

$$x \in D_n \Leftrightarrow (\exists u)\{S(1,u) = 1 \,\&\, S(2,u) = x$$
$$\&\, (\forall i)_{\leqq n}[S(3i,u) = S(L(i),u) + S(R(i),u)]$$
$$\&\, (\forall i)_{\leqq n}[S(3i+1,u) = S(L(i),u) \cdot S(R(i),u)]$$
$$\&\, S(L(n),u) = S(R(n),u)\}.$$

It is clear enough that the predicate on the right-hand side of this equivalence is Diophantine, so it is only necessary to verify the claim:

Let $x \in D_n$ for given x, n. Then there are numbers $t_1, \cdots, t_n$ such that $P_{L(n)}(x, t_1, \cdots, t_n) = P_{R(n)}(x, t_1, \cdots, t_n)$. Choose u (by the sequence number theorem) so that

$$(*) \qquad S(j,u) = P_j(x, t_1, \cdots, t_n), \qquad j = 1, 2, \cdots, 3n+2.$$

Then in particular $S(2,u) = x$ and $S(3i-1,u) = t_{i-1}$, $i = 2, 3, \cdots, n+1$. Thus the right-hand side of the equivalence is true.

Conversely, let the right-hand side hold for given n, x. Set

$$t_1 = S(5,u),\ t_2 = S(8,u),\ \cdots, t_n = S(3n+2,u).$$

Then, (*) must be true. Since $S(L(n),u) = S(R(n),u)$, it must be the case that

$$P_{L(n)}(x, t_1, \cdots, t_n) = P_{R(n)}(x, t_1, \cdots, t_n),$$

so that $x \in D_n$.

Since $D_1, D_2, D_3, \cdots$, gives an enumeration of all Diophantine sets, it is easy to construct a set different from all of them and hence non-Diophantine. That is, define:

$$V = \{n \mid n \notin D_n\}.$$

THEOREM 7.2. *V is not Diophantine.*

Proof. This is a simple application of Cantor's diagonal method. If V were Diophantine, then for some fixed i, $V = D_i$. Does $i \in V$? We have:

$$i \in V \Leftrightarrow i \in D_i; \qquad i \in V \Leftrightarrow i \notin D_i.$$

This is a contradiction.

THEOREM 7.3. *The function $g(n,x)$ defined by:*

$$g(n,x) = 1 \quad \text{if} \quad x \notin D_n,$$
$$g(n,x) = 2 \quad \text{if} \quad x \in D_n,$$

is not recursive.

Proof. If g were recursive then it would be Diophantine (Theorem 6.1), say:

$$y = g(n,x) \Leftrightarrow (\exists y_1, \cdots, y_m) \, [P(n,x,y,y_1,\cdots,y_m) = 0].$$

But then, it would follow that

$$V = \{x \,|\, (\exists y_1,\cdots,y_m) \, [P(x,x,1,y_1,\cdots,y_m) = 0]\}$$

which contradicts Theorem 7.2.

Using Theorem 7.1, write:

$$x \in D_n \Leftrightarrow (\exists z_1, \cdots, z_k) \, [P(n,x,z_1,\cdots,z_k) = 0].$$

where P is some definite (though complicated) polynomial. Suppose there were an algorithm for testing Diophantine equations for possession of positive integer solutions; i.e., *an algorithm for Hilbert's tenth problem*! Then for given n, x this algorithm could be used to test whether or not the equation

$$P(n,x,z_1,\cdots,z_k) = 0$$

has a solution, i.e., whether or not $x \in D_n$. Thus the algorithm could be used to compute the function $g(n,x)$. Since the recursive functions are just those for which a computing algorithm exists, g would have to be recursive. This would contradict Theorem 7.3, and this contradiction proves:

THEOREM 7.4. *Hilbert's tenth problem is unsolvable*!

Naturally this result gives no information about the existence of solutions for any *specific* Diophantine equation; it merely guarantees that there is no single algorithm for testing the class of all Diophantine equations. Also note that:

$$\begin{aligned} x \in V \;&\Leftrightarrow\; \sim (\exists z_1,\cdots,z_k) \, [P(x,x,z_1,\cdots,z_k) = 0] \\ &\Leftrightarrow\; \{(\exists z_1,\cdots,z_k) \, [P(x,x,z_1,\cdots,z_k) = 0] \to 1 = 0\} \\ &\Leftrightarrow\; (\forall z_1,\cdots,z_k) \, [P(x,x,z_1,\cdots,z_k) > 0 \\ &\qquad\qquad \vee \; P(x,x,z_1,\cdots,z_k) < 0] \end{aligned}$$

which shows that if either $\sim$ or unbounded universal quantifiers $(\forall z)$ or implication $(\to)$ are permitted in the language of Diophantine predicates, then non-Diophantine sets will be produced.

It is natural to associate with each Diophantine set a *dimension* and a *degree*; i.e., the *dimension of* S is the least n for which a polynomial P exists for which:

$$(*) \qquad S = \{x \,|\, (\exists y_1,\cdots,y_n) \, [P(x,y_1,\cdots,y_n) = 0]\},$$

and the *degree of* S is the least degree of a polynomial P satisfying (*) (permitting n to be as large as one likes). Now it is easy to see:

THEOREM 7.5. *Every Diophantine set has degree* $\leqq 4$.

Proof. The degree of P satisfying (*) may be reduced by introducing additional variables z_j satisfying equations of the form

$$z_j = y_i y_k$$
$$z_j = y_i^2$$
$$z_j = x\,y_i$$
$$z_j = x^2.$$

By successive substitutions of the z_j's into P its degree can be brought down to 2. Hence the equation is equivalent to a system of simultaneous equations each of degree 2. Summing the squares gives an equation of degree 4.

A less trivial (and more surprising) fact is:

THEOREM 7.6. *There is an integer m such that every Diophantine set has dimension $\leqq m$.*

Proof. Write

$$D_n = \{x \,|\, (\exists y_1, \cdots, y_m)\, [P(x, n, y_1, \cdots, y_m) = 0]\},$$

which is possible by the universality theorem. Then the dimension of D_n is $\leqq m$ for all n.

An interesting example is given by the sequence of Diophantine sets:

$$S_q = \{x \,|\, (\exists y_1, \cdots, y_q)\, [x = (y_1 + 1) \cdots (y_q + 1)]\}.$$

Here S_2 is the set of composite numbers; S_q is the set of "q-fold" composite numbers. It is surely surprising that it is possible to give a Diophantine definition of S_q (for large q) requiring fewer than q parameters (cf. [**19**]).

How large is m, the number of parameters in the universal Diophantine set? A direct calculation using the arguments given here would yield a number around 50. Actually Matiyacevič and Julia Robinson have very recently shown that $m = 14$ will suffice!

The unsolvability of Hilbert's tenth problem can be used to obtain a strengthened form of Gödel's famous incompleteness theorem:

THEOREM 7.7. *Corresponding to any given axiomatization of number theory, there is a Diophantine equation which has no positive integer solutions, but such that this fact cannot be proved within the given axiomatization.*

A rigorous proof would involve a precise definition of "axiomatization of number theory" which is outside the scope of this article. An informal heuristic argument follows:

One uses the given axiomatization to systematically generate all of the theorems (i.e., consequences of the axioms). Among these theorems will be some asserting that some Diophantine equation has no solution. Whenever such is encountered it is placed on a special list called LISTA. At the same time a list, LIST B, is made of Diophantine equations which have solutions. LIST B is constructed by a search procedure, e.g., at the nth stage of the search look at the first n Diophantine equations (in a suitable list) and test for solutions in which each argument is $\leq n$. Thus every Diophantine equation which has positive integer solutions will eventually be placed in LIST B. *If* likewise each Diophantine equation with no solutions would eventually appear in LIST A, then one would have an algorithm for Hilbert's tenth problem. Namely, to test a given equation for possession of a solution simply begin generating LIST A and LIST B until the given equation appears in one list or the other. Since Hilbert's tenth problem is unsolvable, some equation with no solution must be omitted from LIST A. But this is just the assertion of the theorem.

8. Recursively enumerable sets. It is now time to settle the question raised at the beginning: which sets are Diophantine?

DEFINITION. 8.1. *A set S of n-tuples of positive integers is called recursivaty enumerable if there are recursive functions $f(x, x_1, \cdots, x_n)$, $g(x, x_1 \cdot \cdots, x_n)$ such the l:*

$$S = \{\langle x_1, \cdots, x_n \rangle \mid (\exists x) [f(x, x_1, \cdots, x_n) = g(x, x_1, \cdots, x_n)]\}.$$

THEOREM 8.1. *A set S is Diophantine if and only if it is recursively enumerable.*

Proof. If S is Diophantine there are polynomials P, Q with positive coefficients such that:

$$\langle x_1, \cdots, x_n \rangle \in S \Leftrightarrow (\exists y_1, \cdots, y_m) [P(x_1, \cdots, x_n, y_1, \cdots, y_m) = Q(x_1, \cdots, x_n, y_1, \cdots, y_m)]$$

$$\Leftrightarrow (\exists u) [P(x_1, \cdots, x_n, S(1, u), \cdots, S(m, u)) = Q(x_1, \cdots, x_n, S(1, u), \cdots, S(m, u))],$$

so that S is recursively enumerable.

Conversely if S is recursively enumerable there are recursive functions $f(x, x_1, \cdots, x_n)$, $g(x, x_1, \cdots, x_n)$ such that

$$\langle x_1, \cdots, x_n \rangle \in S \Leftrightarrow (\exists x) [f(x, x_1, \cdots, x_n) = g(x, x_1, \cdots, x_n)]$$

$$\Leftrightarrow (\exists x, z) [z = f(x, x_1, \cdots, x_n) \,\&\, z = g(x, x_1, \cdots, x_n)].$$

Thus by Theorem 6.1, S is Diophantine.

9. Historical appendix. The present exposition has ignored the chronological order in which the ideas were developed. The first contribution was by Gödel in his celebrated 1931 paper [**16**]. The main point of Gödel's investigation was the existence of undecidable statements in formal systems. The undecidable statements Gödel obtained involved recursive functions, and in order to exhibit the simple number-theoretic character of these statements, Gödel used the Chinese remainder theorem to reduce them to "arithmetic" form. The technique used is just what is used here in proving Theorem 1.3 (the sequence number theorem) and Theorem 6.1 (in the direction: every recursive function is Diophantine). However without the techniques for dealing with bounded universal quantifiers as discussed in this paper, the best result yielded by Gödel's methods is that every recursive function (and indeed every recursively enumerable set) can be defined by a Diophantine equation preceded by a finite number of existential and bounded universal quantifiers[6]. In my doctoral dissertation (cf. [**5**], [**6**]), I showed that all but one of the bounded universal quantifiers could be eliminated, so that every recursively enumerable set S could be defined as

$$S = \{x \mid (\exists y)\,(\forall k)_{\leqq y}(\exists y_1, \cdots, y_m)\,[P(k, x, y, y_1, \cdots, y_m) = 0]\}.$$

This representation became known as the Davis normal form. (Later R. M. Robinson [**31**], [**32**] showed that in this normal form one could take $m = 4$. More recently Matiyacevič has shown that one can even take $m = 2$. It is known that one cannot always have $m = 0$; whether one can always get $m = 1$ is open.)

Independent of my work and at about the same time, Julia Robinson began her study [**27**] of Diophantine sets. Her investigations centered about the question: *Is the exponential function Diophantine*? The main result was that a certain hypothesis implied that the exponential function was Diophantine. The hypothesis, which became known as the Julia Robinson hypothesis, has played a key role in work on Hilbert's tenth problem. Its statement is simply:

There exists a Diophantine set D such that:

(1) $\langle u, v\rangle \in D$ *implies* $v \leqq u^u$.

(2) *For each* k, *there is* $\langle u, v\rangle \in D$ *such that* $v > u^k$.

The hypothesis remained an open question for about 2 decades. (Actually the set

$$D = \{\langle u, v\rangle \mid v = x_u(2)\ \&\ u > 3\}$$

satisfies (1) and (2) by Lemma 2.19 and is Diophantine by Corollary 3.2, so the truth of Julia Robinson's hypothesis follows at once from the results in this article.) Julia Robinson's proof that this hypothesis implies that the exponential function is Diophantine used the Pell equation. And, the proof that the exponential function is indeed Diophantine given here is closely related to a more recent proof [**28**] by her of this same implication.

In [**27**], Julia Robinson studied also sets and functions which were *exponential Diophantine* (or existentially definable in terms of exponentiation) that is which possess definitions of the form:

$$(\exists u_1, \cdots, u_n, v_1, \cdots, v_n, w_1, \cdots, w_n)\ [P(x_1, \cdots, x_m, u_1, \cdots, u_n, v_1, \cdots, v_n, w_1, \cdots, w_n) = 0$$

$$\&\, u_1 = v_1^{w_1} \& \cdots \&\, u_n = v_n^{w_n}].$$

In particular, the functions $\binom{n}{k}$ and $n!$ were shown by her to be exponential

Diophantine. This is really what is shown in proving (1) and (2) of Theorem 4.1. The present proof of (2) is just hers; the proof of (1) given here is a simplified variant of that in [27]. (It is due independently to Julia Robinson and Matiyasevič.)

The idea of using the Chinese remainder theorem to code the effect of a bounded universal quantifier first occurred in the work of myself and Putnam [7]. In [8], we refined our methods and were able to show, beginning with the Davis normal form, that ***IF*** there are arbitrarily long arithmetic progressions consisting entirely of primes (still an open question), then every recursively enumerable set is exponential Diophantine. In our proof we needed to establish that $h(a, b, y) = \prod_{k=1}^{y} (a + bk)$ is exponential Diophantine, which we did extending Julia Robinson's methods. (The proof given here of (3) of Theorem 4.1 is a much simplified argument found much later by Julia Robinson—cf. [29].) Julia Robinson then showed first how to eliminate the hypothesis about primes in arithmetic progression, and then how to greatly simplify the proof along the lines of Lemma 5.2 of this article. Thus we obtained the theorem of [9] that every recursively enumerable set is exponential Diophantine.

Attention was now focused on the Julia Robinson hypothesis since it was plain that it would imply that Hilbert's tenth problem was unsolvable.

Many interesting propositions were found to imply the Julia Robinson hypothesis.[7]. However the hypothesis seemed implausible to many, especially because it was realized that an immediate and surprising consequence would be the existence of an absolute upper bound for the dimensions of Diophantine sets (cf. Theorem 7.6). Thus in his review [19] Kreisel said concerning the results of [9]: "... it is likely the present result is not closely connected with Hilbert's tenth problem. Also it is not altogether plausible that all (ordinary) Diophantine problems are uniformly reducible to those in a fixed number of variables of fixed degree... ."

The Julia Robinson hypothesis was finally proved by Matiyasevič [23], [24]. Specifically he showed that if we define

$$a_1 = a_2 = 1, \quad a_{n+1} = a_n + a_{n-1}$$

so that a_n is the nth Fibonacci number, then the function a_{2n} is diophantine. Then since, for $n \geqq 3$, as is easily seen by induction,

$$\left(\frac{5}{4}\right)^n < a_n < 2^{n-1},$$

the set

$$D = \{\langle u, v\rangle \mid v = a_{2u} \,\&\, u \geqq 2\}$$

satisfies the Julia Robinson hypothesis. Subsequently, direct diophantine definitions of the exponential function were given by a number of investigators, several of them using the Pell equation as in this article (cf. [3], [4], [14], [18a]). The treatment in §2, 3 is based on Matiyasevič's methods, although the details are Julia Robinson's.

In particular, it was Matiyasevič who taught us how to use results like Lemmas 2.11, 2.12, and 2.22 of the present exposition. (Matiyasevič himself used analogous results for the Fibonacci numbers.)

It was soon noticed (by S. Kochen) that by a simple inductive argument the use of the Davis normal form could now be entirely avoided, as has been done in the present exposition.

Let $\#(P)$ be the number of solutions of the Diophantine equation $P = 0$. Thus $0 \leqq \#(P) \leqq \aleph_0$. Hilbert's tenth problem seeks an algorithm for deciding of a given P whether or not $\#(P) = 0$. But there are many related questions: Is there an algorithm for testing whether $\#(P) = \aleph_0$, or $\#(P) = 1$, or $\#(P)$ is even? I was able to show easily (beginning with the unsolvability of Hilbert's tenth problem) that all of these problems are unsolvable. In fact if

$$A = \{0,1,2,3,\cdots\aleph_0\}$$

and $B \subseteq A$, $B \neq \varnothing$, $B \neq A$, then one can readily show that there is no algorithm for determining whether or not $\#(P) \in B$ (cf. [**15**]).

The fact that no general algorithm such as Hilbert demanded will be forthcoming adds to the interest of algorithms for dealing with special classes of Diophantine equations. Alan Baker and his coworkers [**1**], [**2**] have in recent years made considerable progress in this direction.

Notes

1. These pairing functions (but of course not their being Diophantine) were used by Cantor in his proof of the countability of the rational numbers. J. Roberts and D. Siefkes each corrected an error in my definition of these functions. They, as well as W. Emerson, M. Hausner, Y. Matiyasevič, and Julia Robinson made helpful suggestions.

2. For example, cf. [**25**], pp. 175–180. Matiyasevič used instead the equations $x^2 - xy - y^2 = 1$, $u^2 - muv + v^2 = 1$.

3. The recursive functions are usually defined on the nonnegative integers. This creates a minor but annoying technical problem in comparing the present definition with one in the literature (e.g., cf. [**6**], p. 41; also Theorem 4.2 on p. 51). Thus one can simply note that $f(x_1,\cdots,x_n)$ is recursive in the present sense if and only if $f(t_1+1,\cdots,t_n+1)-1$ is recursive in the usual sense. From the point of view of the intuitive "computability" of the functions involved this doesn't matter at all; one is simply in the position of using the positive integers as a "code" for the nonnegative integers — using $n+1$ to represent n.

4. Inclusion of $S(i,u)$ in this list is redundant. That is, $S(i,u)$ can be obtained using our three operations from the remaining initial functions.

5. The method of proof is Julia Robinson's, [**28**], [**30**]. If one were permitted to use the enumeration theorem in recursive function theory ([**6**], p. 67. Theorem 1.4), the Universality Theorem would follow at once from Theorem 6.1.

6. Actually the result which Gödel stated (as opposed to what can be obtained at once by use of his thceniques) was somewhat weaker. Indeed, the very definition of the class of recursive functions and the perception of their significance came several years later in the work of Gödel, Church, and Tnring. In particular the suggestion that recursiveness was a precise equivalent of the intuitive notion of being computable by an explicit algorithm was made independently by Church and by Turing. And of course it is this identification which is essential in regarding the technical results discussed in this account as constituting a negative solution of Hilbert's tenth problem. (For further discussion and references, cf. [**6**].)

7. For example, I showed ([**13**]) that the Julia Robinson hypothesis would follow from the non-existence of nontrivial solutions of the equation

$$9(u^2 + 7v^2)^2 - 7(x^2 + 7y^2)^2 = 2.$$

The methods used readily show that the same conclusion follows if the equation has only finitely many solutions. Čudnovskii [**4**] claims to have proved that 2^x is diophantine (and hence the Julia Robinson hypothesis) using this equation. Apparently there is a possibility that some of Čudnovskii's work may have been done independently of Matiyasevič — but I have not been able to obtain definite information about this.

References

1. Alan Baker, Contributions to the theory of Diophantine equations: I. On the representation of integers by binary forms, II. The Diophantine equation $y^2 = x^3 + k$, Philos. Trans. Roy. Soc. London Ser. A, 263 (1968) 173–208.

2. Alan Baker, The Diophantine equation $y^2 = ax^3 + bx^2 + cx + d$, J. London Math. Soc., 43 (1968) 1–9.

3. G. V. Čudnovskii, Diophantine predicates (Russian), Uspehi Mat. Nauk, 25 (1970) no. 4 (154), pp. 185–186.

4. ———, Certain arithmetic problems (Russian), Ordena Lenina Akad. Ukrains. SSR, Preprint IM-71-3.

5. Martin Davis, Arithmetical problems and recursively enumerable predicates, J. Symbolic Logic, 18 (1953) 33–41.

6. ———, Computability and Unsolvability, McGraw Hill, New York, 1958.(Dover, 1982).

7. Martin Davis and Hilary Putnam, Reduction of Hilbert's tenth problem, J. Symbolic Logic, 23 (1958) 183–187.

8. ——— and ———, On Hilbert's tenth problem, U. S. Air Force O. S. R. Report AFOSR TR 59–124 (1959), Part III.

9. Martin Davis, Hilary Putnam, and Julia Robinson, The decision problem for exponential Diophantine equations, Ann. Math., 74 (1961) 425–436.

10. Martin Davis, Applications of recursive function theory to number theory, Proc. Symp. Pure Math., 5 (1962) 135–138.

11. ———, Extensions and corollaries of recent work on Hilbert's tenth problem, Illinois J. Math., 7 (1963) 246–250.

12. Martin Davis and Hilary Putnam, Diophantine sets over polynomial rings, Illinois J. Math., 7 (1963) 251–256.

13. Martin Davis, One equation to rule them all, Trans. New York Acad. Sci., Series II, 30 (1968) 766–773.

14. ———, An explicit Diophantine definition of the exponential function, Comm. Pure Appl. Math. 24 (1971) 137–145.

15. ———, On the number of solutions of Diophantine equations, Proc. Amer. Math. Soc., 35 (1972) 552–554.

16. Kurt Gödel, Über formal unentscheidbare Sätze der Principia Mathematica und verwandter Systeme I, Monatsh. Math. und Physik, 38 (1931) 173–198. English translations: (1) Kurt, Gödel, On Formally Undecidable Propositions of Principia Mathematica and Related Systems, Baics

Books, 1962. (2) Martin Davis (editor), The Undecidable, Raven Press, 1965, pp. 5–38. (3) Jean Van Heijenoort (editor), From Frege to Gödel, Harvard University Press, 1967, pp. 596–616.

17. G. H. Hardy and E. M. Wright, An Introduction to the Theory of Numbers, Fourth edition, Oxford University Press, 1960.

18. David Hilbert, Mathematichse Probleme, Vortrag, gehalten auf dem internationalen Mathematiker-Kongress zu Paris 1900. Nachrichten Akad. Wiss. Göttingen, Math. -Phys. Kl. (1900) 253–297. English translation: Bull. Amer. Math. Soc., 8 (1901–1902) 437–479.

18a. N. K. Kosovskii, On Diophantine representations of the solutions of Pell's equation (Russian), Zap. Naučn. Sem. Leningrad. Otdel. Mat. Inst. Steklova, 20 (1971) 49–59.

19. Georg Kreisel, Review of [9]. Mathematical Reviews, 24 (1962) Part A, p. 573 (review number A 3061).

20. Yuri Matiyasevič, The relation of systems of equations in words and their lengths to Hilbert's tenth problem (Russian). Issledovaniya po Konstruktivnoi Matematike i Matematiceskoi Logike II. Vol. 8, pp. 132–144.

21. ———, Two reductions of Hilbert's tenth problem (Russian), *Ibid.*, pp. 145–158.

22. ———, Arithmetic representation of exponentiation (Russian). *Ibid.*, pp. 159–165.

23. ———, Enumerable sets are Diophantine (Russian), Dokl. Akad. Nauk SSSR, 191 (1970) 279–282. Improved English translation: Soviet Math. Doklady, 11 (1970) 354–357.

23a. ———, Diophantine representation of the set of prime numbers (Russian). Dokl. Akad. Nauk SSSR, 196 (1971) 770–773. Improved English translation with Addendum: Soviet Math. Doklady, 12 (1971) 249–254.

23b. ———, Diophantine representation of recursively enumerable predicates, Proc. Second Scandinavian Logic Symp., editor, J. E. Fenstad, North-Holland, Amsterdam, 1971.

23c. ———, Diophantine representation of recursively enumerable predicates, Proc. 1970 Intern. Congress Math., pp. 234–238.

24. ———, Diophantine representation of enumerable predicates (Russian), Izv. Akad. Nauk SSSR, Ser. Mat. 35 (1971) 3–30.

24a. ———, Diophantine sets (Russian), Uspehi Mat. Nauk, 27(1972) 185–222.

25. Ivan Niven and Herbert Zuckerman, An Introduction to the Theory of Numbers, 2nd ed., Wiley, New York, 1966.

26. Hilary Putnam, An unsolvable problem in number theory, J. Symb. Logic, 25 (1960) 220–232.

27. Julia Robinson, Existential definability in arithmetic, Trans. Amer. Math. Soc., 72 (1952) 437–449.

28. ———, Diophantine decision problems, MMA Studies in Mathematics, 6 (1969) [Studies in Number Theory, edited by W. J. LeVeque, pp. 76–116].

29. ———, Unsolvable Diophantine problems, Proc. Amer. Math. Soc., 22 (1969) 534–538.

30. ———, Hilbert's tenth problem, Proc. Symp. Pure Math., 20 (1969) 191–194.

31. Raphael M. Robinson, Arithmetical representation of recursively enumerable sets, J. Symb. Logic, 21 (1956) 162–186.

32. ———, Some representations of Diophantine sets, J. Symb. Logic, forthcoming.

REFERENCES

ADDISON, J. W., and S. C. KLEENE

[1] A Note on Function Quantification, *Proceedings of the American Mathematical Society*, vol. 8, pp. 1002–1006, 1957.

BOONE, WILLIAM W.

[1] Certain Simple, Unsolvable Problems of Group Theory, *Indagationes Mathematicae*, vol. 16, pp. 231–237, 492–497, 1954; vol. 17, pp. 252–256, 571–577, 1955; vol. 19, pp. 22–27, 227–232, 1957.

CARNAP, RUDOLF

[1] "The Logical Syntax of Language," Harcourt, Brace and Company, Inc., New York, 1937.

CHURCH, ALONZO

[1] An Unsolvable Problem of Elementary Number Theory, *American Journal of Mathematics*, vol. 58, pp. 345–363, 1936.

[2] A Note on the Entscheidungsproblem, *The Journal of Symbolic Logic*, vol. 1. pp. 40–41, 1936; Correction, *ibid.*, pp. 101–102.

[2*a*] The Constructive Second Number Class, *Bulletin of the American Mathematical Society*, vol. 44, pp. 224–232, 1938.

[3] "The Calculi of Lambda-conversion," Annals of Mathematics Studies, no. 6, Princeton University Press, Princeton, N.J., 1941; reprinted, 1951.

[4] "Introduction to Mathematical Logic, Part I," Annals of Mathematics Studies, no. 13, Princeton University Press, Princeton, N.J., 1944.

[5] "Introduction to Mathematical Logic," Princeton University Press, Princeton, N.J., vol. 1, 1956; vol. 2, forthcoming.

CHURCH, ALONZO, and S. C. KLEENE

[1] Formal Definitions in the Theory of Ordinal Numbers, *Fundamenta Mathematicae*, vol. 28, pp. 11–21, 1936.

DAVIS, MARTIN

[1] On the Theory of Recursive Unsolvability, doctoral dissertation, Princeton University, 1950.

[2] Relatively Recursive Functions and the Extended Kleene Hierarchy (abstract), *Proceedings of the International Congress of Mathematicians*, vol. 1, p. 723, 1950.

[3] Arithmetical Problems and Recursively Enumerable Predicates, *The Journal of Symbolic Logic*, vol. 18, pp. 33–41, 1953.

[4] An Intrinsic Definition of Recursive Functional (abstract), *Bulletin of the American Mathematical Society*, vol. 63, p. 139, 1957.

FINSLER, PAUL

[1] Formale Beweise und die Entscheidbarkeit, *Mathematische Zeitschrift*, vol. 25, pp. 676–682, 1926.

FRIEDBERG, RICHARD M.

[1] Two Recursively Enumerable Sets of Incomparable Degrees of Unsolvability (Solution of Post's Problem, 1944), *Proceedings of the National Academy of Sciences*, vol. 43, pp. 236–238, 1957.

GÖDEL, KURT

[1] Über formal unentscheidbare Sätze der Principia Mathematica und verwandter Systeme I, *Monatshefte für Mathematik und Physik*, vol. 38, pp. 173–198, 1931.

[2] On Undecidable Propositions of Formal Mathematical Systems, mimeographed lecture notes, Institute for Advanced Study, Princeton, N.J., 1934.

[3] Über die Länge von Beweisen, *Ergebnisse eines mathematisches Kolloquium*, Heft 4, pp. 34–38, 1936.

HALL, MARSHALL, JR.

[1] The Word Problem for Semigroups with Two Generators, *The Journal of Symbolic Logic*, vol. 14, pp. 115–118, 1949.

HARDY, G. H., and E. M. WRIGHT

[1] "An Introduction to the Theory of Numbers," Oxford University Press, New York, 1938.

HILBERT, DAVID

[1] Mathematical Problems, *Bulletin of the American Mathematical Society*, vol. 8, pp. 437–479, 1901–1902.

HILBERT, DAVID, and WILHELM ACKERMANN

[1] "Grundzüge der theoretischen Logik," Springer-Verlag OHG, Berlin, 1928; 2d ed., 1938; reprinted by Dover Publications, New York, 1946; 3d ed., Springer, Berlin, 1949; English translation of 2d ed. under the title "Principles of Mathematical Logic," Chelsea Publishing Company, New York, 1950.

HILBERT, DAVID, and PAUL BERNAYS

[1] "Grundlagen der Mathematik," vol. 1, Springer-Verlag OHG, Berlin, 1934; reprinted, J. W. Edwards, Publisher, Inc., Ann Arbor, Mich., 1944.

[2] "Grundlagen der Mathematik," vol. 2, Springer-Verlag OHG, Berlin, 1939; reprinted, J. W. Edwards, Publisher, Inc., Ann Arbor, Mich., 1944.

KALMÁR, LÁSZLÓ

[1] Über die Axiomatisierbarkeit des Aussagenkalküls, *Acta Scientiarum Mathematicarum*, vol. 7, pp. 222–243, 1934–1935.

[2] Egyszerü példa eldönthetetlen arithmetikai problémára (Ein einfaches Beispiel für ein unentscheidbares arithmetisches Problem) (Hungarian with German abstract), *Matematikai es fizikai lapok*, vol. 50, pp. 1–23, 1943.

[3] On Unsolvable Mathematical Problems, *Proceedings of the 10th International Congress of Philosophy*, pp. 756–758, North-Holland Publishing Company, Amsterdam, 1949.

[4] Eine einfache Konstruktion unentscheidbarer Sätze in formalen Systemen, *Methodos*, vol. 2, pp. 220–226, 1950; English translation, pp. 227–231.

[5] Another Proof of the Gödel-Rosser Incompletability Theorem, *Acta Scientiarum Mathematicarum*, vol. 12, pp. 38–43, 1950.

KLEENE, STEPHEN C.

[1] General Recursive Functions of Natural Numbers, *Mathematische Annalen*, vol. 112, pp. 727–742, 1936.

[2] λ-Definability and Recursiveness, *Duke Mathematical Journal*, vol. 2, pp. 340–353, 1936.

[3] A Note on Recursive Functions, *Bulletin of the American Mathematical Society*, vol. 42, pp. 544–546, 1936.

[3a] On Notation for Ordinal Numbers, *The Journal of Symbolic Logic*, vol. 3, pp. 150–155, 1938.

[4] Recursive Predicates and Quantifiers, *Transactions of the American Mathematical Society*, vol. 53, pp. 41–73, 1943.

[4a] On the Forms of the Predicates in the Theory of Constructive Ordinals, *American Journal of Mathematics*, vol. 66, pp. 41–58, 1944.

[5] A Symmetric Form of Gödel's Theorem, *Indagationes Mathematicae*, vol. 12, pp. 244–246, 1950.

[5a] Recursive Functions and Intuitionistic Mathematics, *Proceedings of the International Congress of Mathematicians*, vol. 1, pp. 679–685, 1950.

[6] "Introduction to Metamathematics," D. Van Nostrand Company, Inc., Princeton, N.J., 1952.

[7] Arithmetical Predicates and Function Quantifiers, *Transactions of the American Mathematical Society*, vol. 79, pp. 312–340, 1955.

[8] On the Forms of Predicates in the Theory of Constructive Ordinals (second paper), *American Journal of Mathematics*, vol. 77, pp. 405–428, 1955.

[9] Hierarchies of Number-theoretic Predicates, *Bulletin of the American Mathematical Society*, vol. 61, pp. 193–213, 1955.

KLEENE, S. C., and EMIL L. POST

[1] The Upper Semi-lattice of Degrees of Recursive Unsolvability, *Annals of Mathematics*, ser. 2, vol. 59, pp. 379–407, 1954.

LINIAL, SAMUEL, and EMIL L. POST

[1] Recursive Unsolvability of the Deducibility, Tarski's Completeness and Independence of Axioms Problems of the Propositional Calculus (abstract), *Bulletin of the American Mathematical Society*, vol. 55, p. 50, 1949.

MARKOV, A. A.

[1] On the Impossibility of Certain Algorithms in the Theory of Associative Systems (Russian), *Doklady Akademii Nauk S.S.S.R.*, n.s., vol. 55, pp. 587–590, 1947; English translation, *Comptes rendus de l'académie des sciences de l'U.R.S.S.*, n.s., vol. 55, pp. 583–586, 1947.

[2] On Some Unsolvable Problems concerning Matrices (Russian), *Doklady Akademii Nauk S.S.S.R.*, n.s., vol. 57, pp. 539–542, 1947.

[3] Impossibility of Certain Algorithms in the Theory of Associative Systems II (Russian), *Doklady Akademii Nauk S.S.S.R.*, n.s., vol. 58, pp. 353–356, 1947.

[4] On the Representation of Recursive Functions (Russian), *Doklady Akademii Nauk S.S.S.R.*, n.s., vol. 58, pp. 1891–1892, 1947.

[5] On the Representation of Recursive Functions (Russian), *Izvestiya Akademii Nauk S.S.S.R.*, n.s., vol. 13, 1949; English translation, translation 54, American Mathematical Society, New York, 1951.

[6] Impossibility of Certain Algorithms in the Theory of Associative Systems (Russian), *Doklady Akademii Nauk S.S.S.R.*, n.s., vol. 77, pp. 19–20, 1951.

[7] Impossibility of Algorithms for Distinguishing Certain Properties of Associative Systems (Russian), *Doklady Akademii Nauk S.S.S.R.*, n.s., vol. 77, pp. 953–956, 1951.

[8] An Unsolvable Problem concerning Matrices (Russian), *Doklady Akademii Nauk S.S.S.R.*, n.s., vol. 78, pp. 1089–1092, 1951.

MARKWALD, W.

[1] Zur Theorie der konstruktiven Wohlordnungen, *Mathematische Annalen*, vol. 127, pp. 135–149, 1954.

MOSTOWSKI, ANDRZEJ

[1] On Definable Sets of Positive Integers, *Fundamenta Mathematicae*, vol. 34, pp. 81–112, 1947.

[2] On a Set of Integers Not Definable by Means of One-quantifier Predicates, *Annales de la société polonaise de mathématique*, vol. 21, pp. 114–119, 1948.

[3] A Classification of Logical Systems, *Studia Philosophica*, vol. 4, pp. 237–274, 1951.

MUCNIK, A. A.

[1] Negative Answer to the Problem of Reducibility of the Theory of Algorithms (Russian), *Doklady Akademii Nauk S.S.S.R.*, n.s., vol. 108, pp. 194–197, 1956.

MYHILL, JOHN

[1] Three Contributions to Recursive Function Theory, *Proceedings of the 11th International Congress of Philosophy*, vol. 14, pp. 50–59, North-Holland Publishing Company, Amsterdam, 1953.

[2] Creative Sets, *Zeitschrift für mathematische Logik und Grundlagen der Mathematik*, vol. 1, pp. 97–108, 1955.

MYHILL, JOHN, and J. C. SHEPHERDSON

[1] Effective Operations on Partial Recursive Functions, *Zeitschrift für mathematische Logik und Grundlagen der Mathematik*, vol. 1, pp. 310–317, 1955.

NOVIKOFF, P. S.

[1] On the Algorithmic Unsolvability of the Word Problem in Group Theory (Russian), *Akademiya Nauk S.S.S.R. Matematicheskii Institut Trudy*, no. 44, Moscow, 1955.

PÉTER, RÓZSA

[1] "Rekursive Funktionen," Akadémiai Kiadó, Budapest, 1951.

POST, EMIL L.

[1] Finite Combinatory Processes—Formulation I, *The Journal of Symbolic Logic*, vol. 1, pp. 103–105, 1936.

[2] Formal Reductions of the General Combinatorial Decision Problem, *American Journal of Mathematics*, vol. 65, pp. 197–215, 1943.

[3] Recursively Enumerable Sets of Positive Integers and Their Decision Problems, *Bulletin of the American Mathematical Society*, vol. 50, pp. 284–316, 1944.

[4] A Variant of a Recursively Unsolvable Problem, *Bulletin of the American Mathematical Society*, vol. 52, pp. 264–268, 1946.

[5] Note on a Conjecture of Skolem, *The Journal of Symbolic Logic*, vol. 11, pp. 73–74, 1946.

[6] Recursive Unsolvability of a Problem of Thue, *The Journal of Symbolic Logic*, vol. 12, pp. 1–11, 1947.

[7] Degrees of Recursive Unsolvability: Preliminary Report (abstract), *Bulletin of the American Mathematical Society*, vol. 54, pp. 641–642, 1948.

[8] Absolutely Unsolvable Problems and Relatively Undecidable Propositions—Account of an Anticipation, unpublished, 1941.

RICE, H. G.

[1] Classes of Recursively Enumerable Sets and Their Decision Problems, *Transactions of the American Mathematical Society*, vol. 74, pp. 358–366, 1953.

ROBINSON, JULIA

[1] General Recursive Functions, *Proceedings of the American Mathematical Society*, vol. 1, pp. 703–718, 1950.

[2] Existential Definability in Arithmetic, *Transactions of the American Mathematical Society*, vol. 72, pp. 437–449, 1952.

ROBINSON, RAPHAEL M.

[1] Primitive Recursive Functions, *Bulletin of the American Mathematical Society*, vol. 53, pp. 925–942, 1947.

[2] Recursion and Double Recursion, *Bulletin of the American Mathematical Society*, vol. 54, pp. 987–993, 1948.

[3] Arithmetical Representation of Recursively Enumerable Sets, *The Journal of Symbolic Logic*, vol. 21, pp. 162–186, 1956.

Rosenbloom, Paul C.

[1] "Elements of Mathematical Logic," Dover Publications, New York, 1951.

Rosser, Barkley

[1] Extensions of Some Theorems of Gödel and Church, *The Journal of Symbolic Logic*, vol. 1, pp. 87–91, 1936.

Shapiro, Norman

[1] Degrees of Computability, *Transactions of the American Mathematical Society*, vol. 82, pp. 281–299, 1956.

Skolem, Thoralf

[1] Begründung der elementaren Arithmetik durch die rekurrierende Denkweise ohne Anwendung scheinbarer Veränderlichen mit unendlichem Ausdehnungsbereich, *Skrifter utgitt av Det Norske Videnskaps-Akademi i Oslo*, I, no. 6b, 1923.

Spector, Clifford

[1] Recursive Well-orderings, *The Journal of Symbolic Logic*, vol. 20, pp. 151–163, 1955.

[2] On Degrees of Recursive Unsolvability, *Annals of Mathematics*, vol. 64, pp. 581–592, 1956.

Tarski, Alfred

[1] Der Wahrheitsbegriff in den formalisierten Sprachen, *Studia Philosophica*, vol. 1, pp. 261–405, 1936; English translation in "Logic, Semantics, Metamathematics," Oxford University Press, New York, 1956.

Tarski, Alfred, Andrzej Mostowski, and Raphael M. Robinson

[1] "Undecidable Theories," Studies in Logic and the Foundations of Mathematics, North-Holland Publishing Company, Amsterdam, 1953.

Turing, A. M.

[1] On Computable Numbers, with an Application to the Entscheidungsproblem, *Proceedings of the London Mathematical Society*, ser. 2, vol. 42, pp. 230–265, 1936–1937; Correction, *ibid.*, vol. 43, pp. 544–546, 1937.

[2] Computability and λ-Definability, *The Journal of Symbolic Logic*, vol. 2, pp. 153–163, 1937.

[3] Systems of Logic Based on Ordinals, *Proceedings of the London Mathematical Society*, ser. 2, vol. 45, pp. 161–228, 1939.

[4] The Word Problem in Semigroups with Cancellation, *Annals of Mathematics*, ser. 2, vol. 52, pp. 491–505, 1950.

INDEX

Page references for a notion relative to a set A will be found under the corresponding unrelativized notion. For example, page references for *A-simple sets* will be found under *simple sets.*

A CATALOG OF SELECTED

DOVER BOOKS

IN SCIENCE AND MATHEMATICS

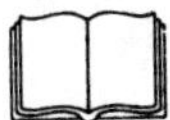

A CATALOG OF SELECTED

DOVER BOOKS

IN SCIENCE AND MATHEMATICS

THE ELECTROMAGNETIC FIELD, Albert Shadowitz. Comprehensive undergraduate text covers basics of electric and magnetic fields, builds up to electromagnetic theory. Also related topics, including relativity. Over 900 problems. 768pp. 5⅝ × 8¼. 65660-8 Pa. $18.95

FOURIER SERIES, Georgi P. Tolstov. Translated by Richard A. Silverman. A valuable addition to the literature on the subject, moving clearly from subject to subject and theorem to theorem. 107 problems, answers. 336pp. 5⅜ × 8½. 63317-9 Pa. $8.95

THEORY OF ELECTROMAGNETIC WAVE PROPAGATION, Charles Herach Papas. Graduate-level study discusses the Maxwell field equations, radiation from wire antennas, the Doppler effect and more. xiii + 244pp. 5⅜ × 8½. 65678-0 Pa. $6.95

DISTRIBUTION THEORY AND TRANSFORM ANALYSIS: An Introduction to Generalized Functions, with Applications, A.H. Zemanian. Provides basics of distribution theory, describes generalized Fourier and Laplace transformations. Numerous problems. 384pp. 5⅜ × 8½. 65479-6 Pa. $9.95

THE PHYSICS OF WAVES, William C. Elmore and Mark A. Heald. Unique overview of classical wave theory. Acoustics, optics, electromagnetic radiation, more. Ideal as classroom text or for self-study. Problems. 477pp. 5⅜ × 8½. 64926-1 Pa. $12.95

CALCULUS OF VARIATIONS WITH APPLICATIONS, George M. Ewing. Applications-oriented introduction to variational theory develops insight and promotes understanding of specialized books, research papers. Suitable for advanced undergraduate/graduate students as primary, supplementary text. 352pp. 5⅜ × 8½. 64856-7 Pa. $8.95

A TREATISE ON ELECTRICITY AND MAGNETISM, James Clerk Maxwell. Important foundation work of modern physics. Brings to final form Maxwell's theory of electromagnetism and rigorously derives his general equations of field theory. 1,084pp. 5⅜ × 8½. 60636-8, 60637-6 Pa., Two-vol. set $21.90

AN INTRODUCTION TO THE CALCULUS OF VARIATIONS, Charles Fox. Graduate-level text covers variations of an integral, isoperimetrical problems, least action, special relativity, approximations, more. References. 279pp. 5⅜ × 8½. 65499-0 Pa. $7.95

HYDRODYNAMIC AND HYDROMAGNETIC STABILITY, S. Chandrasekhar. Lucid examination of the Rayleigh-Benard problem; clear coverage of the theory of instabilities causing convection. 704pp. 5⅜ × 8¼. 64071-X Pa. $14.95

CALCULUS OF VARIATIONS, Robert Weinstock. Basic introduction covering isoperimetric problems, theory of elasticity, quantum mechanics, electrostatics, etc. Exercises throughout. 326pp. 5⅜ × 8½. 63069-2 Pa. $8.95

DYNAMICS OF FLUIDS IN POROUS MEDIA, Jacob Bear. For advanced students of ground water hydrology, soil mechanics and physics, drainage and irrigation engineering and more. 335 illustrations. Exercises, with answers. 784pp. 6⅛ × 9¼. 65675-6 Pa. $19.95

CATALOG OF DOVER BOOKS

GEOMETRY OF COMPLEX NUMBERS, Hans Schwerdtfeger. Illuminating, widely praised book on analytic geometry of circles, the Moebius transformation, and two-dimensional non-Euclidean geometries. 200pp. 5⅝ × 8¼.
63830-8 Pa. $8.95

MECHANICS, J.P. Den Hartog. A classic introductory text or refresher. Hundreds of applications and design problems illuminate fundamentals of trusses, loaded beams and cables, etc. 334 answered problems. 462pp. 5⅜ × 8½. 60754-2 Pa. $9.95

TOPOLOGY, John G. Hocking and Gail S. Young. Superb one-year course in classical topology. Topological spaces and functions, point-set topology, much more. Examples and problems. Bibliography. Index. 384pp. 5⅝ × 8¼.
65676-4 Pa. $9.95

STRENGTH OF MATERIALS, J.P. Den Hartog. Full, clear treatment of basic material (tension, torsion, bending, etc.) plus advanced material on engineering methods, applications. 350 answered problems. 323pp. 5⅜ × 8½. 60755-0 Pa. $8.95

ELEMENTARY CONCEPTS OF TOPOLOGY, Paul Alexandroff. Elegant, intuitive approach to topology from set-theoretic topology to Betti groups; how concepts of topology are useful in math and physics. 25 figures. 57pp. 5⅜ × 8½.
60747-X Pa. $3.50

ADVANCED STRENGTH OF MATERIALS, J.P. Den Hartog. Superbly written advanced text covers torsion, rotating disks, membrane stresses in shells, much more. Many problems and answers. 388pp. 5⅜ × 8½. 65407-9 Pa. $9.95

COMPUTABILITY AND UNSOLVABILITY, Martin Davis. Classic graduate-level introduction to theory of computability, usually referred to as theory of recurrent functions. New preface and appendix. 288pp. 5⅜ × 8½. 61471-9 Pa. $7.95

GENERAL CHEMISTRY, Linus Pauling. Revised 3rd edition of classic first-year text by Nobel laureate. Atomic and molecular structure, quantum mechanics, statistical mechanics, thermodynamics correlated with descriptive chemistry. Problems. 992pp. 5⅜ × 8½. 65622-5 Pa. $19.95

AN INTRODUCTION TO MATRICES, SETS AND GROUPS FOR SCIENCE STUDENTS, G. Stephenson. Concise, readable text introduces sets, groups, and most importantly, matrices to undergraduate students of physics, chemistry, and engineering. Problems. 164pp. 5⅜ × 8½. 65077-4 Pa. $6.95

THE HISTORICAL BACKGROUND OF CHEMISTRY, Henry M. Leicester. Evolution of ideas, not individual biography. Concentrates on formulation of a coherent set of chemical laws. 260pp. 5⅜ × 8½. 61053-5 Pa. $6.95

THE PHILOSOPHY OF MATHEMATICS: An Introductory Essay, Stephan Körner. Surveys the views of Plato, Aristotle, Leibniz & Kant concerning propositions and theories of applied and pure mathematics. Introduction. Two appendices. Index. 198pp. 5⅜ × 8½. 25048-2 Pa. $7.95

THE DEVELOPMENT OF MODERN CHEMISTRY, Aaron J. Ihde. Authoritative history of chemistry from ancient Greek theory to 20th-century innovation. Covers major chemists and their discoveries. 209 illustrations. 14 tables. Bibliographies. Indices. Appendices. 851pp. 5⅜ × 8½. 64235-6 Pa. $18.95